Ultrasonic Spectral Analysis for Nondestructive Evaluation

Ultrasonic Spectral Analysis for Nondestructive Evaluation

Dale W. Fitting
and
Laszlo Adler

Ohio State University
Columbus, Ohio

Springer Science+Business Media, LLC

Library of Congress Cataloging in Publication Data

Fitting, Dale.
  Ultrasonic spectral analysis for nondestructive evaluation.

  Bibliography: p.
  Includes indexes.
  1. Non-destructive testing. 2. Ultrasonic testing. 3. Spectrum analysis. I. Adler,
Laszlo, 1932-   . Title.
TA417.4. F54   620.1'1274   80-14991

ISBN 978-1-4613-3128-5   ISBN 978-1-4613-3126-1 (eBook)
DOI 10.1007/978-1-4613-3126-1

© 1981 Springer Science+Business Media New York
Originally published by Plenum Press, New York in 1981.
Softcover reprint of the hardcover 1st edition 1981

Acknowledgments

The authors are grateful to Battelle Memorial Institute for financial support of this work. Special thanks are extended to Mr. G. J. Posakony, Battelle Northwest Laboratories Program Monitor, for his initiation and continuous interest in this problem.

Appreciation is also extended to the following:

Nondestructive Testing Laboratory, Metals & Ceramics Division, Oak Ridge National Laboratory—especially R. W. McClung, Hugh Whaley (now at Babcock & Wilcox, Lynchburg, Virginia), K. V. Cook, and Bill Simpson—who initiated our interest in ultrasonic spectroscopy;

Ultrasonics Group, The University of Tennessee—especially M. A. Breazeale and T. K. Bolland—for their suggestions and cooperation;

Maxine Martin for her editorial assistance and excellent typing of the manuscript.

Gale Slutski and Kerstin Kleber for transferring our ideas into the illustrations.

Last but not least, we would like to acknowledge the scientists, both here and abroad, whose contributions to the field of ultrasonic spectral analysis made this work possible.

The work contained in this book was performed while the authors were associated with the University of Tennessee, Knoxville.

Ohio State University

Dale W. Fitting
Laszlo Adler

Contents

Introduction

Ultrasonic spectroscopy is the study of ultrasonic waves resolved into their Fourier frequency components. Since many material properties manifest themselves as amplitude or phase changes in ultrasonic waves used to interrogate a specimen, ultrasonic spectroscopy has proven to be quite valuable.

Initially, investigations were carried out using narrowband ultrasonic instrumentation. Later, the value of a broadband ultrasonic pulse was recognized. Analysis of the pulse was found to give information simultaneously over a wide range of frequencies. The first system to utilize multifrequency (broadband) ultrasonic pulses was built by Gericke (1960-059)* at the Army Research Laboratory. His system used a contact transducer operated in the pulse–echo mode. Gericke also appears to have been the first to use the term *ultrasonic spectroscopy*. In analogy with visual inspection by white light, he postulated that additional information could be obtained corresponding to the "color" of the waves after interaction with the material. Gericke attempted to obtain "white ultrasound" by shock-exciting a damped ceramic transducer. His results indicated qualitatively that the ultrasonic spectrum of an echo contains information related to the configuration of a void in a solid and to the microstructure (grain size) of a material.

A few years later, Whaley and Cook (005) at Oak Ridge National Laboratory built the first immersion system for spectrum analysis. Whaley and Adler (036) extended spectrum analysis to a multitransducer system, thus introducing pitch-catch and through-transmission ultrasonic spectroscopy. In a number of model experiments (012, 036, 037, 038, 039) Whaley and Adler demonstrated that quantitative information could be obtained from the spectrum of the received ultrasonic signal. Their interference model was the first analytical approach for relating size and orientation of simple discontinuities to the frequency spectrum.

About the same time as work in the United States was going on, several groups of investigators in England began developing ultrasonic spectroscopic systems. Brown and Lloyd at City University in London (026, 030, 180, 192, 245) began to utilize spectroscopy for the study of laminated structures and for defect characterization. In addition, they began development of broadband transducers. Note should be made of the two symposia dealing with ultrasonic spectroscopy which were held at City University—one in 1970 (268–272), the next in 1976 (241–250).

During the last few years, additional groups of investigators in the United States have become involved with ultrasonic spectroscopy. Much of this research has been stimulated by a program on quantitative nondestructive testing supported by

*See Chapter 5, Abstracted Bibliography, for an explanation of the referencing system.

the Advanced Research Projects Agency (ARPA), the Air Force Materials Laboratory (AFML), and operated through the Rockwell Science Center. Theoretical investigations were funded, as well as basic experimental research and applications-oriented studies.

At the Rockwell Science Center, Tittmann and Elsley (004, 133, 379, 409) have studied scattering of ultrasonic waves from voids. Their unique experimental system permits accumulation of data on the angular as well as the frequency dependence of scattering. Additional work by Tittmann and others at Rockwell (181) has been directed toward the development of improved ultrasonic standards. Studies of waves diffracted by fluid-filled cavities were carried out at Cornell by Pao and Sachse (087, 091, 149, 176).

Ultrasonic spectroscopy has been applied to the evaluation of adhesive bonds by Alers (200, 432, 459, 466) at Rockwell, Rose and Meyer (020, 034, 118) at Drexel, and Chang, Couchman, and Yee (177, 218, 256) at General Dynamics.

Not all investigations have centered on experimental work. Several theoreticians have recently developed theories for elastic wave diffraction by voids. These theories provide the basis for relating the frequency dependence of the scattered signal to defect characteristics. In the long-wavelength limit Gubernatis, Domany, and Krumhansl (304, 314, 318) used the Born approximation, Datta (325, 333) employed a matched asymptotic method, and Pao and Varaden (327, 329, 330) as well as Waterman (326, 328) used a scattering matrix technique to investigate the frequency-dependent nature of ultrasonic scattering. Achenbach, Gautesen, and McMaken (323, 324) used elastodynamics and ray theory to obtain analytically the diffraction field in the short-wavelength limit. For this same regime, Adler and Lewis (14, 145) applied Keller's theory to study scattering from planar cracks. Inversion techniques developed by Bleistein and Cohen (427, 455) and Mucciardi *et al.* (119, 126, 447, 454) provide a means for relating spectral features to flaw characteristics.

Adler, Cook, and Simpson at Oak Ridge National Laboratory have introduced several applications of ultrasonic spectroscopy, which include wall thickness, phase, and attenuation measurements (005, 044). Investigations are being made of phase spectroscopy by Nabel (073, 232) in Germany, and by Simpson, Adler, and Elsley in the United States. An important contribution to the understanding of ultrasonic spectroscopy was provided by Simpson (002, 013), who used a Fourier-transform model to explain features in the diffracted wave spectra.

Correlations of the ultrasonic spectrum with strength-related properties have been made theoretically by Rice and Budiansky (234), and experimentally by Vary (334, 488, 489) and Elsley, Richardson, and Thompson (004).

Ultrasonic studies of surface roughness and periodicity by Quentin, Jungman, deBilly, and Cohen-Tenoudji (138, 139, 140) have shown the value of information contained in the spectra. Szabo (211) and Richardson and Tittmann (321, 403) have developed inversion techniques for deducing subsurface gradients from surface wave dispersion measurements.

Papadakis at Ford Motor Company has performed thorough investigations of the application of ultrasonic spectroscopy to attenuation measurements (016, 023, 212) and microstructure (021, 031, 186).

In preparing this report we noted a paucity of discussion concerning the ultrasonic spectroscopic system. In Chapter 2 we have reviewed the operation of a spectroscopic device by way of a system model.

Chapter 3 outlines some of the applications of ultrasonic spectroscopy which have been identified to date. In addition to the use of spectroscopy for defect characterization and assessment of adhesive bonds, spectroscopy has proven useful for: determining surface properties and subsurface gradients, monitoring corrosion, deducing strength-related properties, investigating the microstructure of a material as well as the measurement of frequency-dependent attenuation, and velocity.

During our work on this report a questionnaire was mailed to over one hundred leading scientists and engineers, asking their opinion of the current state-of-the-art of ultrasonic spectroscopy. Their comments are summarized in Chapter 4. Also included is a listing of investigators worldwide who are active in ultrasonic spectroscopy.

An extensive bibliography with subject and author indexes forms Chapter 5. Time and space did not permit mentioning in Chapters 2 and 3 all references to the literature. Additional information may be found by referring to the subject index.

We hope that this report will serve to move ultrasonic spectroscopy closer to the solution of relevant problems and encourage research in this vital area.

Ultrasonic Spectroscopic Systems

Introduction

This chapter of the report concerns itself with the systems (and their components) which are used for ultrasonic spectroscopy. We begin by introducing the
concept of a linear time-invariant model for the spectroscopic system. Through the
use of this model one may investigate, in detail, the operation of system components
and also arrive at the overall system performance.

Each element of the system is treated separately, then as a part of one of three
basic subsystems and finally as a complete instrument. Examples of systems presently in use are given and the advantages and shortcomings of each type are
discussed.

System Model

Although there are many possible designs for an ultrasonic spectrum analysis
system, each contains provisions for (1) generating ultrasound, (2) receiving a
portion of the ultrasound which has interacted with the material under study, and
(3) analyzing the received wave to determine the magnitude (and sometimes phase)
of the ultrasound at a number of frequencies. Figure 1 illustrates the major
components of a generalized spectroscopic system.

In this common system configuration (Figure 1), an electrical waveform generated by the transmitter is applied to the transmitting transducer. Conversion of the
electrical energy into mechanical energy occurs within the transducer, producing an
ultrasonic wave. As the wave propagates through the material being studied,
interactions of the ultrasonic energy with the material alter the amplitude, phase,
and direction of the wave. A receiving transducer intercepts a portion of the
ultrasonic energy and conversion occurs from mechanical to electrical energy.
Because the electrical signal is usually small, an amplifier is used to increase its
amplitude. The purpose of the analysis system, which follows the amplifier, is to sort
out the magnitude and timing of the ultrasonic interactions (within the material) and
present amplitude and phase spectra.

The operation of the system is probably best analyzed by first examining each
of its components in detail and then noting its effect on system performance. A
model proposed by Frederick and Seydel (001, 009) is a reasonable starting place for
our analysis. Each component of the system is considered a linear time-invariant

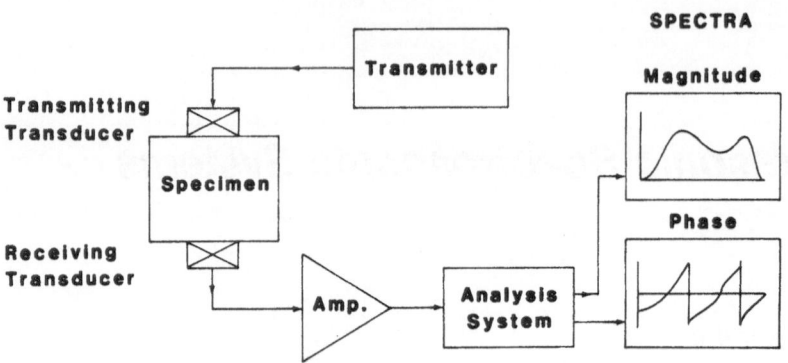

Figure 1. Generalized ultrasonic spectroscopy system.

(LTI) system. Superposition holds for systems which are linear. That is, if input 1 to the system produces output 1 and input 2 produces output 2, the output of the system in response to simultaneous inputs of 1 and 2 may be found by summing (in the appropriate manner) output 1 and output 2. Time invariance implies that the output for a given input is independent of the time the input was applied to the system. Although analysis of ultrasonic spectroscopes as linear time-invariant systems is not always appropriate, it does provide a good basis for assessing the performance of system components.

The behavior of a LTI system is completely described by its impulse response (in the time domain) or its frequency response (in the frequency domain). The two descriptions of system response are equivalent and are linked by the following relations:

$$H(f) = \int_{-\infty}^{\infty} h(t) e^{-j2\pi ft}\, dt \tag{1}$$

$$h(t) = \frac{1}{2\pi} \int_{-\infty}^{\infty} H(f) e^{j2\pi ft}\, df \tag{2}$$

where $h(t)$ is the impulse response and $H(f)$ is the frequency response (transfer function) [t is time, f is frequency, and $j = (-1)^{1/2}$]. The responses $h(t)$ and $H(f)$ are said to form a Fourier transform pair.

It is important to recognize that $H(f)$ is a complex quantity. That is,

$$H(f) = \mathrm{Re}[H(f)] + j\,\mathrm{Im}[H(f)] \tag{3}$$

The magnitude spectrum is defined as

$$|H(f)| = \left\{ \mathrm{Re}^2[H(f)] + \mathrm{Im}^2[H(f)] \right\}^{1/2} \tag{4}$$

and the phase spectrum is

$$\angle H(f) = \tan^{-1} \frac{\mathrm{Im}[H(f)]}{\mathrm{Re}[H(f)]} \tag{5}$$

For a linear time-invariant system the input and output are related as shown in Figure 2. The output $o(t)$ is found by convolving the input $i(t)$ with the impulse response of the system $h(t)$. That is,

$$o(t)=i(t)*h(t)=\int_{-\infty}^{\infty} i(t')h(t-t')\,dt' \tag{6}$$

Convolution may also be carried out in the frequency domain, and the mathematics (once the transfer functions are known) is reduced to a multiplication of two complex numbers. The output is

$$O(f)=\mathrm{Re}[O(f)]+j\mathrm{Im}[O(f)]=I(f)H(f) \tag{7a}$$

$$O(f)=\mathrm{Re}[I(f)]\,\mathrm{Re}[H(f)]-\mathrm{Im}[I(f)]\,\mathrm{Im}[H(f)]$$

$$+j\{\mathrm{Re}[I(f)]\,\mathrm{Im}[H(f)]+\mathrm{Re}[H(f)]\,\mathrm{Im}[I(f)]\} \tag{7b}$$

At each frequency the output is just the product of the input and system frequency response at that frequency.

Figure 3 is a block diagram of an ultrasonic spectroscope modeled as an LTI system. The symbols for the impulse and frequency response given for each component in Figure 3 will be used throughout this report. Although the impulse response is important, we will be dealing almost entirely with frequency responses since frequency is the relevant parameter in spectral analysis. Notice that for most of the components, the signal of interest is an electrical waveform. However, for an important section of the system the relevant signal is an ultrasonic wave propagating through a coupling medium and the material under study.

In general, the time-domain representation of the signal is that which is monitored; the analysis subsystem provides the transformation to the frequency domain. The natural selection of the independent variable through system components where an ultrasonic wave propagates is distance (z). Distance may be converted to time if the velocity of wave travel is known. System components in which ultrasonic waves propagate are shown with impulse responses in terms of

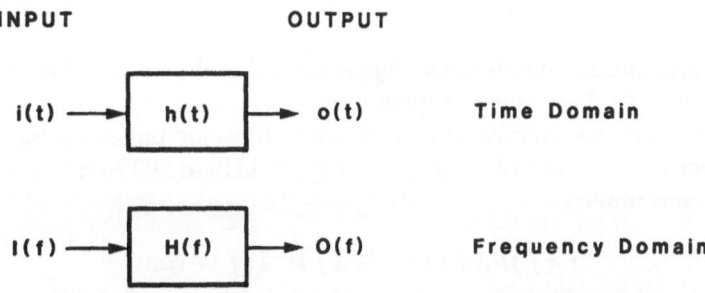

Figure 2. Equivalent representations of a linear time-invariant system.

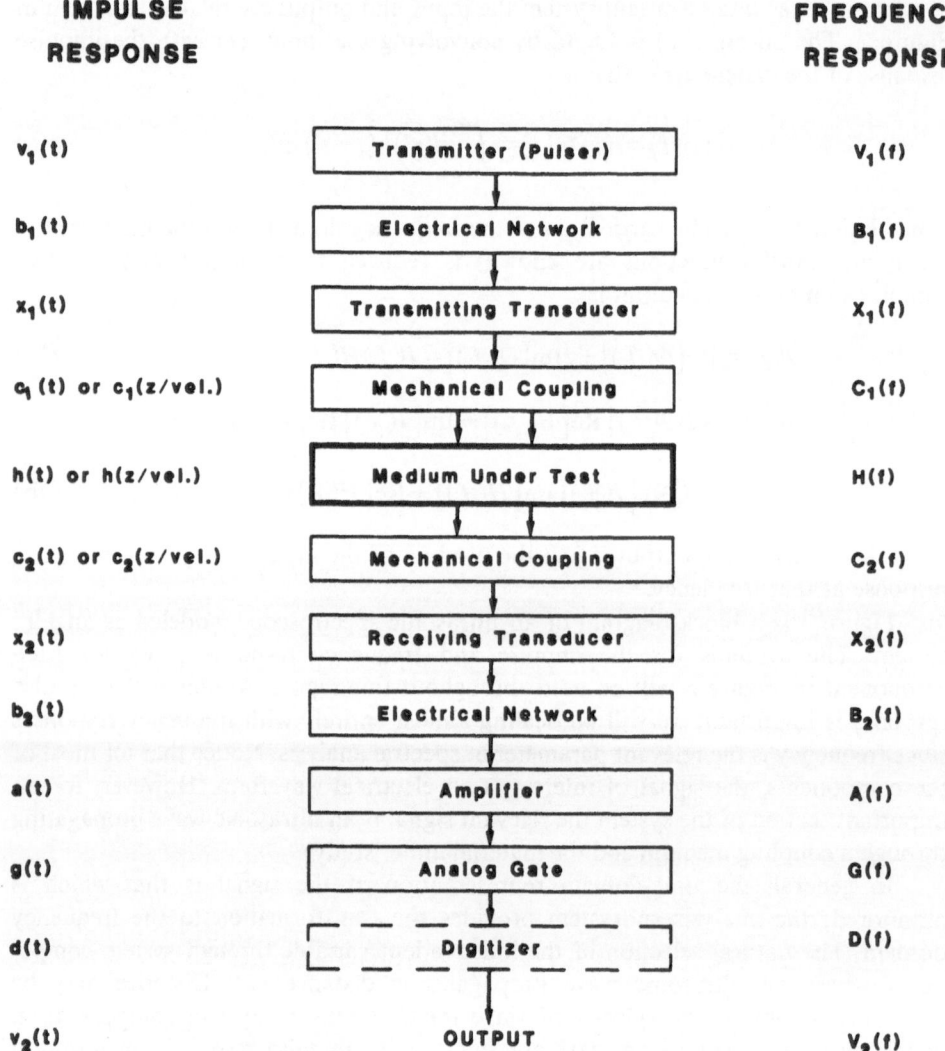

IMPULSE RESPONSE

FREQUENCY RESPONSE

Impulse Response	Block	Frequency Response
$v_1(t)$	Transmitter (Pulser)	$V_1(f)$
$b_1(t)$	Electrical Network	$B_1(f)$
$x_1(t)$	Transmitting Transducer	$X_1(f)$
$c_1(t)$ or $c_1(z/vel.)$	Mechanical Coupling	$C_1(f)$
$h(t)$ or $h(z/vel.)$	Medium Under Test	$H(f)$
$c_2(t)$ or $c_2(z/vel.)$	Mechanical Coupling	$C_2(f)$
$x_2(t)$	Receiving Transducer	$X_2(f)$
$b_2(t)$	Electrical Network	$B_2(f)$
$a(t)$	Amplifier	$A(f)$
$g(t)$	Analog Gate	$G(f)$
$d(t)$	Digitizer	$D(f)$
$v_2(t)$	OUTPUT	$V_2(f)$

Figure 3. Elements of an ultrasonic spectroscopic system modeled as a linear, time-invariant system.

time-varying quantities and distance. Again recall that the transfer functions [as well as the output, $V_2(f)$] are complex quantities.

An ideal spectroscopic system would emit ultrasonic pulses whose spectrum is uniform over a large range of frequencies (e.g., 20 kHz to 500 MHz). For the model in Figure 3 this implies

$$V_1(f)B_1(f)X_1(f)C_1(f) = T(f) = \text{const} \tag{8}$$

for the frequency range. A receiving section whose response is uniform over the same

Figure 4. Spectroscopic system for the determination of a medium's transfer function. (Transmitter and receiver are fully characterized. $V_2(f)$ is the measured output.)

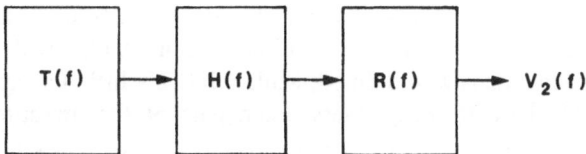

Figure 4. Spectroscopic system for the determination of a medium's transfer function. (Transmitter and receiver are fully characterized. $V_2(f)$ is the measured output.)

frequency range is also desirable. That is,

$$C_2(f)X_2(f)B_2(f)A(f)\cdots = R(f) = \text{const} \tag{9}$$

If the ideal system just described were used to test a material, with impulse response $h(t)$, the output of the spectrum analysis system $[V_2(f)]$ would be $H(f)$ multiplied by a constant. However, such an ideal spectroscopic system is not strictly realizable over the wide frequency range mentioned earlier.

Fortunately the LTI model provides a method of correcting for the nonuniform frequency response in nonideal systems, if the transfer functions of the system components are known. Assume an ultrasonic detector is available whose response (versus frequency) is known. Then if the detector is placed such that it receives the ultrasonic wave produced by the transmitting transducer, the detector output will be (or can be corrected to be) $T(f)$. Coupling the transmission section of the spectroscope through a lossless, nondispersive medium $[H(f)=1]$ to the receiver will produce output $V_2(f)$. The frequency responses of these components are related by

$$T(f)H(f)R(f) = V_2(f) \tag{10a}$$

or

$$R(f) = \frac{V_2(f)}{T(f)H(f)} \tag{10b}$$

Since $H(f)=1$

$$R(f) = \frac{V_2(f)}{T(f)} \tag{11}$$

The system transmitter and receiver, thus fully characterized by their frequency responses, can be used to analyze materials having an unknown transfer function $[H(f)]$.

Suppose spectrum $T(f)$ was transmitted into the medium under test and $V_2(f)$ was the output of the spectroscopic system (Figure 4). The transfer function of the test medium is found by deconvolution. This process (although possible in the time domain) is most easily carried out in the frequency domain, where

$$H(f) = \frac{V_2(f)}{T(f)R(f)} \tag{12}$$

In the sections of this chapter which follow, the frequency response of each of the system components will be examined. One will notice that the behavior of some elements is well defined, while relatively little has appeared in the literature in terms of characterizing others. Each part of the measurement system will be examined separately. These discussions will be followed by observations on the interaction of components to form subsystems, while the discussion of complete systems will close this chapter of the report.

Transmitter

The transmitter produces an electrical waveform of sufficient amplitude to excite the transmitting transducer. The wave shape (and frequency content) produced by the transmitter in an ultrasonic spectroscopic system is dictated by the type of circuitry employed. The time and frequency domain representations of several ideal and realizable transmitter wave shapes are given in Table 1.

Table 1
Time and Frequency-Domain Representations of Commonly Used Waveforms

Time domain	Frequency domain

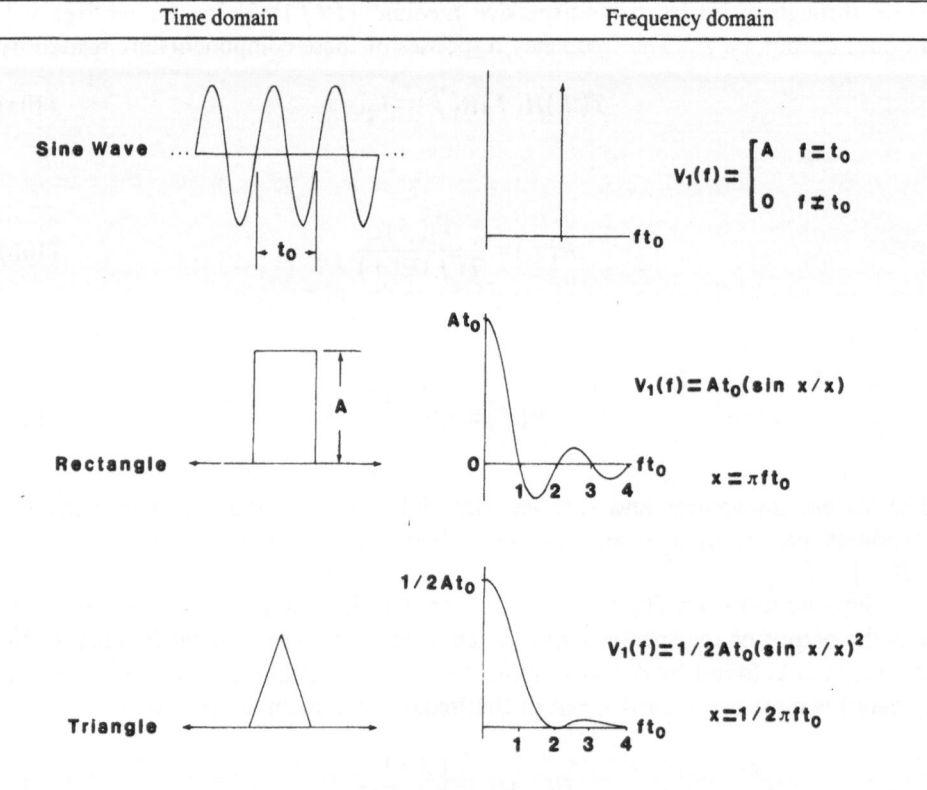

Sine Wave

$$V_1(f) = \begin{bmatrix} A & f = t_0 \\ 0 & f \neq t_0 \end{bmatrix}$$

Rectangle

$$V_1(f) = At_0(\sin x/x)$$

$$x = \pi f t_0$$

Triangle

$$V_1(f) = 1/2\,At_0(\sin x/x)^2$$

$$x = 1/2\pi f t_0$$

Table 1 *(continued)*

Time domain	Frequency domain

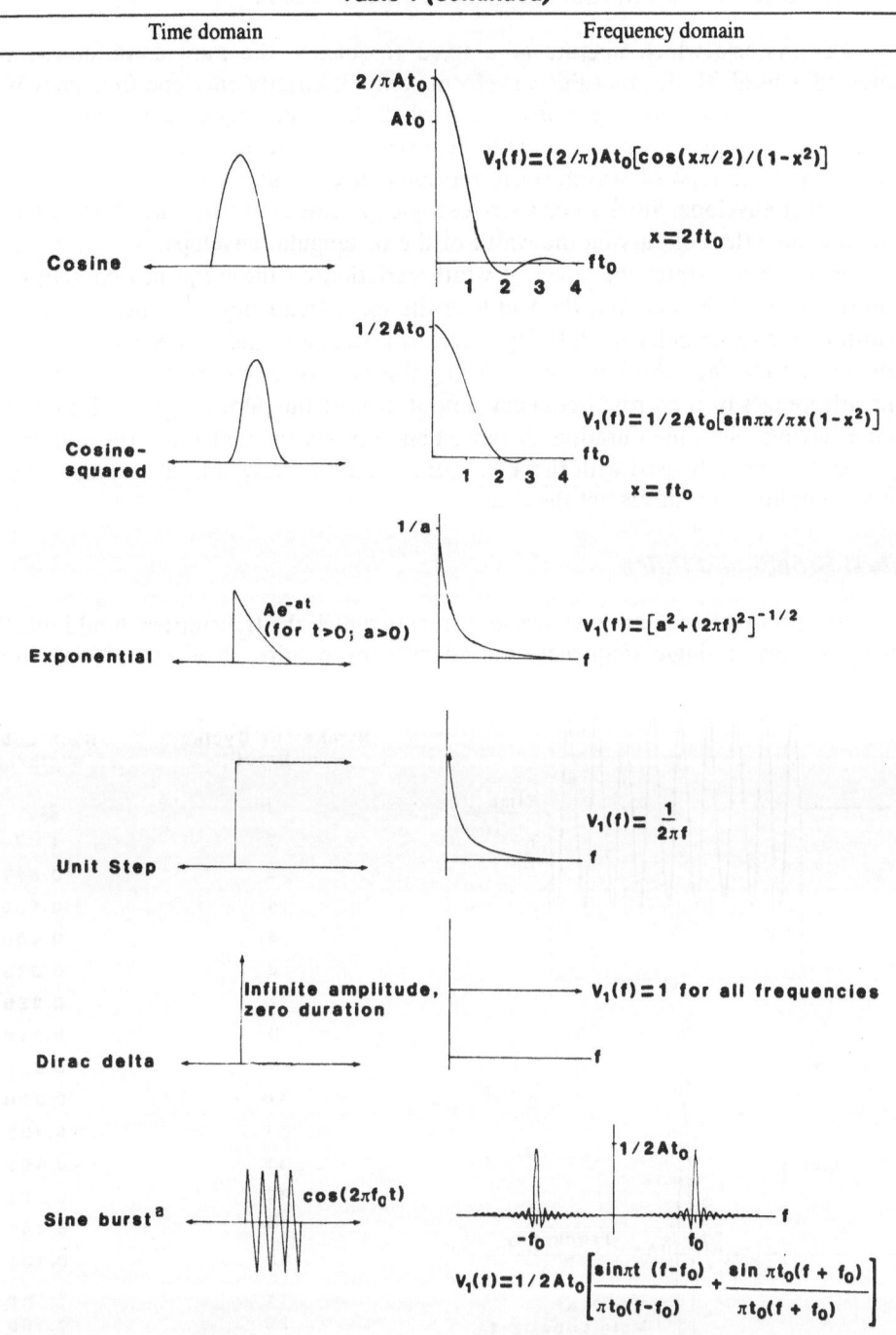

$V_1(f) = (2/\pi)At_0[\cos(x\pi/2)/(1-x^2)]$

$x = 2ft_0$

$V_1(f) = 1/2At_0[\sin\pi x/\pi x(1-x^2)]$

$x = ft_0$

$V_1(f) = [a^2 + (2\pi f)^2]^{-1/2}$

$V_1(f) = \dfrac{1}{2\pi f}$

$V_1(f) = 1$ for all frequencies

$V_1(f) = 1/2At_0\left[\dfrac{\sin\pi t\ (f-f_0)}{\pi t_0(f-f_0)} + \dfrac{\sin\ \pi t_0(f+f_0)}{\pi t_0(f+f_0)}\right]$

[a]The spectrum of a sinusoidal burst is the same as that of a rectangular pulse, but shifted by f_0 (frequency of the sine wave). In practice, the portion of $V_1(f)$ at negative frequencies is accounted for by doubling $V_1(f)$ for positive frequencies.

Continuous-Wave Sinusoid and Sinusoidal Burst

For systems which operate at a fixed frequency, the pure continuous-wave sinusoid is used. Notice that this waveform contains strictly only one frequency if it is infinitely extended in time. For cases in which the sinusoid exists for only a finite length of time, the frequency domain representation includes more than a single frequency. This type of waveform is represented as a sine wave modulated by a rectangular envelope. Since some spectroscopic systems emit sine-wave bursts, let us examine the effect of varying the width of the rectangular envelope.

Figure 5 illustrates the effect of width variation on the frequency content of a sinusoidal burst. Notice that the width of the main frequency lobe increases as the width of the rectangular modulating waveform decreases and is independent of the sine wave frequency. An investigator, using this type of electrical pulse, must weigh the advantages of a narrow frequency output against the poor range resolution of a pulse having long time duration. A pulse length of six to eight times the sinusoidal period is commonly used with the assumption that a narrowband ultrasonic wave is produced; however, this is not the case.

Ideal Broadband Pulse

As opposed to a continuous-wave system in which the transmitter should ideally produce only a single frequency, pulsed systems require an electrical waveform

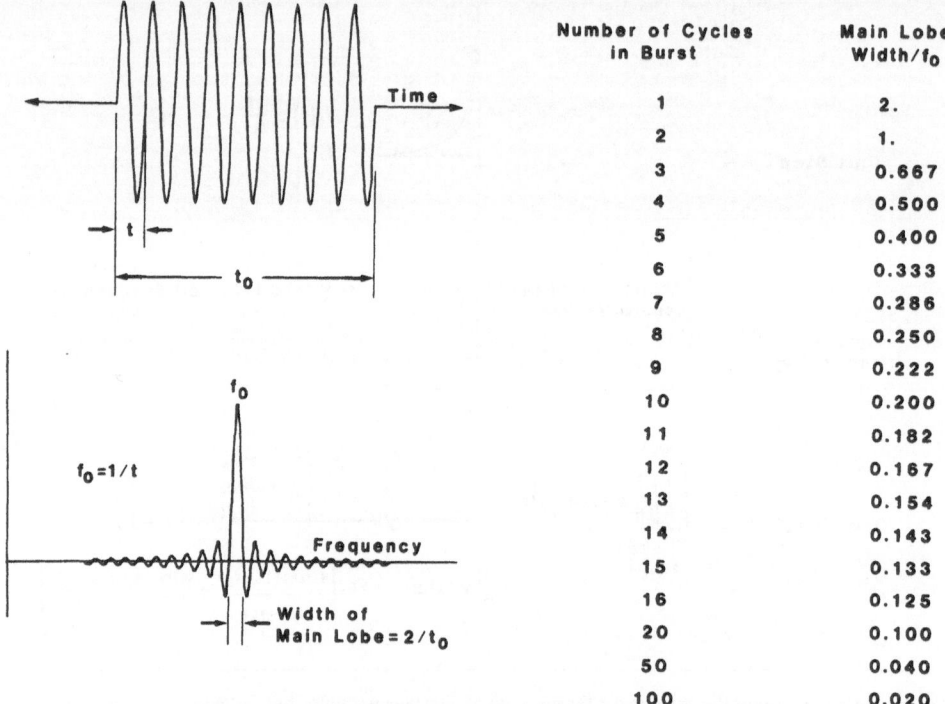

Number of Cycles in Burst	Main Lobe Width/f_0
1	2.
2	1.
3	0.667
4	0.500
5	0.400
6	0.333
7	0.286
8	0.250
9	0.222
10	0.200
11	0.182
12	0.167
13	0.154
14	0.143
15	0.133
16	0.125
20	0.100
50	0.040
100	0.020

Figure 5. Effect of pulse duration of a sinusoid upon the spectrum.

containing a wide range of frequencies. An instantaneous voltage spike of infinite amplitude contains uniform magnitude of all frequencies. Such a waveform (termed the Dirac delta function) certainly cannot be realized. However, pulses having very short duration and large amplitudes can be produced. Reference to Table 1 indicates the spectrum contained in a short rectangular pulse contains a reasonably wide range of frequencies.

Rectangular Pulse

Gericke (006–008, 062) described the use of a pulser which produces a narrow rectangular pulse. Although the amplitude spectrum is far from uniform (refer to Table 1), a spectrum with minima and maxima may in fact be desirable because of nonuniformities in the response of other elements of the transmission subsystem. Gericke (061) found negative rectangular pulses of 50-V amplitude and 0.1-μsec duration to be adequate for his work.

Other Pulse Shapes

Reference to Table 1 indicates that pulses of shapes other than rectangular could be used for the electrical exciting waveform. In fact, a triangular, cosine, or cosine-squared pulse has a bandwidth greater than that of a rectangular pulse having the same duration (t_0); however, the circuitry required to produce them is a bit more complex. A method for producing a triangular driving waveform has been described by Dixon (157). The cosine pulse is used in some medical ultrasonic equipment. A cosine-squared pulse has apparently not been used.

Single Transition

A pulse with a single transition to ground is attractive because of superior range resolution and also the large bandwidth possible. The bandwidth (BW) and transition time (t_r) of the step are related by

$$BW = 1/2t_r \qquad (13)$$

For a transition time of 5 nsec, the pulse would have a bandwidth of 100 MHz.

Engineers at Panametrics (1973-168) investigated using single-step transitions for transducer excitation. They used a commercially available pulser with a maximum amplitude output of ± 50 V, and a transition time of 7 nsec. They found indeed that the pulse yielded an improvement in bandwidth; however, several problems were encountered. Cable effects (discussed in the next section) degraded the transition time and thus affected bandwidth. Also, the voltage output from the pulser was considered to be too low. Furthermore, the ultrasonic pulse shape (in the material under investigation) deteriorated appreciably for propagation distances of only a few millimeters.

In the time since 1973, considerable advances have been made in components for electrical switching. Step transitions of several hundred volts within nanoseconds are now possible (Cheney *et al.*, 155). The problem encountered in degradation of ultrasonic pulse shape appears to be a manifestation of frequency-dependent attenuation (and dispersion) in the sample. The troublesome effects of cable capacitance may be overcome as will be noted in the next section.

Exponential Pulse

An often used scheme for pulsing an ultrasonic transducer is to momentarily connect it to a high-voltage supply, then let it return toward zero. The type of pulse produced is a unit step with an exponential decay. The type of circuitry used to produce this type of waveform is diagrammed in Figure 6. The switch in early equipment was a thyratron. In modern equipment a silicon-controlled rectifier (SCR) or avalanche transistor is used. As will be shown later, the transducer and electrical coupling are a capacitive load on the pulser. Because of this, the current waveform is simply the derivative of the voltage waveform. Transducer excitation occurs only during the transition of the excitation pulse with no appreciable excitation during the exponential decay.

Arbitrary Pulse Shapes

Pulser waveforms of shapes other than sinusoids, impulses, square waves, or step transitions have not been investigated. A method for producing waveforms of arbitrary shape has been reported by Hill (240, 255). Two transmitters based on Hill's method are shown in Figure 7.

In the step approximation circuit (Figure 7a) a digital memory contains binary-coded samples of the waveform to be produced. The clock and binary counter address successive words stored in memory. Output of the digital-to-analog converter is the voltage waveform stored digitally in memory. When the output voltages must vary over a wide range, the resolution of the D/A converter and number of bits in the digital word should be increased to minimize departures (between successive samples) from the waveform desired.

An alternate to increasing the digital word length is to utilize the piecewise approximation circuit shown in Figure 7b. The operational amplifier at the output of

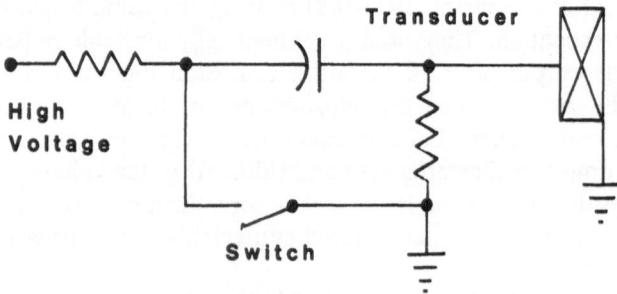

Figure 6. Example of circuit used to produce a unit-step pulse with an exponential decay.

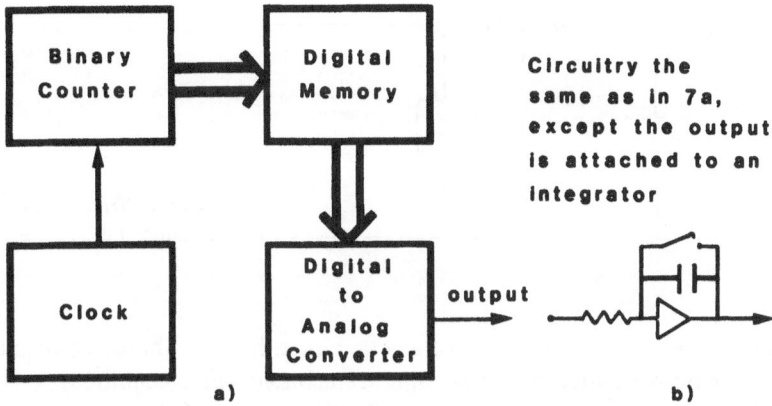

Figure 7. Digital generation of a pulse of arbitrary shape. (a) Step approximation circuit; (b) piecewise linear approximation circuit (after Hill, 257).

the D/A is configured as an integrator. Words stored in the digital memory of this system represent the difference in voltage between successive samples of the desired waveform.

In either of the two systems described, the output would be connected through a low-pass filter to an amplifier, the amplifier being required to raise the output to several hundred volts. Although the equipment used by Hill restricted its use to low frequencies, other clocks, counters, memories, and D/A converters are available for use at ultrasonic frequencies.

A more complex technique for producing pulses of selected shape will be introduced later in this chapter. The components required in this technique are similar to those in the upgrading of Hill's method to higher frequencies. The properties of these components will be discussed at that time.

Electrical Coupling Network, I

The electrical waveform produced by the transmitter must be coupled from the transmitter to the transducer. Whether this coupling is as simple as a coaxial cable or as complex as a compensating filter, it will have a decided effect on the ultrasonic spectrum. An important component, the electrical coupling, seems to have been often overlooked by many when designing a spectroscopic system. It certainly deserves some attention.

Coaxial Cable

The most common electrical coupling network is a coaxial cable, several meters in length. An equivalent circuit of a coaxial cable is shown in Figure 8. The parameters R (resistance/meter), L (inductance/m), G (conductivity/m), and C

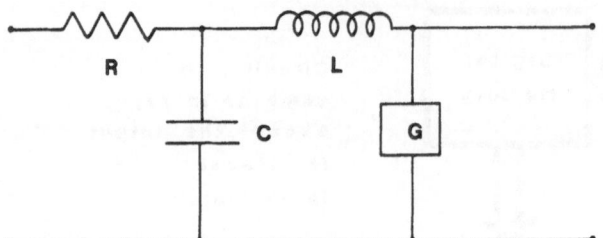

Figure 8. Transmission-line representation of coaxial cable.

(capacitance/m) are not lumped parameters, but rather distributed constants. Analysis of such a network must be made using transmission-line techniques.

The transfer function of the transmission line in Figure 8 is

$$B_1(f) = e^{-\gamma l} \tag{14}$$

where γ (called the propagation constant) is defined as

$$\gamma = [(R + j2\pi f L)(G + j2\pi f C)]^{1/2} \tag{15}$$

and l is the length of the cable. The propagation constant is often written in terms of its real and imaginary parts, as

$$\gamma = \alpha + j\beta \tag{16}$$

where for radio frequencies at which

$$R \ll j2\pi f L \quad \text{and} \quad G \ll j2\pi f C \tag{17}$$

$$\alpha \cong \tfrac{1}{2}[R(C/L)^{1/2} + G(L/C)^{1/2}] \tag{18}$$

and

$$\beta \cong 2\pi f(LC)^{1/2} \tag{19}$$

The constants R, L, G, and C of a coaxial cable depend on its physical characteristics (size of conductors, size of insulator) and on its electrical characteristics (dielectric constant of the insulator). The loss (energy dissipation in the dielectric) in most common coaxial cable is proportional to frequency. Small-diameter cable can significantly degrade pulse rise time because of the large losses at high frequencies.

As was noted previously, coaxial cable introduces a delay as the electrical signal passes along the cable. The velocity of propagation is dependent on cable properties and may be determined from the relation

$$\text{velocity} = 2\pi/\beta = 1/[f(LC)^{1/2}]$$

An important consideration in using coaxial cable (in addition to realizing its capability for limited bandwidth transmission) is the importance of properly matching impedances of both ends of the cable. That is, the impedances of the pulser, cable, and ultrasonic transducer must be the same, or alternatively impedance-matching networks should be employed between components. The penalty for improper impedance matching is a reflection of part of the incident energy. Energy reflected from the end of the cable can combine with the original waveform to change its shape (and spectrum). For electrical pulses of duration longer than the two-way transit time along the connecting cable, reflections can cause problems. As an example, the two-way transmission time of a pulse along a 3-ft length of RG-58A/U cable is 9.2 nsec. For pulse durations of greater than 9.2 nsec and partial reflection from the end of the cable, the original pulse and reflection will interfere. The interference can be constructive or destructive depending on the cable length.

The vector ratio of the reflected voltage to the incident voltage at the end of a mismatched line is given by

$$\text{reflection coefficient} = \frac{Z_L - Z_0}{Z_L + Z_0} \tag{20}$$

where Z_L is the impedance of the load and Z_0 is the characteristic impedance of the cable. Z_0 in terms of the cable's electrical parameters is

$$Z_0 = \left(\frac{R + j2\pi f L}{G + j2\pi f C} \right)^{1/2} \tag{21}$$

For the conditions given in (17), this simplifies to

$$Z_0 = (L/C)^{1/2} \tag{22}$$

approximately a pure resistance.

Since the impedance of most ultrasonic transducers is largely capacitive reactance, a broadband impedance-matching network should be utilized. The alternative to incorporating matching networks at the ends of the cable is to simply eliminate the cable altogether. The pulse would be mounted directly on the transducer. Myers *et al.* (239) constructed such a pulser, and it did indeed produce an ultrasonic pulse with wide bandwidth.

One may choose to experimentally determine the cable transfer function. One end of the cable is connected to a variable-frequency sine wave generator, the other end being terminated by a resistance equal to the characteristic impedance of the cable. The peak-to-peak voltages at the generator (V_i), and termination (V_0) as well as the time shift (t_s) between the two signals are recorded at each frequency (f). The magnitude and phase angle of the transfer function are given by

$$|B_1(f)| = V_0/V_i \tag{23}$$

$$\angle B_1(f) = (t_s/T)2\pi \text{ (radians)} \tag{24}$$

or

$$\angle B_1(f) = (t_s/T)(360°) \quad \text{(degrees)}. \tag{25}$$

where T is the period of the ultrasonic wave (f^{-1}).

Response Equalization Networks

Rather than use a simple coaxial cable to connect the transmitter and trans-ducer, a more complex network may be employed. The purpose of this component is to alter the spectrum of the energy from the transmitter in such a way that the ultrasonic output from the transmitting transducer is reasonably uniform over a wide range of frequencies. These networks are discussed later in this chapter after the frequency response of the transducer has been introduced.

Transmitting Transducer

The transmitting transducer converts electrical energy, supplied by the pulser, into mechanical vibratory energy—ultrasound. A wide variety of transducer types exist; however, the discussion which follows will center on piezoelectric (and ferroelectric) devices. Other types of electromechanical transducers such as elec-tromagnetic devices cannot provide the high frequencies or wide bandwidths re-quired for ultrasonic spectroscopy. Electrostatic and acoustoelectric transducers have properties which make them attractive for use in spectroscopic systems, and will be mentioned later.

Frequency Response

One possible approach to the problem of determining a transducer's transfer characteristics is to begin from first principles and consider the mathematics relating the electrical and elastic properties of the piezoelectric materials. A number of authors have taken this approach.

Consider a piezoelectric disk of thickness l and with surface area of the circular face equal to S. The disk, much larger in diameter than thickness, is assumed to be made of a material having no electrical conductivity and which produces no ultrasonic attenuation. The transducer is operated in a thickness expander mode to produce longitudinal ultrasonic waves propagating in the $-x$ and $+x$ directions.

The acoustic impedance of the transducer is

$$Z_1 = \rho_1 c_1$$

where ρ_1 is the density and c_1 is the ultrasonic velocity. Subscripts 2 and 3 are assigned to the material properties, respectively, behind ($-x$) and in front ($+x$) of the transducer. Additionally, the two bounding media are considered to be infinitely extended. Because only longitudinal waves are considered, the problem becomes one

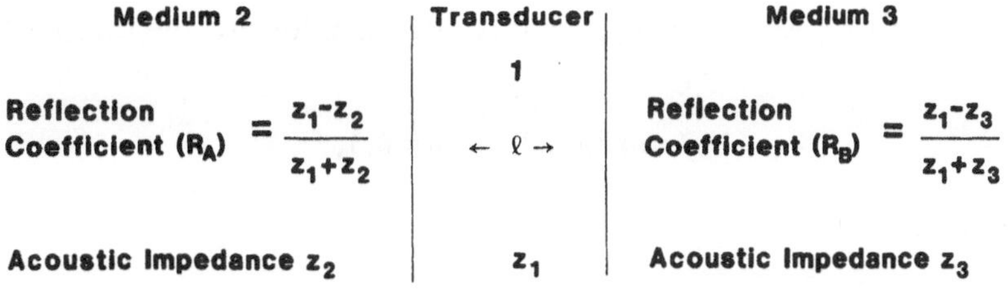

Medium 2	Transducer	Medium 3
	1	
Reflection Coefficient (R_A) $= \dfrac{z_1-z_2}{z_1+z_2}$	$\leftarrow \ell \rightarrow$	Reflection Coefficient (R_B) $= \dfrac{z_1-z_3}{z_1+z_3}$
Acoustic Impedance z_2	z_1	Acoustic Impedance z_3

Figure 9. Transducer bounded by two media. The reflection coefficients at each boundary are shown.

dimensional. The amplitude reflection coefficients (from transducer to bounding media) are given by the formulas in Figure 9.

The derivation proceeds by considering the equation describing the direct piezoelectric effect:

$$T = C\frac{\partial u}{\partial x} - hD \tag{26}$$

where T is the stress on a face perpendicular to the x axis, u is the particle displacement in the x direction, D is the electric displacement, C is the elastic stiffness at constant electric displacement in the x direction, and h is the piezoelectric coefficient at constant strain. Once an initial (Dirac delta) excitation, $D\delta(t)$, occurs, the disk's behavior is determined solely by the propagation characteristics of the ultrasound in the transducer and the loading at its faces. The complete analysis of this situation (Lakestani, Baboux, and Fleischmann, 223) leads to the following transfer function (in Laplace notation) for ultrasonic waves radiating into medium 2:

$$X_1(s) = \frac{-h(1+R_A)(1-e^{-t's})(1-R_Be^{-t's})}{2Z_1(1-R_AR_Be^{-2t's})} = \frac{P(s)}{V_1(s)} \tag{27}$$

where t' is the transit time of an ultrasonic pulse through the transducer (i.e., l/c_1), $P(s)$ and $V_1(s)$ are the Laplace transforms of the pressure (force/area) and driving voltage, respectively, and s is the Laplace operator.

Let us examine the frequency-dependent part of $X_1(f)$. The frequency response may be written

$$X_1(f) = K_1 \frac{(1-e^{-j2\pi ft'})(1-R_Be^{-j2\pi ft'})}{1-R_AR_Be^{-j4\pi ft'}} \tag{28}$$

The magnitude of $X_1(f)$ is

$$|X_1(f)| = K_1 \frac{2|\sin(2\pi ft'/2)|(1+R_B^2-2R_B\cos 2\pi ft')^{1/2}}{(1+R_A^2R_B^2-2R_AR_B\cos 4\pi ft')^{1/2}} \tag{29}$$

Notice that $|X_1(f)|$ is periodic (period $=2\pi/t'$). Additionally, the transfer function goes to zero for

$$2\pi f = n(2\pi/t'), \qquad n=0,1,2,\dots \tag{30}$$

that is, at frequencies

$$f = n/t' = nc_1/l \tag{31}$$

In addition to these minima, $X_1(f)$ has maxima, where

$$2\pi f = (2n+1)\pi/t' \tag{32}$$

$$f = (2n+1)/2t' \tag{33}$$

Thus the transducer demonstrates a frequency response with variations. The amplitude of the modulation is determined by the reflection coefficients at the transducer boundaries.

Figure 10 contains plots of transducer response [as calculated using Eq. (28)] for several boundary conditions. Notice that the response broadens for appropriate backing.

For the case when the reflection coefficients R_A and R_B are the same ($=R$), the expression for $X_1(f)$ simplifies considerably:

$$X_1(f) = \frac{1-e^{-j2\pi ft'}}{1+Re^{-j2\pi ft'}} \tag{34}$$

and

$$|X_1(f)| = \frac{2|\sin(2\pi ft'/2)|}{(1+R^2+2R\cos 2\pi ft')^{1/2}} \tag{35}$$

A plot of the magnitude of the transducer response for $R_A=R_B=0.9$ is given in Figure 10.

When $R_B=0$ (matched backing, i.e., $Z_1=Z_2$), the frequency response has its widest bandwidth. $X_1(f)$ becomes

$$X_1(f) = 1-e^{-j2\pi ft'} \tag{36}$$

and

$$|X_1(f)| = 2|\sin(2\pi ft'/2)| \tag{37}$$

Theoretical and experimental investigations have been performed to assess the effects of backing and loading on the frequency response of a piezoelectric transducer. Information concerning this subject may be found in the publications of Kossoff (282), Sittig (284), and Simanski, Pouliquen, and Defebvre (191).

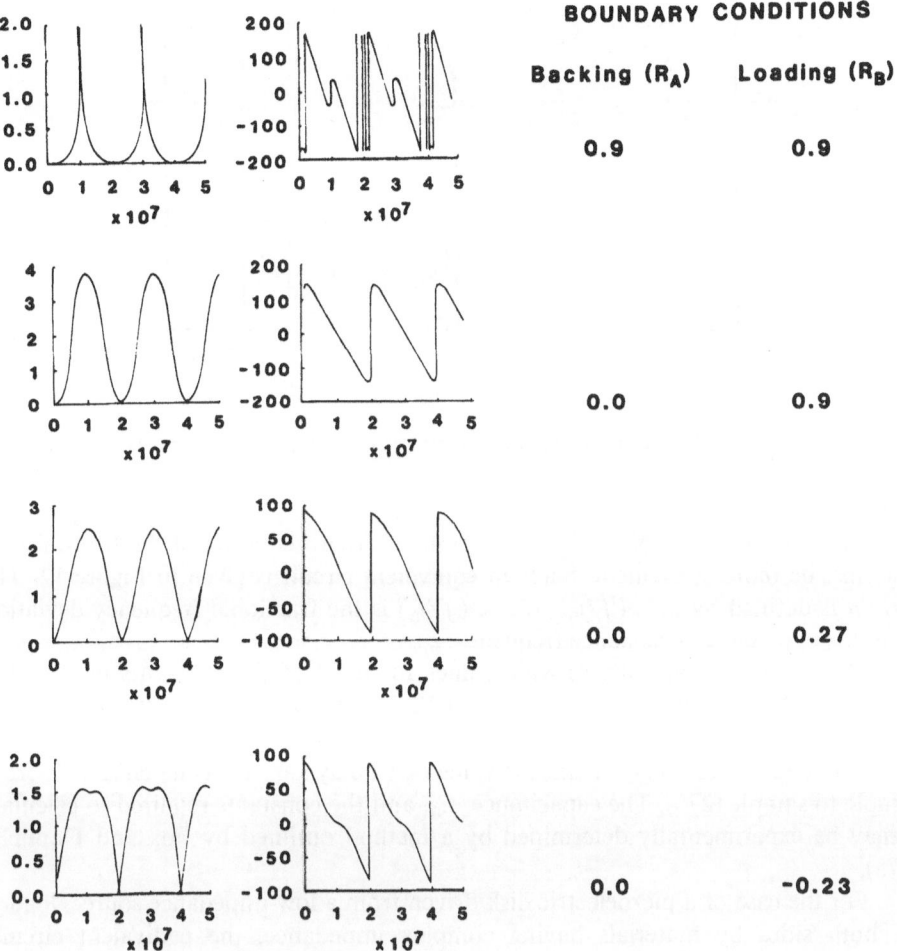

Figure 10. Frequency response of a transducer disk for a variety of backing and loading materials.

Equivalent-Circuit Representation

Because the transducer is driven by an electrical generator, it is often conveni-ent to represent the transducer as an electrical network (Mason, 273). The derivation of the equivalent circuit begins with the differential equation [Eq. (26)] which links the electrical and mechanical properties of piezoelectric materials. This relation, combined with an expression denoting oscillatory movements of particles in the medium (wave propagation) leads to a differential equation, part of which is identical to that describing propagation in a transmission line. Careful consideration of terms in the equation leads one to associate them with the voltage applied to the transducer, $v_1(t)$, and the resultant forces produced at the boundaries of a piezoelec-tric disk (F_f is the front face and F_b is the back face of the disk). Figure 11 presents the equivalent-circuit representation of a piezoelectric transducer. C_0 is the clamped capacitance, and the characteristic mechanical impedance of the piezoelectric disk is denoted Z_1.

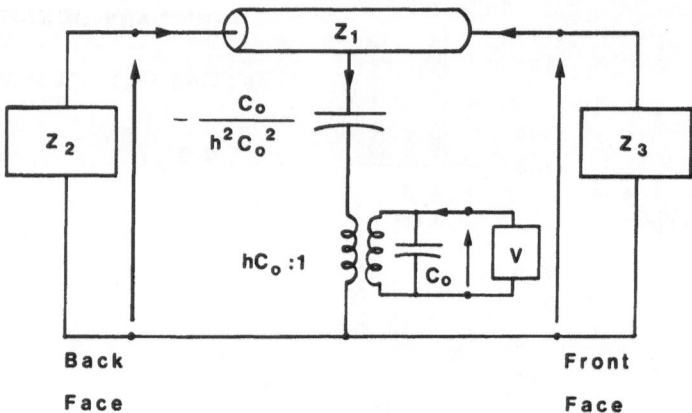

Figure 11. Equivalent-circuit of a piezoelectric transducer (after Mason, 273).

In some cases, use of the lumped parameter representation of the transmission line may be more convenient. Such an equivalent circuit is given in Figure 12. The term α is defined by $\alpha = \pi(f/f_0)$. where (f/f_0) is the fractional frequency deviation from the open-circuit resonance frequency, f_0.

Manufacturers typically provide values for most of the constants used in the equivalent circuits given in Figures 11 and 12. The terms in the transformer ratio, h and C_0, are usually unknown. The piezoelectric constant, h, or parameters necessary to evaluate it, is seldom given and the value of C_0 may vary by as much as 60% from sample to sample (275). The capacitance, C_0, and the constants required to calculate h may be experimentally determined by a method outlined by Fox and Donnelly (275).

For the case of a piezoelectric disk driven from a low-impedance source, loaded on both sides by materials having complex impedance, the equivalent circuits proposed by Kossoff (282) may be used. Near the resonant frequency of the disk or when $Z_3 = Z_2$, the impedance of the branch containing the tangent terms (Figure 12) approaches infinity. In such a case the circuit is considerably simplified (Figure 13).

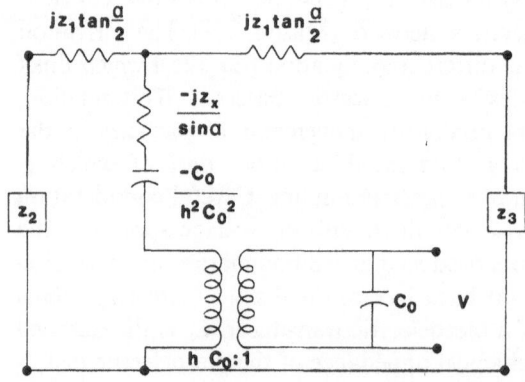

Figure 12. Equivalent circuit of a piezoelectric transducer with a lumped parameter representation of the transmission line (after Kossoff, 282).

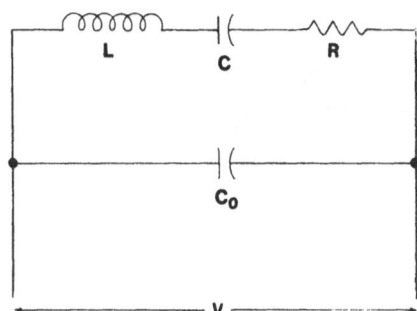

Figure 13. Equivalent circuit of a piezoelectric transducer, valid for frequencies near resonance.

The equivalent circuits shown in Figures 11–13 are exact assuming that (1) there are no losses in the dielectric (piezoelectric disk), (2) attenuation of the ultrasonic pulses in the transducer is negligible, (3) the bounding media are infinitely extended, (4) the lateral extent of the disk is much larger than the ultrasonic wavelength, and (5) there is no coupling to other vibratory modes. If it appears that all of interest has been assumed away, we would like to allay these fears by noting that although the exact calculation of frequency response from fundamental constants using the models above is not strictly exact, their derivation is valuable. As with the mathematical models of transducers which were described earlier, the transfer characteristics of the equivalent circuits may be found and parametric investigations performed.

Although thickness mode transducers have been stressed in this review, this should not be taken to indicate that significant work has not been done to characterize other types of transducers. One is referred to the work of Mason (273) and Berlincourt *et al.* (274) for equivalent circuits of transducers operated in any of several modes to excite longitudinal or transverse waves.

Derivation of Frequency Response from the Equivalent Circuit

Kazys and Lukosevicius (355) analyzed an equivalent circuit of a piezoelectric transducer and found that the transfer function could be represented as the product of two functions:

$$X_1(f) = X_1'(f) X_1''(f) \tag{38}$$

The first term $[X_1'(f)]$ describes the transfer characteristics of a transducer with a small "electromechanical coupling factor," driven by a voltage source. This function is identical to that given in Eq. (27). The second term $[X_1''(f)]$ accounts for the influence of the negative capacitance $(-C_0)$ and the source impedance of the driving generator (Z_{V_1}):

$$X_1''(f) = \frac{1}{j2\pi f C_0 [Z_{V_1}(f) + Z_{IN}(f)]} \tag{39}$$

where $Z_{IN}(f)$ is the electrical impedance at the input of the piezoelectric transducer, defined by

$$Z_{IN}(f)=\frac{1}{j2\pi fC_0}$$

$$\times\left[1-\frac{1}{j2\pi fC_0Z_1}\frac{(K_1+K_2+2)-4e^{-j2\pi ft'}-(K_1+K_2-2)e^{-j4\pi ft'}}{(1+K_1+K_2+K_1K_2)-(1-K_1-K_2+K_1K_2)e^{-j\pi ft'}}\right]$$

(40)

where $K_1=Z_3/Z_1$, $K_2=Z_2/Z_1$, and $t'=l/c_1$.

As the source impedance of the driving generator (Z_{V_1}) decreases, the distortions in the frequency response caused by $X_1''(f)$ decrease. For the case of a current-driving source, the negative capacitance in the equivalent circuit has no influence on the frequency response, except for a constant term. The response may be expressed as

$$X_1(f)=\frac{-h(1+R_A)S}{Z_1(Z_3+Z_1)}\frac{(1-e^{-j2\pi ft'})(1-R_Be^{-j2\pi ft'})}{1-R_AR_Be^{-j4\pi ft'}}$$

(41)

Mechanical Pulse Propagation

Consider a voltage impulse applied to the circuit in Figure 11. The electrical pulse is transformed into two mechanical pulses originating at the faces of the transducer disk and which travel along the transmission line. Reflections occur at the boundaries of the line, the magnitude depending on the acoustic impedance mismatch. A wave train (Figure 14) of reverberations is produced, with a portion of the ultrasonic energy transferred to the bounding media each time the wave encounters the interface.

It is the interference of the waves originating from the back and front (and their reflections within the transducer) which create the wide variation of response versus frequency for piezoelectric transmitters. By eliminating reflections at the backing,

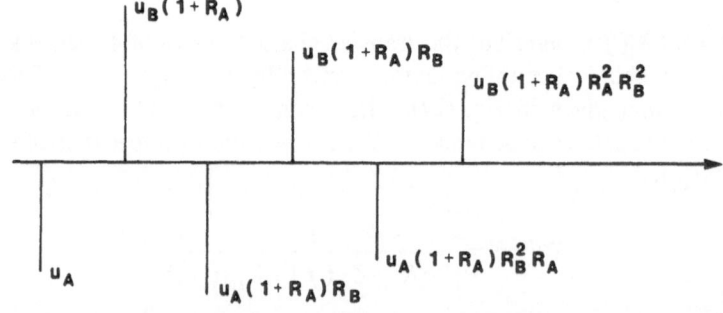

Figure 14. Theoretical response of a transmitting transducer to an electrical impulse.

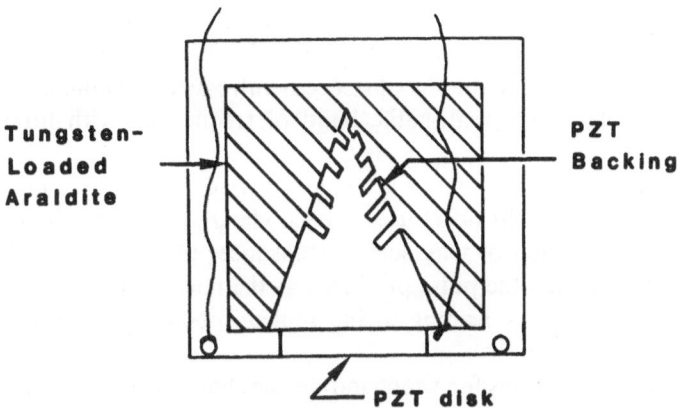

Tungsten-Loaded Araldite

PZT Backing

PZT disk

Figure 15. Wideband transducer construction (after Brown and Weight, 192).

some of the interferences, which produce peaked response at resonance, are eliminated; this occurs, however, at the expense of decreased coupling of energy into ultrasound at that frequency. Notice that in order to reduce reflections to zero, the backing must not only be acoustically matched, but should also be dissipative such that no ultrasonic waves are returned to the boundary.

A practical approach to removal of the interfering back pulse is to back the transducer with a piece of material physically equivalent to the transducer. For piezoelectric ceramics, this would be an unpoled chunk of the ceramic shaped as shown in Figure 15.

With matched backing alone, all sources of interfering waves have not been eliminated. The mechanical pulse originating at the back surface of the disk and traveling toward the front will interfere with the wave initially emitted from the front surface. The effect of these two pulses on the spectrum of the emitted ultrasound may be found by taking the Fourier transform of two impulses of opposite polarity which are spaced in time by t'. The amplitude spectrum varies sinusoidally with frequency. The minima in the spectrum occur at intervals of

$$f_{min} = 1/t' \qquad (42)$$

If the mechanical pulses arising from the front and back boundaries of the transducer are of equal amplitude, then the modulation of the spectrum will be complete (i.e., the response at minima will reach zero). For front and back pulses of unequal amplitude the modulation will be less than 100% (the response at minima is not zero). A number of techniques for decreasing or eliminating the back surface pulse have been proposed.

Broadband Transmitting Transducers

Perhaps the simplest approach to alleviating the interfering pulse from the backing boundary is to make the thickness of the disk larger. The ultrasonic waves returning from the back surface would be separated a considerable time period from the waves emitted by the front surface. Alternatively, the thickness of the transducer could be made so small that the maxima occur infrequently enough that measurements may be made over a reasonable band of frequencies with little change in response.

A design intended to decrease backwall effects proposed by Lloyd and mentioned by Brown and Weight (192) uses a back electrode smaller than that on the front face. The explanation of why such a transducer performs so well is lacking; however, it is thought that the pulse produced at the back boundary is smaller than that produced at the front—leading to incomplete modulation and hence a wideband response.

Several other possibilities for wideband devices have been conceived. One early device was constructed of several piezoelectric elements of differing thickness, cemented together (Sittig, 285; Lloyd, 264). Each of the disks, which are pulsed sequentially, produces an ultrasonic spectrum modulated at different frequencies and having different resonant frequencies. Figure 16 illustrates the concept. The spectra produced by each disk are made to overlap. If the responses are added the overall response is reasonably uniform. Although the spectrum emitted by any one disk is limited, the spectrum produced by sequentially pulsing the composite covers a wide range of frequencies. Lloyd notes that the construction of composite transducers is complicated by the close tolerances on the compliance and thickness of the Araldite bonding the disks together.

Another transducer configuration has been examined by Lloyd (269). It is shown in Figure 17. Slots are machined on the active face of the transducer. Then the peaks and valleys are electroded. On transmission the acoustic pulse arising from the back (valley) electrodes travels, not toward the front, but backward through the matching backing.

Mitchell (272) has explored an interesting approach to producing ultrasound over a wide band of frequencies. He considers a material in which the piezoelectric properties vary through the thickness of a plate transducer. The purpose of this gradient is, again, to eliminate the pulse arising from the back face of the transducer. It can be shown that if the distribution of piezoelectric stress, $hD(x)$, is known, then

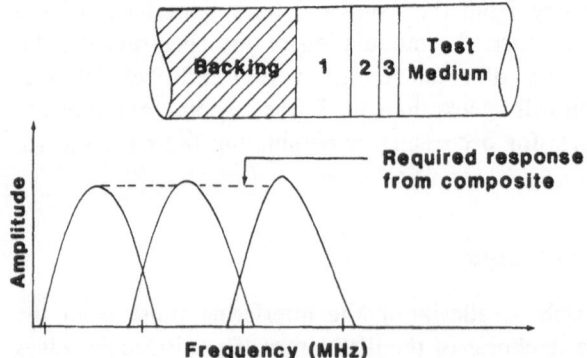

Figure 16. Structure of a composite transducer and its response characteristics (after Lloyd, 264).

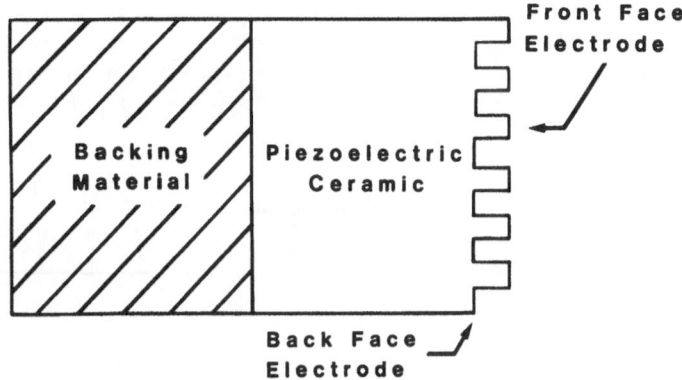

Figure 17. Wideband transducer proposed by Lloyd (269).

the stress wave output may be expressed as

$$T(x,t) = -j\frac{k}{2}\left[\int hD(x)e^{-j2\pi fx}\,dx\right]e^{j(2\pi ft-kx)} \tag{43}$$

where $k=2\pi/\lambda$. Assuming there is an acoustic mismatch at both boundaries of the transducer, the amplitude spectrum of the stress wave is seen to be modulated by the Fourier transform of the gradient of the distribution $hD(x)$:

$$\int \frac{\partial(hD)}{\partial x}e^{-j2\pi fx}\,dx = jk\mathcal{F}(hD(x)) \tag{44}$$

To illustrate the effect of various gradients of piezoelectric properties on the spectrum, Figure 18 has been plotted.

Figure 18a demonstrates that a conventional transducer with piezoelectric stress (hD) constant throughout its length has a gradient represented by two delta functions of the opposite polarity. Such a gradient has the familiar sinusoidally modulated spectrum. However, if the piezoelectric material is made to have the distribution of Figure 18b, the bandwidth is considerably broadened. A similar situation exists (Mitchell, 272) for a transducer having a concave back face (Figure 18c). The shape has the effect of spreading out the piezoelectric stress and eliminating any abrupt changes. Again the effect is to broaden the spectrum of emitted ultrasound.

The technique of controlling transducer frequency response by changing the hD distribution seems promising; however, creation of the proper gradient presents a problem. Several techniques have been proposed to create the required distributions:

1. Controlled heating of one face of a piezoelectric ceramic to provide a variation of depolarization with depth.
2. Variation of h by control of the polarizing current during ceramic poling.
3. Gradation of h by vacuum deposition of ceramic material.
4. Variation of D by diffusing conductivity centers into the transducers.

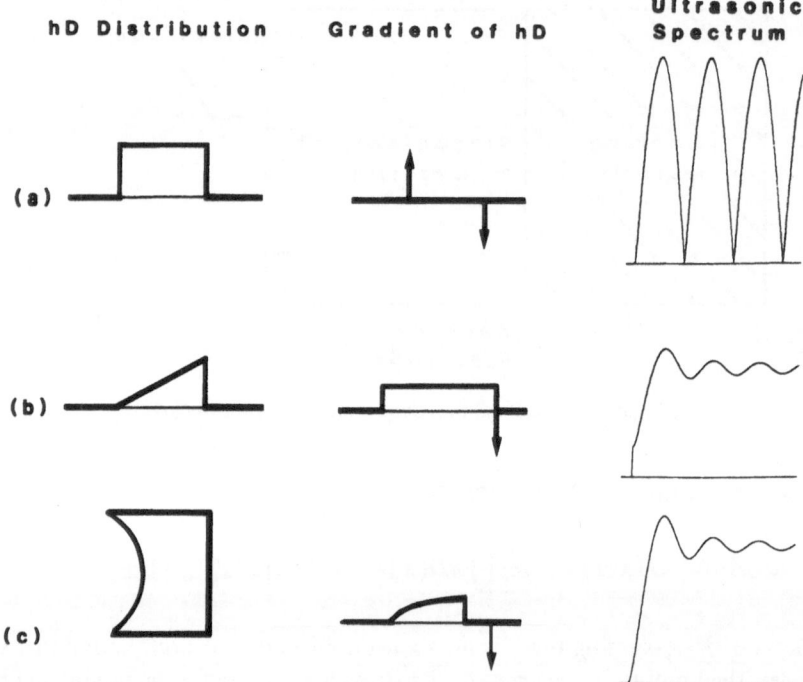

Figure 18. Effect of gradients in *hD* on the spectrum of emitted ultrasound. (a) Gradient in conventional transducer, (b) gradient in wideband device, (c) gradient in transducer with curved back surface.

5. Shaping the rear electrode as in Figure 18c.
6. Possibly utilizing a smaller back electrode to take advantage of the fringing electric field (because the field lines are not parallel to the transducer thickness, *h* varies with direction and *D* is changed).

Apparently, the difficulties in producing transducers of this type are enough to have limited their use, since little has appeared in the literature since the early 1970s.

Yet another wideband transducer of unusual design was conceived and built by van der Pauw (261). Figure 19 illustrates the design of his interdigital transducer. Normally, such a device is used to generate surface waves; however, significant differences exist so as to make this unit capable of producing bulk waves. Alternate electrodes are given opposite charges during poling. Subsequent electrical pulsing of the same electrode pairs produces an excitation of the transducer's thickness mode of vibration, with no interfering pulses from the back wall. The mechanism of operation is thought to be due to an exponential variation of *h* with depth, or in Lloyd's analysis, a transducer folded back on itself so that pulses from the front and "rear" are emitted in the same direction. The device also appears to be capable of transduction to produce both bulk and surface waves. Morgan (085) demonstrated its use to excite broadband (0.5–10 MHz) surface waves.

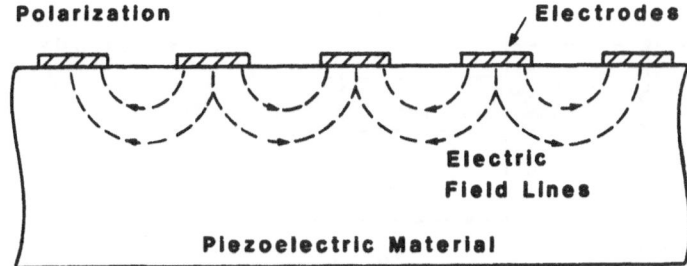

Figure 19. Interdigital transducer for excitation of broadband longitudinal and surface wave pulses (after vander Pauw, 261).

Kazhis and Lukoshevichyus (279) describe a transducer capable of generating longitudinal ultrasonic pulses containing a very broad band of frequencies. The construction of this device (Figure 20) again exploits the warping of the electric field in a nonuniform manner to reduce (and almost eliminate) the rear wall pulse.

Electrostatic Transducers

In addition to piezoelectric transmitting transducers, electrostatic devices should also be considered. Summaries of important properties of these devices are given in Mason (273), Mercier (82), Legros and Lewiner (288), and Cantrell and Breazeale (353). Although the efficiency of conversion of electrical to mechanical energy in electrostatic drivers is substantially lower than that for piezoelectric devices, the wide bandwidth characteristics make them attractive.

Consider a capacitive driver constructed as shown in Figure 21. Layer 2 is a dielectric film with permittivity of $\varepsilon = \varepsilon_r \varepsilon_0$. This material is assumed to exhibit no piezoelectric or electrostrictive properties. Layers 1 and 3 between the electrodes and dielectric are thin air spaces trapped as a result of irregularities on the surfaces of the electrodes and the film.

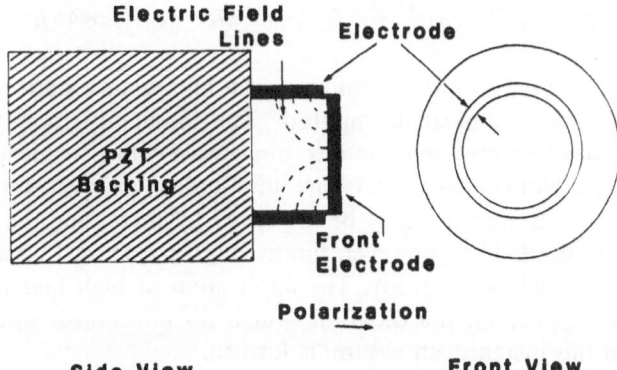

Figure 20. Construction of a wideband transducer, with the "rear" electrode wrapped around the side (after Kazhis and Lukoshevichyus, 279).

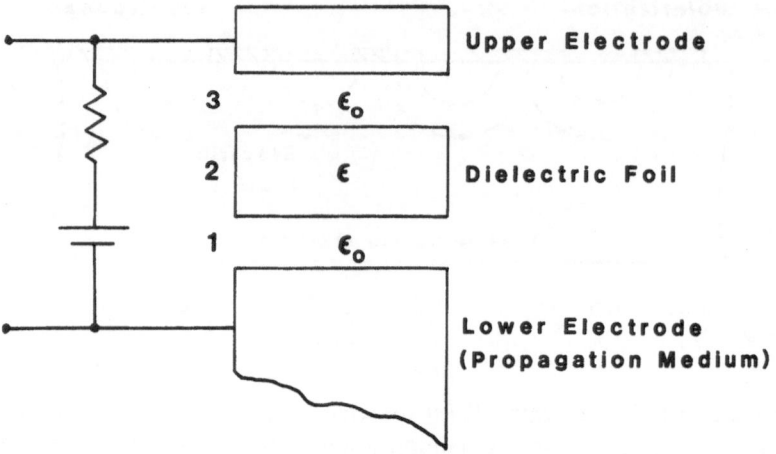

Figure 21. An electrostatic transducer.

The electrostatic transducer may be operated in several modes. If no electrical bias is supplied to the device, and a radio frequency (rf) voltage is applied to the electrodes, an electrical field $\varepsilon_1 = e_1 \sin 2\pi ft$ is produced near the lower electrode. The electrostatic pressure produced on this surface is

$$p = \tfrac{1}{4}\varepsilon_0 e_1^2 (1 - \cos 4\pi ft) \tag{45}$$

Thus an ultrasonic wave with frequency twice that of the driving voltage is generated in the material. Because ε_0 is small the amplitude of the ultrasonic waves will be small, even for large voltages.

A capacitive driver may also be operated with a high-voltage dc bias. The field the bias produces, E_1, is superimposed on the field ε_1, generated by the applied rf potential. For this mode of operation, the pressure on the lower electrode becomes

$$p = \tfrac{1}{2}\varepsilon_0 \left(E_1^2 + \tfrac{1}{2}e_1^2 \right) + \varepsilon_0 E_1 e_1 \sin 2\pi ft - \tfrac{1}{4}\varepsilon_0 e_1^2 \cos 4\pi ft \tag{46}$$

The first term in Eq. (46) is unimportant to us as it represents a static pressure, while the third term gives a wave of small amplitude. The dominant term in the expression is the second term, which represents an ultrasonic wave with a frequency the same as the driving voltage. Notice that the pressure amplitude is independent of frequency. Efficiency of such a transmitter may be improved by increasing the bias field E_1. The magnitude of the field is, however, limited by the electrical breakdown of the dielectric and its mechanical rigidity. The application of high bias fields also can cause charges to appear on the dielectric which do not vanish when the bias is disconnected. In this instance, an electret is formed.

The biasing voltage on an electret may be reversed after electricification of dielectric. In such an instance the electrical field produced at the lower electrode (E_1) is the sum of the field produced by the "permanent" charges on the dielectric and the field produced by the bias. Because E_1 is large the transducer efficiency is high. In addition, since the fields due to the bias and electret are opposed, the field in the dielectric is the difference of the two fields and chances of electrical breakdown are decreased.

Unfortunately, charges placed on a dielectric making it an electret are not permanent, and as the transducer ages efficiency decreases. Nevertheless, electrets have been used at ultrasonic frequencies (289).

Electromagnetic Acoustic Transducers

An electromagnetic acoustic transducer (EMAT) consists of a conducting coil and a magnet such that, when placed near the surface of a metal and driven by a current at the desired frequency, ultrasonic waves are produced. Transduction occurs through the interaction of the induced eddy currents and the magnetic field. Reciprocally, the same transducer produces a voltage in response to ultrasonic waves which set the metal surface in motion.

Unique features of EMATs include contactless (and thus reproducible) operation, ability to excite selected wave modes (longitudinal, shear, and surface), and the capability of scanning the angle of the longitudinal and shear waves simply by changing the frequency of the rf driving signal. However, the transduction efficiency is lower than that of piezoelectric devices. Additionally, a tradeoff must be made between bandwidth and efficiency.

The following authors describe the particulars of EMATs and their associated electronics: Maxfield (369, 395, 438), Moran (381, 439), Thomas (396), Thompson and Fortunko (406, 440), and Szabo (437).

Transmitting Subsystem

Foster and Hunt (230) made an experimental comparison of the performance of several of the wideband transmitting transducers described previously in this report. Figure 22 demonstrates their apparatus. The pulser (an Aerotech ultrasonic transducer analyzer, UTA) produced a 185-V step excitation having a rise time of less than 25 nsec. A wideband receiving transducer was used for measurement of the amplitude of the emitted ultrasonic pulse. The received waveforms for three types of transmitters are reproduced in Figure 23. We have computed phase and amplitude spectra for each.

Matching Components to Optimize Response

As was described earlier in the report, the frequency content of the ultrasonic pulse is determined by several components—the pulser, the transducer, and the electrical connection between the two. The bandwidth of the ultrasonic pulse may be broadened by an appropriate choice of the electrical pulse used to excite the transducer or selection of an appropriate electrical network to connect the pulser and transducer.

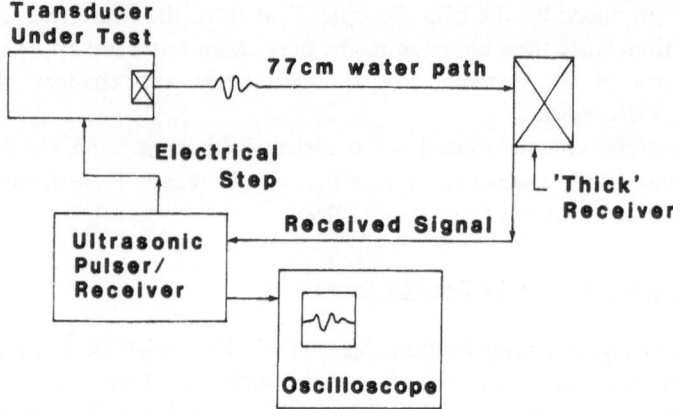

Figure 22. Experimental apparatus for measurement of wideband transducer response (after Foster and Hunt, 230).

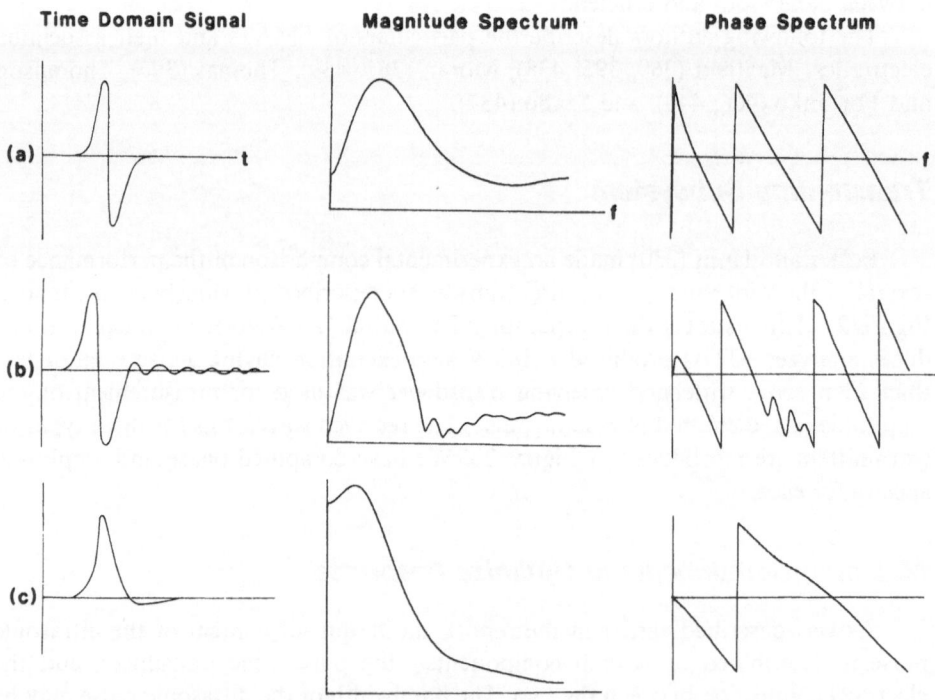

Figure 23. Performance of several prototypic wideband transducers. (a) Transducer with thick, matched backing, (b) transducer shown in Figure 15, (c) transducer shown in Figure 20 (in part, after Foster and Hunt, 230).

Gericke (116) realized the importance of matching the pulse waveform and the transducer. The frequency response of conventional piezoelectric transducers demonstrates approximately equally spaced maxima and minima [Eqs. (31) and (33) and Figure 10]. Maxima occur at the fundamental resonant frequency and its odd (3, 5, 7, ...) harmonics, while the minima occur at the even (2, 4, 6, ...) harmonics. Figure 24 illustrates how the electrical pulse may be "matched" to the transducer. The first curve represents the frequency response of the pulser $V_1(f)$. The second curve is the transducer response as determined by a method described by Gericke (50). The last plot is the loop response of pulser, transmitting transducer, and receiving transducer. The transducer was operated in pulse–echo mode and was coupled to a specimen which did not appreciably affect the spectrum (aluminum plate with plane parallel surfaces). The pulser is "matched" to the transducer by varying the pulse width until a maxima in pulser response is made to coincide with a mininma in transducer response. For the example, in Figure 24, the third maximum in $V_1(f)$ is positioned near the frequency of the second minimum of the transducer response. Considering that the third plot is a loop response (twice through the transducer, first sending then receiving) the uniformity demonstrated in the region of matching is reasonably good.

Response Equalization Networks. The bandwidth of an ultrasonic transducer may be broadened by selection of an appropriate response for a network in the

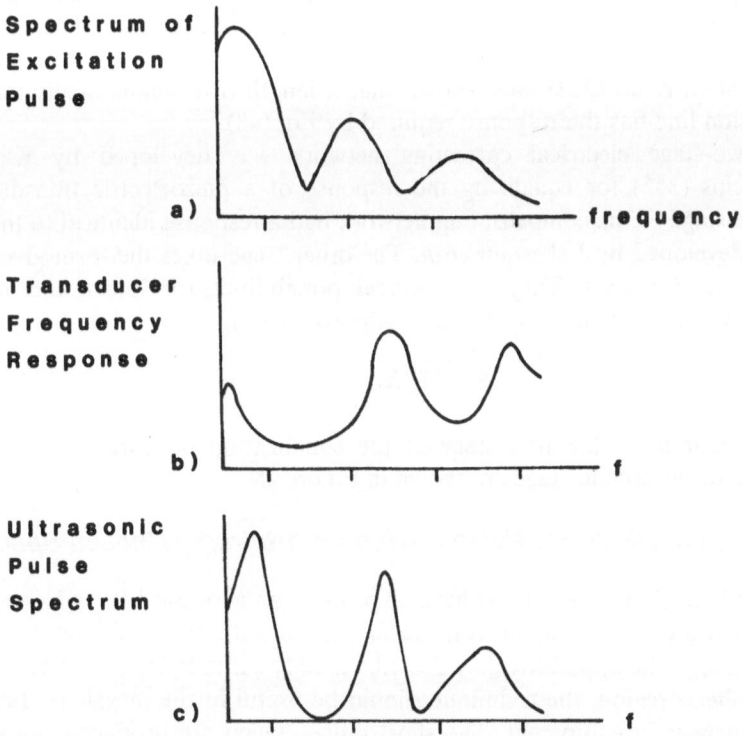

Figure 24. Example of matching pulser and transducer. Spectra of (a) rectangular excitation pulse, (b) transducer response, and (c) pulse–echo, "loop" response (after Gericke, 116).

transmitting subsystem. Although the network may be implemented acoustically, as $C_1(f)$, or electrically, as $B_1(f)$, the latter is usually chosen because the parameters are more easily controlled. The product of the network response and the transducer response should be constant over the frequency range desired:

$$B_1(f)X_1(f)=K_{b1} \quad \text{(a constant)}$$

or

$$C_1(f)X_1(f)=K_{b1} \tag{47}$$

Synthesis of the network begins by considering the response of the transducer [Eq. (28)]. The response of the equalization network is the reciprocal of that of the transducer,

$$B_1(f) \quad \text{or} \quad C_1(f)=\frac{K_{b1}}{X_1(f)} \tag{48}$$

Thompson (238) used a low-pass filter to correct the response of a low-frequency (20–80-kHz) piezoelectric transducer. The order and cutoff frequency of the filter were chosen so the filter skirt had a decreasing response which approximated the increasing response of the transducer. A high-pass filter may be more appropriate for higher-frequency transducers, where the response decreases as frequency increases above resonance.

Lakestani *et al.* (223) have shown that a length of asymmetrically terminated transmission line has the response required by Eq. (48).

A two-stage electrical correcting network was developed by Kazys and Lukosevicius (355) for equalizing the response of a piezoelectric transducer [Eq. (38)]. One stage of the compensating network had a response identical to that of the network developed by Lakestani *et al.* The other stage takes the form given by the reciprocal of Eq. (39). They give several possibilities for linear and nonlinear electrical networks which have responses approximating

$$1/X_1''(f)$$

A block diagram of the first stage of the equalization network and a nonlinear realization of the second stage are shown in Figure 25.

Producing an Ultrasonic Pulse Having an Arbitrarily Chosen Spectrum

A method for driving a transducer to produce an acoustic pulse which is shorter than the natural period of the transducer was introduced by Winter and Pereira (190). Although the apparatus they describe is only suitable for pulses band limited to the kilohertz region, the technique should be useful in the megahertz range with suitable changes in equipment. The short pulses which are produced can be of an arbitrarily chosen shape. For an appropriate selection of shape, the pulse could be made to contain a very wide range of frequencies.

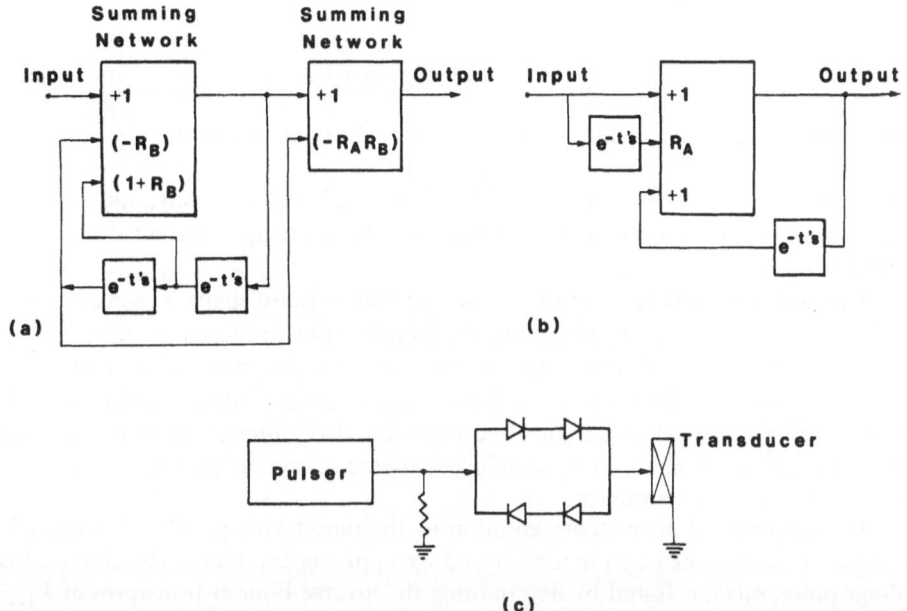

Figure 25. Transducer-response equalization network stages compensating for the transducer response given in: (a) Eq. (28), for the case of a symmetrically loaded transducer (loading and backing are the same); (b) Eq. (28), for the case of an asymmetrically loaded transducer; and (c) Eq. (38), a nonlinear réalization (after Kazys and Lukosevicius, 355).

If the transducer can be considered linear and time invariant, the system of pulser, transducer, and cable may be represented as in Figure 26. For the discussion which follows the cable and coupling effects (with the transducer radiating into a given material) are lumped into $x_1(t)$. Here $v_1(t)$ is the electrical pulse shape used to excite the transducer [having impulse response $x_1(t)$] producing the ultrasonic pulse $u(t)$. The ultrasonic pulse shape $u(t)$ may be found from the convolution of the impulse responses for the pulser and transducer:

$$u(t)=v_1(t)*x_1(t)=\int_{-\infty}^{\infty} v_1(t')x_1(t-t')\,dt' \tag{49}$$

Equivalently, in the frequency domain

$$U(f)=V_1(f)X_1(f) \tag{50a}$$

We will also need the following two forms of Eq. (50a):

$$X_1(f)=U(f)/V_1(f) \tag{50b}$$

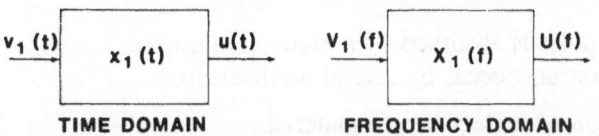

Figure 26. Linear, time-invariant representation of the transmitting subsystem.

and

$$V_1(f) = U(f)/X_1(f) \tag{50c}$$

The frequency response of the transducer, $X_1(f)$, may be determined by applying a known $v_1(t)$ and recording the resulting $u(t)$ (see Figure 27). The frequency-domain representations are found and Eq. (50b) is applied. Alternatively, the frequency response of the transducer may be determined by utilizing a system described by Gericke (50).

The next step requires that the output ultrasonic pulse shape be selected. After the shape is chosen, it is digitized and the Fourier transform computed to give the desired $U(f)$. For a wideband output, the output pulse shape is of little consequence; however, a desirable frequency-domain representation would be a flat response [$U(f)$=const] across the frequency band of interest. In order to avoid ringing in the time domain $U(f)$ should be tapered to zero at the edges of the band of constant frequency response.

The frequency-domain representation of the pulser voltage, $V_1(f)$, required to produce the wideband $U(f)$ is then found by applying Eq. (50c). The shape of the voltage pulse, $v_1(t)$, is found by determining the inverse Fourier transform of $V_1(f)$. If the representation of the pulse shape is stored digitally, it may be read out through a digital-to-analog (DAC) converter (and low-pass filter, for smoothing) to produce the required $v_1(t)$. References 240 and 257 contain useful information for generation of waveforms from digitally stored information.

As was mentioned before, the equipment utilized by Winter and Pereira is only suitable for transducers at frequencies in the kilohertz range. However, by utilizing fast digital memory (such as emitter-coupled logic) and a video DAC the technique could be used for frequencies up to approximately 40 MHz. Since the voltage level output by the DAC is only 10 V, an amplifier is required. The slew rate limitations of the amplifier may limit the useful frequency of the device.

This method of broadening the band of ultrasonic frequencies emitted by a transducer has an advantage over frequency-compensating filters applied to the output of the receiving transducer. Incorporation of the compensating network in the transmitting rather than the receiving subsystem allows the signal-to-noise ratio to be made fairly uniform over the useful frequency range. If the time-averaged amplitude of noise in the spectroscopic system is reasonably constant over the frequency range of interest, then the signal-to-noise ratio (S/N) varies as the transducer output changes. Signal-to-noise is high at frequencies corresponding to peaks in transducer response and S/N is low for frequencies where minima occur. By making the ultrasonic output spectrum uniform, signal-to-noise is made uniform.

Laser Generation of Ultrasonic Waves

When a high-energy laser pulse is absorbed by a material, ultrasonic waves are produced. Stress wave generation may occur by several mechanisms:

1. Thermoelastic expansion of rapidly heated material.

2. Vaporization or ionization (the stress is caused by recoil momentum of the vaporized material).
3. Nonuniform radiation intensity distribution causes a spatially nonuniform heating of the medium.
4. Dielectric breakdown (in liquids).
5. Electrostriction (transparent media).

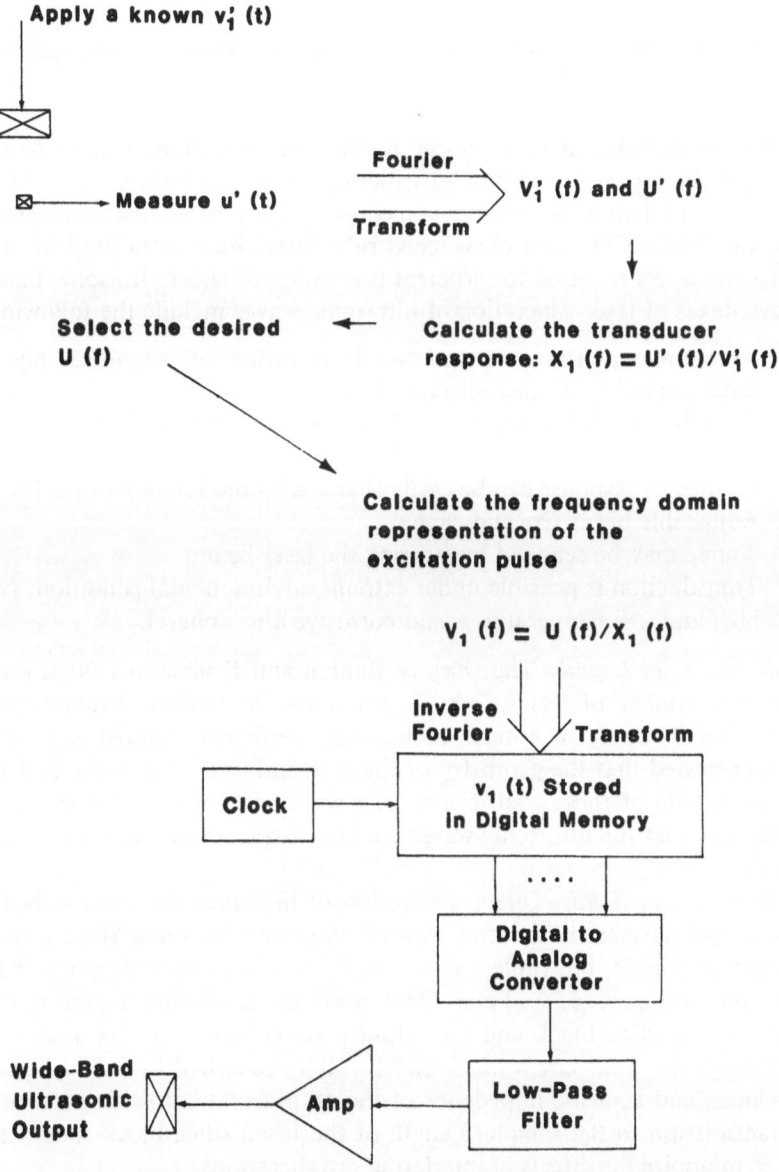

Apply a known v'_i (t)

Measure u' (t)   Fourier Transform $\longrightarrow$ **V'_i (f) and U' (f)**

Select the desired U (f) $\longleftarrow$ **Calculate the transducer response: X_1 (f) = U' (f)/V'_i (f)**

Calculate the frequency domain representation of the excitation pulse

$$V_1 \text{ (f)} = U \text{ (f)}/X_1 \text{ (f)}$$

Inverse Fourier Transform

Clock

v_1 (t) Stored in Digital Memory

· · · ·

Digital to Analog Converter

Low-Pass Filter

Amp

Wide-Band Ultrasonic Output

Figure 27. A Technique for broadbanding the output of a transmitting transducer.

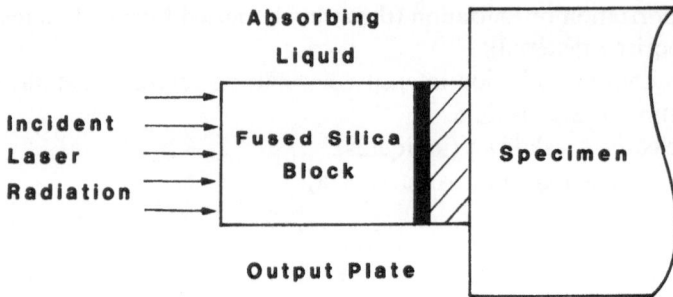

Figure 28. Laser generation of stress waves in a solid, utilizing a constrained absorbing liquid (after Felix, 382).

The pressure pulse shape produced by the laser radiation depends on the laser pulse shape, the spatial intensity distribution of the radiation, and the optical attenuation coefficient of the absorbing medium. High-power pulsed dye, Q-switched neodymium–YAG, CO_2, and Q-switched ruby lasers have been used to attain the high intensity levels required for efficient generation of short ultrasonic pulses.

Advantages of laser generation of ultrasonic waves include the following:

1. Amplitudes of the waves produced are orders of magnitude higher than those possible with piezoelectric devices;
2. Wide bandwidth of ultrasonic frequencies (extending into the microwave frequency region);
3. Frequency response can be easily changed by modulation of the laser pulse;
4. Adjustable beam size and shape;
5. Source may be scanned by steering the laser beam;
6. Transduction is possible under extreme environmental conditions (vacuum, high and low temperatures, and corrosive atmosphere).

Bulk Waves in Liquids. The work of Bunkin and Komissarov (486) provides a theoretical treatment of "sound-wave" generation in liquids. Experiments using liquids having a variety of optical attenuation coefficients (Sigrist and Kneubühl, 497) demonstrated that the geometry of the laser-induced stress wave is dependent on the magnitude of these coefficients. The work of Carome, Moeller, and Clark (364) indicates that the ultrasonic waves contain frequency components in excess of 2400 MHz.

Bulk Waves in Solids. Direct absorption of high-intensity laser pulses by the material causes surface destruction. Absorbers placed on the surface increase the stress wave amplitude, but (unless fairly thick) do not prevent destruction (Rocha, Griffen, and Thomas, 477). Felix (382) used an absorbing liquid sandwiched between a fused silica block and an output plate (Figure 28). The laser radiation passes through the transparent block and is totally absorbed by a thin liquid layer. The thickness and acoustic impedance of the output plate are chosen to maximize stress transmission to the sample. Length of the fused silica block should be large enough to minimize the effects of interfering reverberations.

von Gutfeld and Melcher (365) and von Gutfeld (476) continued the investigations of thermoelastic expansion of constrained surfaces and found that elastic energy could be increased by as much as 56 dB with respect to that produced on an unconstrained surface.

Surface Waves in Solids. In addition to producing bulk waves, the absorption of laser radiation by a solid gives rise to surface waves. Since the frequencies of the elastic waves which are produced are components of the power spectrum of the laser pulse, modulation of the laser can be used to alter the spectrum. One can also enhance the generation of certain frequencies by using a spatially periodic distribution of laser radiation (by using a mask). The period of the slot spacing in the mask is made equal to the surface elastic wavelength corresponding to the frequency desired.

Radiation Coupling, I

The effect of material intervening between the transducer face and the region of interest in the specimen is considered in this section. The mechanisms of interaction of the ultrasound with the material in which the wave propagates are important, as is the ultrasonic field. The interactions of interest are those giving rise to frequency-dependent phase changes (dispersion) and amplitude losses (attenuation).

Mechanical Coupling Layers

For ultrasonic spectroscopic examination of the bulk properties of a material (e.g., attenuation, velocity) there is usually a single coupling layer between the transmitting transducer and the material. When performing an examination of selected regions (e.g., site of defect), the effects of the material intervening between the transducer and the region of interest must be considered as well as the transducer coupling. The coupling layer alone will be considered in the next few paragraphs.

In the case of "contact" testing, a thin layer of liquid is interposed between transducer and specimen. Immersion testing, on the other hand, is carried out using a much longer ultrasonic path length in the coupling media. The analysis of the three-layer problem (transducer-coupling media sample) must take into account the reflections at both boundaries, attenuation in the coupling layer, delay time of the pulse traveling through the liquid media, and the effect of possible reverberations in the coupling layer.

A thorough treatment of wave propagation in layers may be found in the book by Brekhovskikh (296). Expressions for the amplitude transmission coefficient (T_{13}) and the amplitude reflection coefficient (R_{13}) derived in this reference for the three-layer problem, are reproduced here. For a plane wave incident from layer 3,

$$R_{13} = \frac{R_{23} + R_{12}\exp(j2k_2d)}{1 + R_{12}R_{23}\exp(j2k_2d)} \tag{51}$$

$$T_{13} = \frac{4Z_1Z_2}{(Z_1+Z_2)(Z_2+Z_3)} \frac{1}{\exp(-jk_2d) + R_{12}R_{23}\exp(jk_2d)} \tag{52}$$

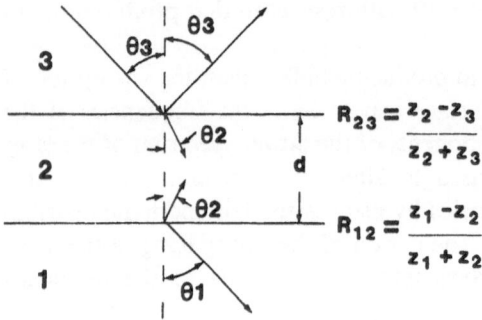

Figure 29. Designation of parameters for an acoustic impedance matching layer.

where the Z_i are normal acoustic impedances, and the reflection coefficients R_{12} and R_{23} are given in Figure 29. Note that

$$Z_i = \frac{\rho_i c_i}{\cos \theta_i} \tag{53}$$

reduces to the characteristic impedance

$$Z_i = \rho_i c_i$$

when the angle (θ_i) between the wave normal and normal to the surface is zero (normal incidence). Attenuation in the coupling layer is accounted for by a complex k_2.

The effects of the transducer wear plate, surface roughness, and gas trapped in the coupling layer are discussed by Lutsch (224).

Canella (498) describes some of the problems inherent in contact testing: resonance and phase changes in the coupling layer, and "wettability" of the surface.

Acoustic Impedance Matching. One commonly thinks one is at the mercy of the coupling and the situation is assessed only to determine the effects it has produced. However, the presence of coupling layers may be used to advantage if appropriate concern is taken with the choice of thickness and acoustic impedance of the layers. As may be inferred from Brekovshikh's work, a single coupling layer produces a response with resonances and antiresonances. Such a response, nonuniform, is far from ideal. Broadband acoustic impedance matching networks are required. Several authors have considered this problem. Kossoff (282) showed that quarter-wave matching at both the backing and load considerably widens bandwidth. Impedance matching by the use of several thin layers was considered by Highmore (183). The changes in response of intermediate matching layers, over a range of temperature, were investigated by Kazys and Domarkas (354). Shibayama, Matsunaka, and Sato (195) used matching layers with log-thickness taper to achieve efficient matching.

Intrinsic Energy Losses—Scattering and Absorption

The energy loss of an ultrasonic wave may be broadly categorized as geometrical or intrinsic. The former classification includes reflection and refraction, waveguide effects, and beam spreading. Intrinsic losses include scattering and absorption. Also included in this latter category are energy loss mechanisms due to: magneto-elastic interactions, interactions with conduction electrons in metals, interactions with paramagnetic and nuclear spin systems, electrical interaction phenomena in piezoelectric and semiconductor crystals, interaction with thermal phonons, and dislocation damping.

Consider a plane wave traveling in the x direction, with particle displacements given by

$$u = u_0 \exp\left[j(2\pi ft - kx)\right] \tag{54}$$

where u_0 is the maximum amplitude of the wave, t is time, and k is the wave number. Attenuation may be introduced into Eq. (54) by letting the wave number be a complex number. That is,

$$k = k_0 + j\alpha(f) \tag{55}$$

where k_0 and α are real, and α is defined as the linear, amplitude attenuation coefficient (units of length^{-1}).

The effects of the material may be considered by comparing the particle displacements at two points in the ultrasonic field (x_1 and x_2). For plane waves,

$$c_1(f) = u_1/u_2 = \exp\left[jk(x_2 - x_1)\right] = \exp(jkl) \tag{56}$$

where l is the difference in ultrasonic path length.

For a lossless material, the magnitude of the wave remains unchanged as it propagates:

$$|c_1(f)| = 1 \tag{57}$$

However, the phase does change with distance and with frequency:

$$\angle c_1(f) = \tan k_0 l = \tan \frac{2\pi f}{c} l \tag{58}$$

For a lossless dispersive material (velocity a function of frequency)

$$\angle c_1(f) = \tan \frac{2\pi f}{c(f)} l \tag{59}$$

An attenuating material alters the magnitude as well as phase spectrum:

$$c_1(f) = \exp\left\{j[k_0 + j\alpha(f)]l\right\} \tag{60}$$

$$|c_1(f)| = \exp\left[-\alpha(f)l\right] \tag{61}$$

The phase spectrum is given by Eq. (59).

The attenuation coefficient incorporates both the losses due to scattering, $\alpha_s(f)$, and those due to absorption, $\alpha_a(f)$:

$$\alpha(f)=\alpha_s(f)+\alpha_a(f) \tag{62}$$

Geometrical Energy Loss—The Ultrasonic Field

Let us begin by investigating the frequency-dependent nature of the ultrasonic field produced by a plane, circular emitter. Consider a disk located at the origin oriented with the normal to its faces aligned with the x axis. The transducer, driven from a sinusoidal voltage source at frequency f_0, behaves as a pistonlike source. The radiation field produced by such a source is the same field as would be produced by a plane wave diffracted by a circular opening having the same radius, a, as the disk. This diffraction problem, solved by Rayleigh and Sommerfeld, predicts that the pressure p at a point which is a distance x from the aperture and at a radius r from the axis at time t will be (Mercier, 083)

$$p(r,x,t)=\frac{jkP_0}{x}\exp\left[j(2\pi ft-kx)\right]\exp\left[-\frac{jk}{2x}(r^2+a^2)\right]a^2$$

$$\times\int_0^1\exp\left[\frac{jka^2}{2x}(1-w^2)\right]J_0\frac{kraw}{x}w\,dw \tag{63}$$

P_0 is the pressure amplitude at the opening, w is r'/a, and J_0 is a Bessel function. The variable r' represents the radius of a point in the aperture. The integral "sums" all contributions to the field point from all points on the source. The frequency dependence of this equation is obvious in that the wave number ($k=2\pi f/c$) occurs in every term.

The magnitude of $p(r,x,t)$ gives the pressure at the point. To determine the mean pressure on a surface, the function p is integrated over the area. The value of the integral divided by the surface area gives the mean pressure.

For points on the axis of the transducer, the integral in Eq. (63) simplifies considerably. The pressure on axis and in the near field is given by

$$p(0,x,t)=jP_0\exp\left[j(2\pi ft-kx)\right]\exp\left(-j\frac{ka^2}{2x}\right)\left[\sin\frac{ka^2}{2x}-j\left(\cos\frac{ka^2}{2x}-1\right)\right] \tag{64}$$

Plots of amplitude (pressure) and phase given by Eqs. (64) and (66) for two frequencies are contained in Figure 30. Notice that the field along the axis demonstrates appreciable changes as a function of frequency.

In addition to the radiation field varying along the axis of the transmitter, Eq. (63) indicates that appreciable changes occur for variation in distance off axis (r). In the far field of the ultrasonic beam, beginning at

$$x_{ff}=\frac{4a^2-(c/f)^2}{4(c/f)} \tag{65}$$

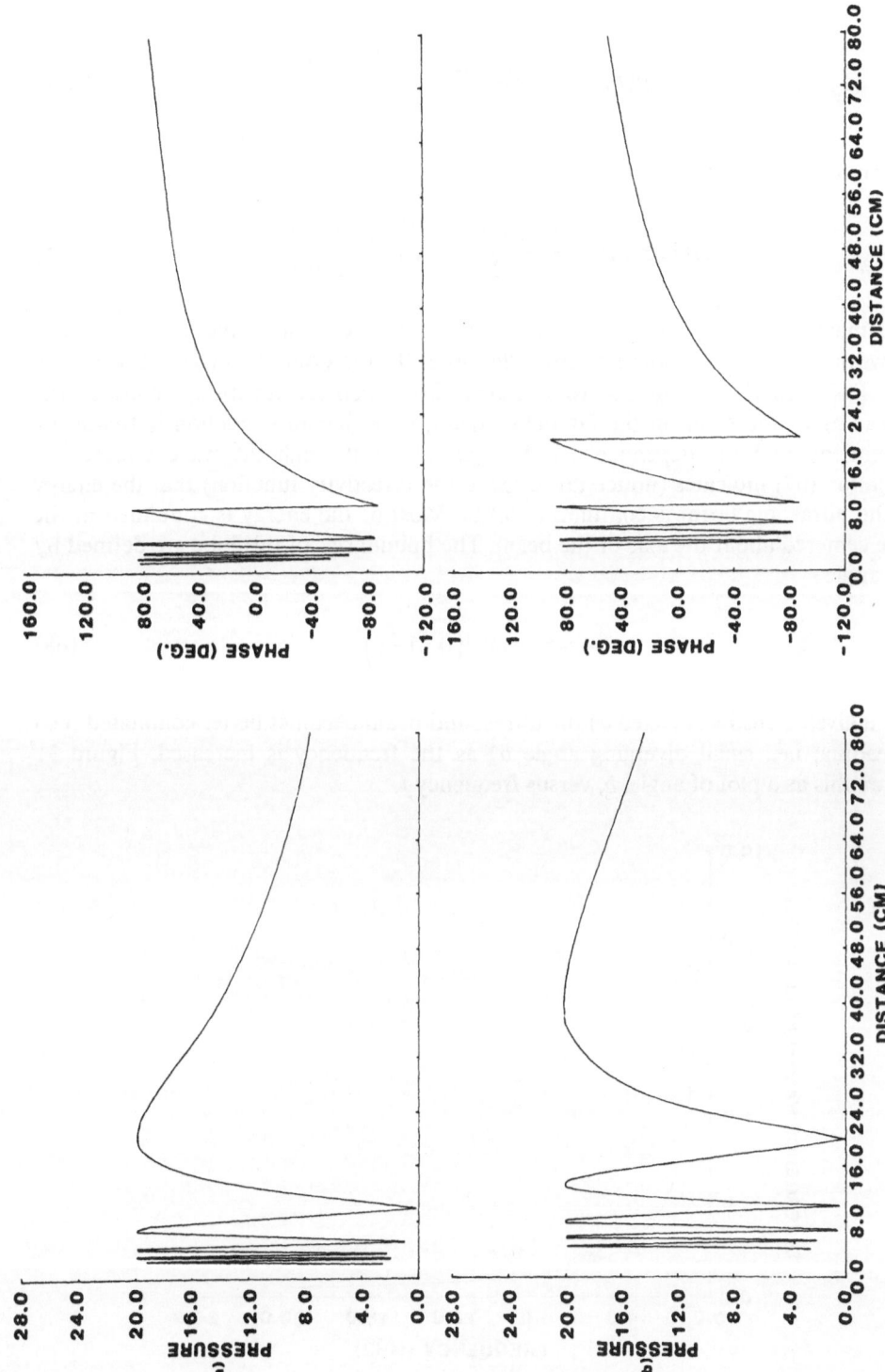

Figure 30. Pressure and phase variations along the axis of a circular transducer. Frequency: (a) 7.5 MHz (b) 15 MHz. (0.5-in.-diameter transducer in water.)

the ultrasonic field is given by

$$p(r,z,t)=jP_0e^{j(2\pi ft-kx)}\frac{a}{r}J_1\frac{kra}{x} \tag{66}$$

Commonly, p is written as

$$p(r,z,t)=\frac{jP_0ka^2e^{j(2\pi ft-kx)}}{2x}\frac{2J_1(ka\sin\phi)}{ka\sin\phi} \tag{67}$$

The magnitude of $jP_0ka^2e^{2\pi ft-kx}/2x$ is the pressure on-axis and ϕ is the angle between the x axis and the line from the origin to the point of interest at (r,ϕ,x).

The second term in Eq. (67), called the directivity function, describes the spreading of the beam in the far field. Again, note that this function is frequency dependent, and the dependence is brought about through the wave number, k. Equation (67) indicates (notice the zeros in the directivity function) that the energy of the ultrasonic beam is confined to lobes. Most of the energy is contained in the lobe centered about the axis of the beam. The boundaries of this lobe are defined by the angle

$$\phi=b=\sin^{-1}\left(0.61\frac{c}{af}\right) \tag{68}$$

For a given transducer (fixed a) the ultrasound beam becomes better collimated (less spreading, i.e., small diverging angle b) as the frequency is increased. Figure 31 shows this as a plot of angle, b, versus frequency f.

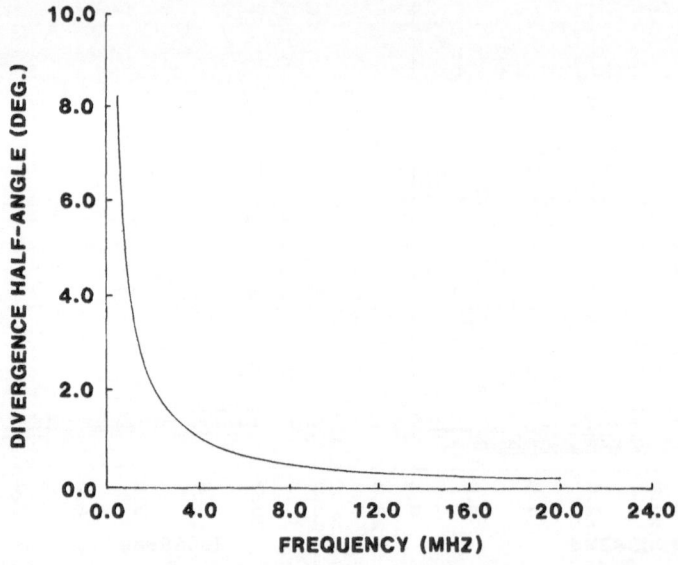

Figure 31. Divergence half-angle b as a function of frequency ($c=1.5\times10^5$ cm/sec; $a=1.27$ cm).

As has been noted in the preceding discussion, the shape of the radiation field for a continuous-wave exciting source is strongly dependent on frequency and source dimensions (ka). For the case of excitation of the transducer by voltage transients, the problem of discerning the radiation field is somewhat more complex than that with cw excitation. One possible approach has been advanced by Stephanishen (280), Freedman (278), and Robinson, Lees, and Bess (281). In this method, the transducer impulse response is made to vary throughout the acoustic field. The transient pressure is found by a linear-systems approach—the impulse response at the point is convolved with the velocity function transmitted at the face of the transducer. Beaver (292) utilized this time domain technique to calculate the near field of piston radiators for the case of several types of transient excitation. Figure 32 shows the pressure profiles calculated for a transducer disk excited by typical piston-velocity pulses. For sine-wave burst excitation the pressure profiles tend toward those produced with cw excitation, as the length of the burst is increased. Transient excitation is seen to produce pressure profiles which do not have the ripples characteristic of cw excitation.

A second method for finding the radiation field utilizes a Fourier-transform approach (Foster and Hunt, 230). The pressure pulse at the transducer face is decomposed into its Fourier components. Rayleigh–Sommerfeld diffraction theory is applied separately to each of the cw components. The radiation field is then found by the superposition of the fields from all components (Papadakis and Fowler, 294). Proper phase relationships must be kept for correctly reassembling the components.

Application of this frequency-domain method leads to the following expression for the pressure on-axis. Letting $P(f) = \{p(0,0,t)\}$, and $\mathscr{F}$ represent Fourier transformation, then

$$p(0, x, t) = \mathscr{F}^{-1}\left(P(f)\left\{ \exp jk\left[x - (x^2 + a^2)^{1/2}\right] - 1 \right\} \right) \qquad (69)$$

Figure 33 presents a comparison of the on-axis pressure for a transducer for continuous and transient excitation. The profile produced by driving the transducer with a short pulse is free from near-field variations, and in the far-field approaches the profile computed for cw excitation.

The Fourier-transform method may also be applied to calculations of off-axis pressure variations in the far field. The pressure field is

$$p(r, x, t) = \mathscr{F}^{-1}\left[P(f)\frac{ka^2}{2\,jx}\frac{2J_1(2\pi far/cx)}{2\pi far/cx} \right] \qquad (70)$$

Plots (230) of the off-axis field demonstrate a beam more peaked toward the central axis than under conditions of cw excitation, and having no side lobes. Experimental measurements confirm the applicability of using Eqs. (69) and (70) for predicting the transient behavior of piezoelectric transducers.

The intrinsic energy loss or redistribution mechanisms of reflection, refraction, and waveguide effects will not be discussed here. The reader is referred to any complete book on theoretical acoustics.

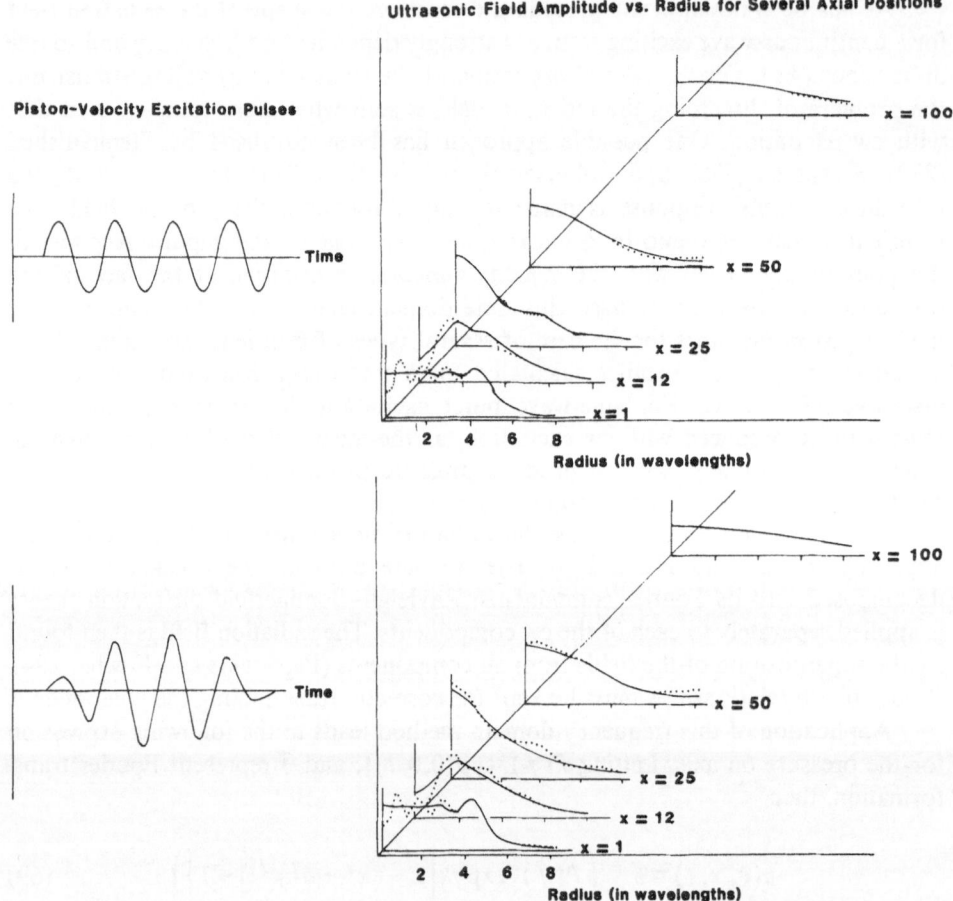

Figure 32. Sonic-field amplitude profiles vs. radius for several axial distances (transducer radius = 5λ, x in units of wavelength). Dotted curves represent the amplitude profiles for cw excitation (after Beaver, 292).

Material under Investigation

The sole purpose of a spectroscope is to determine the properties of the material or system being tested. Implicit in the assumption that useful information can be derived from the output of an ultrasonic spectroscopic system is that the relevant properties of the material are, or can be' related to, properties which are frequency dependent. This point cannot be overemphasized. If the desired properties cannot be related to functions of frequency, then ultrasonic spectroscopy is of no use.

The next chapter of this report describes the many applications of ultrasonic spectroscopy identified thus far. Each use of spectroscopy has arisen because a definite relation (theoretical or experimental) has been found between the frequency response and an important material property.

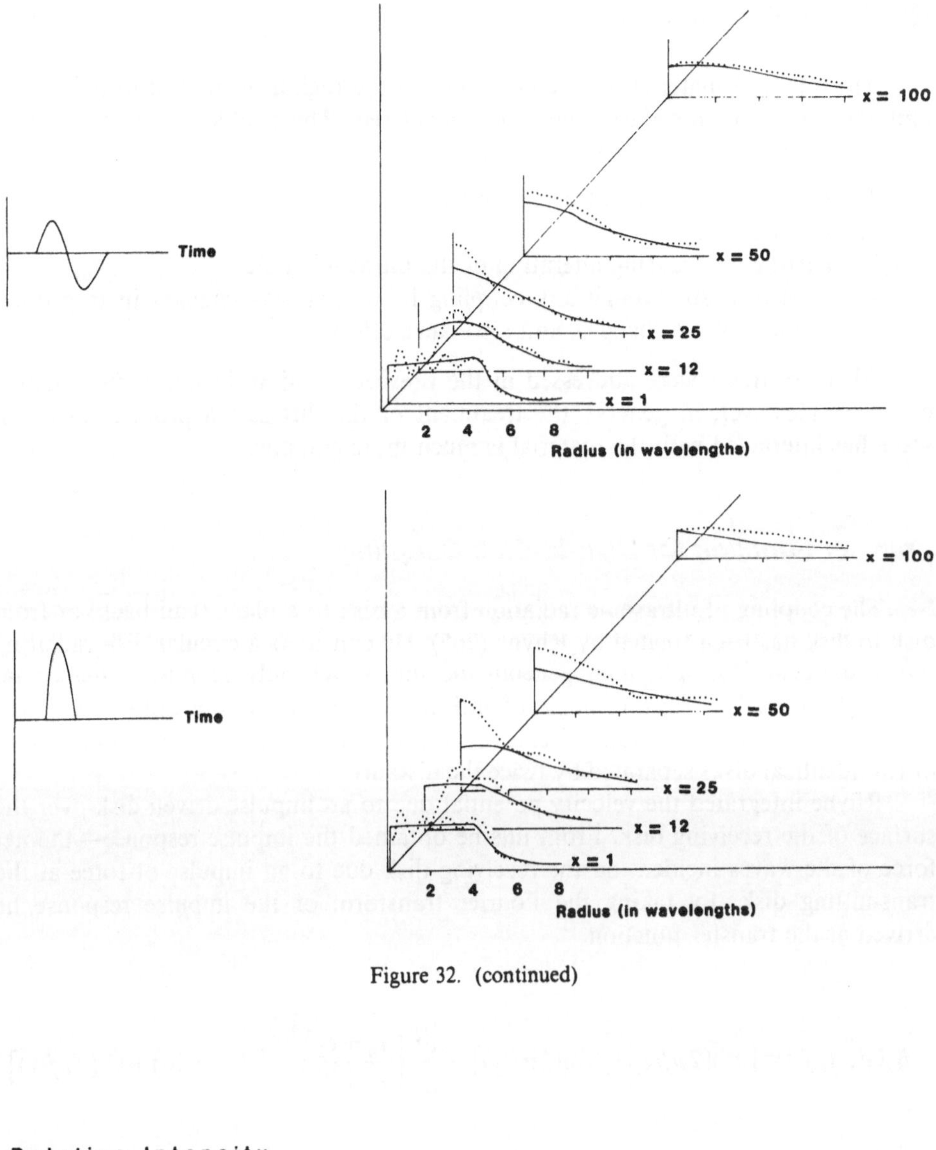

Time

Radius (in wavelengths)

x = 100

x = 50

x = 25

x = 12

x = 1

2 4 6 8

Time

Radius (in wavelengths)

x = 100

x = 50

x = 25

x = 12

x = 1

2 4 6 8

Figure 32. (continued)

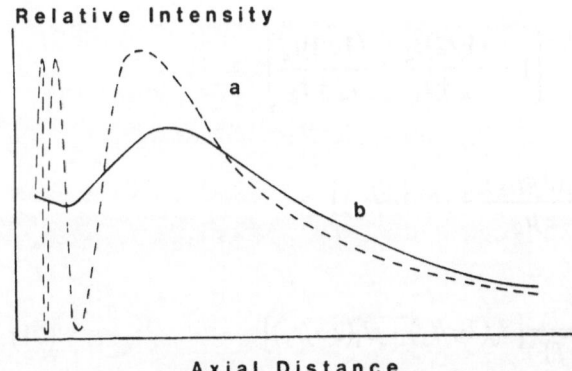

Relative Intensity

a

b

Axial Distance

Figure 33. On-axis pressure variation for a transducer with (a) continuous and (b) transient excitation (after Foster and Hunt, 230).

Radiation Coupling, II

The effect of material intervening between the region of interest in the sample and the receiving transducer must be considered. The problems which must be contended with are as follows:

1. Attenuation of the ultrasonic waves.
2. Dispersion effects.
3. Diffraction, causing alteration of the ultrasonic field.
4. Transmission through any coupling layers (causing changes in transducer loading, phase changes, and resonance effects).

All these items were addressed in the previous section dealing with radiation coupling. However, in general, the treatment of the diffraction problem after the wave has interacted with the material is much more complex.

Transfer Function for Disk-to-Disk Coupling

The coupling of ultrasonic radiation from a disk to a plane (and back) or from disk to disk has been treated by Rhyne (265). He considers a circular disk radiating into a uniform, isotropic, homogeneous medium which only supports longitudinal waves. A perfectly reflecting plane of infinite extent reflects the ultrasonic waves back to the disk (Figure 34a). An equivalent problem (shown in Figure 34b) is that of two identical disks separated by twice the distance.

Rhyne integrated the velocity potential due to an impulse-driven disk over the surface of the receiving disk. From this he obtained the impulse response—the net force of the waves incident on the receiving disk due to an impulse of force at the transmitting disk. By taking the Fourier transform of the impulse response he arrived at the transfer function

$$B_2(d, a, f) = [\cos(2\pi ft_3) + j\sin(2\pi ft_3)] - \frac{c^2}{a^2}\left(\frac{t_4 + t_2}{t_4 + t_1}\right)^{1/2}[J_0(2\pi ft_3) + jJ_1(2\pi ft_3)]$$

$$+ \frac{c^2}{a^2}t_3^2\left(\frac{t_4 + t_2}{t_4 + t_1}\right)^{1/2}\left[1 - \frac{(1/2)t_4}{t_4 + t_1} + \frac{(1/2)t_4}{t_4 + t_2}\right]$$

$$\times \left(J_0(2\pi ft_3) - \frac{J_1(2\pi ft_3)}{2ft_2} + jJ_1(2\pi ft_3)\right)$$

$$+ \sum_{n=2}^{\infty} a_n t_3 j^n \frac{\partial^n}{\partial(2\pi j)^n}[J_0(2\pi ft_3) + jJ_1(2\pi ft_3)] \qquad (71)$$

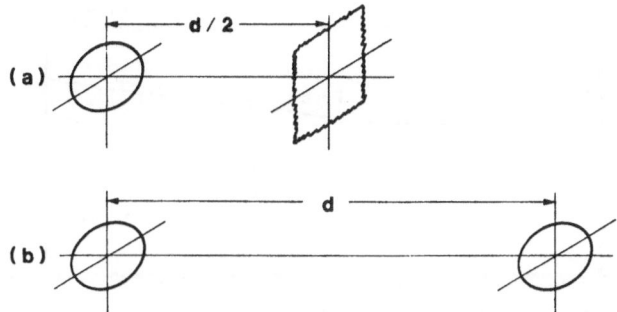

Figure 34. Two identical radiation coupling problems. (a) Coupling from disk to plane and back to the disk. (b) Coupling between two identical disks (after Rhyne, 265).

where $J_0(\cdots)$, $J_1(\cdots)$ are Bessel functions of the first kind, and

$$t_1 \equiv d/c \tag{72a}$$

$$t_2 \equiv \left(\frac{4a^2 + d^2}{c^2} \right)^{1/2} \tag{72b}$$

$$t_3 \equiv \tfrac{1}{2}(t_2 - t_1) \tag{72c}$$

$$t_4 \equiv \tfrac{1}{2}(t_2 + t_1) \tag{72d}$$

The magnitude and phase of the frequency response for fixed a (0.3175 cm) and d (3 cm) are plotted versus frequency in Figure 35. The radiation coupling filter acts as a high-pass filter, having zero transmission at zero frequency and with transmission approaching a constant at higher frequencies.

By plotting the transfer function versus distance, d, for fixed a and frequency (Figure 36), one obtains a "diffraction" correction which can be applied to measured

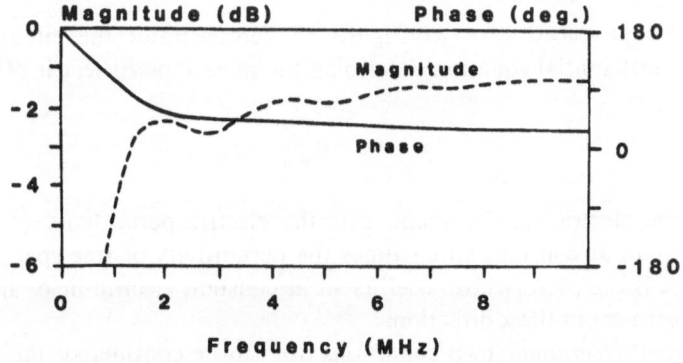

Figure 35. Transfer function of radiation coupling for fixed transducer radius ($a=0.3175$ cm) and distance ($d=3$ cm) for varying frequency (after Rhyne, 265).

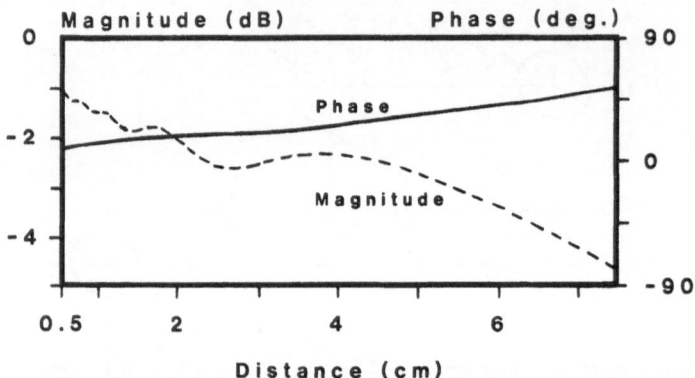

Figure 36. Transfer function of radiation coupling for fixed transducer radius ($a=0.3175$ cm) and frequency ($f=2.25$ MHz) for varying distance (d) (after Rhyne, 265).

attenuation values to account for beam-spreading (diffraction) losses as well as losses due to phase variations across the receiver surface.

Receiving Transducer

The receiving transducer converts incoming ultrasonic waves to an electrical signal. As was the case for transmitting transducers, attention in this section of the report will center on piezoelectric devices.

Piezoelectric Receivers

Let us again consider a piezoelectric disk of thickness l and surface area of one face equal to s. The disk, much larger in diameter than thickness, is assumed to be made of a material having no electrical conductivity. Additionally, the transducer material does not attenuate the ultrasonic wave. Longitudinal ultrasonic waves (from medium 2) are incident on the disk with the wave front parallel to the front circular face.

Figure 9 reproduced here as Figure 37 demonstrates the situation to be analyzed. The differential equation describing the inverse piezoelectric effect is

$$D=\varepsilon E+e\frac{\partial u}{\partial x} \tag{73}$$

where D is the electric displacement, ε is the electric permittivity ($=K_s\varepsilon_0$, i.e., dielectric constant at constant strain times the permittivity of free space), E is the electric field, e is the piezoelectric coefficient at constant electric field, and u is the particle displacement in the x direction.

Open-Circuit Operation. Two conditions need to be considered: the transducer terminated by (1) an open or (2) a short circuit. With $D=0$ (open circuit), it is straightforward (148, 223, 236, 297) to show that the output voltage (x_2) depends

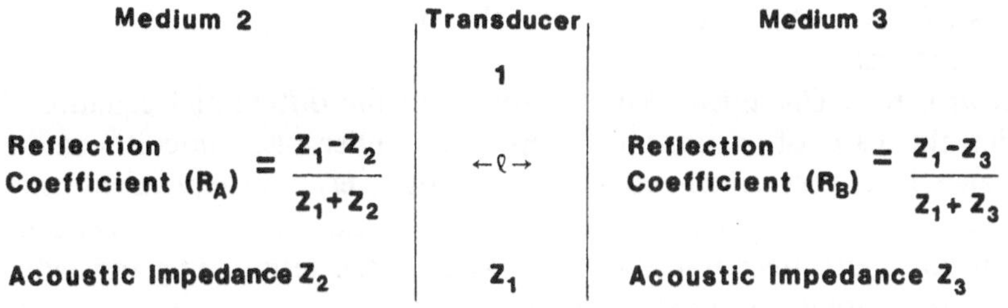

Medium 2	Transducer	Medium 3
	1	
Reflection Coefficient $(R_A) = \dfrac{Z_1 - Z_2}{Z_1 + Z_2}$	$\leftarrow \ell \rightarrow$	Reflection Coefficient $(R_B) = \dfrac{Z_1 - Z_3}{Z_1 + Z_3}$
Acoustic Impedance Z_2	Z_1	Acoustic Impedance Z_3

Figure 37. Transducer bounded by two media. The reflection coefficients at each boundary are given.

only on the particle displacements u_A and u_B at the disk faces:

$$x_2 = \frac{e}{\varepsilon}(u_B - u_A) \tag{74}$$

For thick transducers (i.e., l/c_1 greater than the ultrasonic pulse duration), $u_B = 0$ until the pulse reaches the back surface of the disk, and

$$x_2(t) = -\frac{e}{\varepsilon} u_A = -\frac{e}{\varepsilon} \int \left(\frac{\partial u}{\partial t}\right)_A dt \tag{75}$$

that is, the output voltage is proportional to the *integral* of the force on the transducer's front face.

If an ultrasonic pulse creates a particle velocity delta function at the front face (A) of the transducer, the transfer function of the receiver (in Laplace notation) is

$$X_2(s) = -\frac{e(1 - R_A)(1 - e^{-t's})(1 - R_B e^{-t's})}{s\varepsilon(1 - R_A R_B e^{-2t's})} = \frac{V(s)}{U(s)} \tag{76}$$

where t' is the transit time of the ultrasonic pulse in the transducer (i.e., l/c_1), V_2 is the voltage between the faces of the transducer, and U is the particle velocity of the incident wave.

One should compare this expression with Eq. (27), the transfer function of the transmitting transducer. They are identical, except for the $1/s$ term in the response of the receiver. This integration may be eliminated if the transducer output is passed through an electronic differentiating network. When this is done, the frequency response relations given in the transmitting transducer section may be used for the receiver. Again, the frequency response of the transducer is dependent on the acoustic impedances of the loading and sample media, and on the thickness of the transducer disk.

Kazys and Lukosevicius (355) note the response of an open-circuited receiver is not influenced by the negative capacitance in the equivalent circuit. They give the receiver response in terms of the response of a transmitting transducer. That is,

$$X_2(f) = \frac{S}{j2\pi f(Z_3 + Z_1)C_0} X_1'(f) \tag{77}$$

where $X'_1(f)$ is given in Eq. (27) and S is the surface area of one face of the piezoelectric disk.

Short-Circuit Operation. We now return to the differential equation (73) and consider the case of a transducer operated under short-circuit conditions (or terminated in a very small resistance). Alternatively, the equivalent circuit of a piezoelectric transducer (Figure 11) may be analyzed. For time $<l/c_1$ the backing material does not affect transducer response. If the terminating resistance, R_e, is small the response is governed solely by

$$-\frac{C_0\varepsilon^2}{e^2 C_0^2} \quad \text{and} \quad Z_1 + Z_2$$

When the time constant of this combination is large compared with the transit time of the pulse through the transducer disk, the voltage across R_e will closely follow the force applied to the front of the disk (Redwood, 297). The output voltage, however, will be quite small. As R_e is made smaller and smaller, the voltage representation of force becomes better, but the signal amplitude gets ever smaller.

If the backing is perfectly matched to the transducer, then there will be no ultrasonic pulse reverberations. The special case described above for times $0 < t < l/c_1$ can be extended for all times.

A number of wideband transducers, discussed in the section on transmitting transducers, were tested by Foster and Hunt (230) as part of a receiving subsystem. The broadband transmitters also acted as broadband receivers. The response covered a wide range of frequencies (broadband) but was not uniform. Although the frequency response was not flat, the $[(\sin x)/x]$-type response observed with commercially available transducers was not evident. Rather, a gentle transition from peak to minima occurred (refer to Figure 23—the spectral responses as receivers are essentially those shown for transmitting response).

Acoustoelectric Transducer

Often the phase-sensitive characteristics of piezoelectric detectors are a disadvantage. The difficulties arise because the output of a piezoelectric transducer depends on the vector average of the incident ultrasonic waves. For ultrasonic measurements in nonparallel, rough, or inhomogeneous material, an acoustoelectric receiver (Heyman, 363, and Heyman and Cantrell, 185) is preferred. In an acoustoelectric converter, the acoustic wave couples to charge carriers which are "dragged" in the direction of wave propagation. The unbalance of charge carriers creates an electric field which is dependent on the flux and not on the ultrasonic phase.

Figure 38 demonstrates the advantage of an acoustoelectric converter (AEC) for making attenuation measurements in an inhomogeneous material. The differences in path lengths for the ultrasonic rays give rise to phase changes. For a piezoelectric detector phase cancellation occurs, producing a smaller output voltage than expected and giving an abnormally high calculated attenuation. Although the effects may be lessened by making the piezoelectric transducer small, one sacrifices directionality. The AEC, a power detector, responds to incident flux and gives true attenuation.

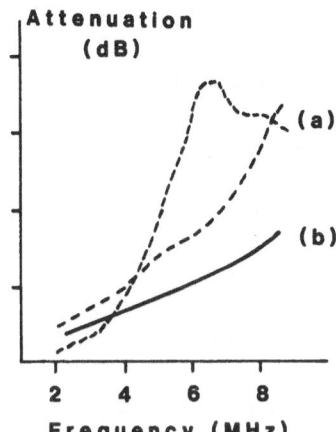

Figure 38. Phase cancellation effects in a piezoelectric detector (a) (which give abnormally high attenuation measurements) may be eliminated by using an acoustoelectric converter (b) (after Busse *et al.*, 467).

Capacitive Receiver

Capacitive detectors have the desirable characteristic of a flat frequency response. Another attractive feature is the elimination of the coupling layer between transducer and sample. However, the very small voltages produced by these devices, in response to an ultrasonic wave, require low-noise electronics and often some type of signal averaging.

A capacitive receiver and associated electronics have been described by Gauster and Breazeale (352) and Cantrell and Breazeale (353). The sensitivity of this device is such that displacement amplitudes on the order of 10^{-12} m can be detected. The bandwidth of the device, apparently limited only by the preamplifier, spans the range 1–150 MHz. The frequency response is flat from 3 to 70 MHz.

Electrical Coupling Network, II

As is the case for the electrical connection between the pulser and transmitting transducer, the electrical network interposed between the receiving transducer and amplifier can have a profound effect on the displayed spectrum. Again, even the simplest connection, a coaxial cable, can alter the frequency response quite appreciably.

Since the response of the receiving transducer is dependent upon the resistance of the electrical termination, one must decide on the operational mode for the receiver before progressing with the network design. If the transducer is to be terminated with a low resistance, then a coaxial cable having this characteristic impedance should be directly connected to the receiver. The cable should be terminated at the amplifier with a resistance equal to the characteristic impedance of the cable. Proper termination eliminates reflections at the ends of the cable which cause peaks in frequency response.

If a thick receiving transducer is terminated with a high resistance, and a coaxial cable is used to transfer the signal to the amplifier, a broadband impedance-matching network should be used between the high resistance and cable. Additionally, a differentiating network is required to remove the integration term from the receiver response.

Differentiator

The differentiator transfer function (in Laplace notation),

$$B_2(s)=K_d s \qquad \text{(where } K_d \text{ is a constant)} \tag{78}$$

is not physically realizable (there is one more zero than pole in the transfer function); however, one may be satisfied with the approximation shown in Figure 39a. That is, the high-pass filter function

$$B_2(s)=\frac{K_{hp}s f_0 f_1}{(s+f_0)(s+f_1)} \tag{79}$$

could be used. Indeed, the ideal function of Eq. (78) is never needed in practice, since the differentiation is performed on signals restricted to a finite frequency range. The function [Eq. (79)] will block direct current, differentiate signals with spectral content well below $f=f_0$, and pass frequencies between f_0 and f_1. Differentiation of signals is a notoriously noisy operation, but a judicious choice of f_0 will help minimize these effects. An operational amplifier implementation of the transfer function in Eq. (79) is given in Figure 39b.

Low-Pass Filter

Later in this report (the analysis section) it will be shown that in some cases (when the signal is to be sampled) the upper frequency should be limited. A low-pass filter (LPF) is utilized for this purpose. Although the ideal LPF (Figure 40a) is not realizable, a close approximation can be constructed (Figure 40b). The filter can be designed with active devices or R–L–C components. An operational-amplifier

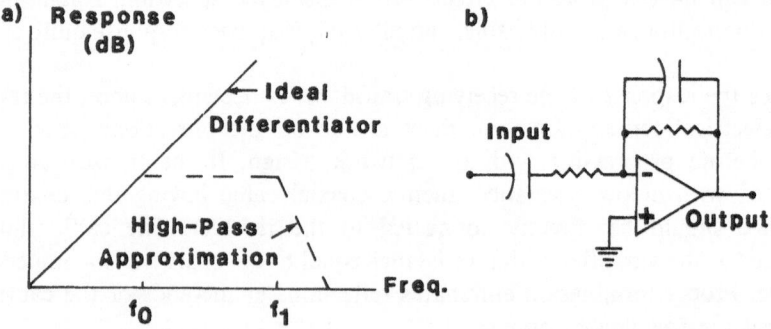

Figure 39. Differentiator and its high-pass filter approximation. (a) Frequency response; (b) op-amp implementation of high-pass approximation.

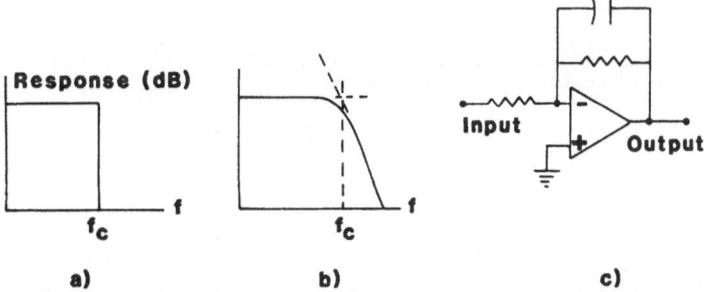

Figure 40. Low-pass filter. (a) Ideal frequency response; (b) frequency response of a realizable filter; (c) op-amp implementation of first-order LPF.

implementation of the LPF is given in Figure 40c. Additional stages may be added to improve cutoff characteristics.

Amplifier

Since the electrical signals produced by the receiving transducer are low in amplitude, and the signal-processing electronics requires higher voltages, an amplifier is often incorporated into the system. The gain of the device,

$$A(f) = \frac{\text{output voltage}}{\text{input voltage}}$$

is a complex quantity, thus having a frequency response with phase as well as amplitude variations.

If no reactive circuit elements were utilized, there were no stray reactances, and the parameters of the active devices (transistors, etc.) were not frequency dependent, then the amplifier's frequency response would extend uniformly from dc to infinite frequencies. However, blocking and bypass capacitors limit the response at low frequencies, while stray reactances and characteristics of the active devices limit performance at high frequencies. Analysis of the frequency response begins by representing the amplifier as a lumped-parameter equivalent circuit. All of the blocking and bypass capacitors are included. Additional elements are added to the circuit to represent stray capacitances.

The frequency response of an amplifier stage is essentially that of a bandpass filter:

$$A(f) = A_{\text{mid}} \frac{j(f/f_h)}{1 + j(f/f_h)} \frac{1}{1 + j(f/f_l)} \tag{80}$$

where A_{mid} is the midfrequency gain, and f_l and f_h are, respectively, the low and high cutoff frequencies. The amplitude and phase spectra for this type of response are

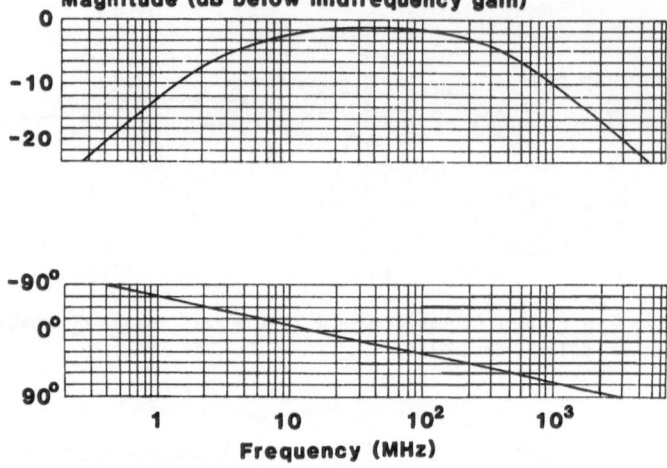

Figure 41. Frequency response of a single-stage amplifier.

shown in Figure 41. Although the high-frequency cutoff normally proves a limitation, it can be used to advantage in systems employing a digitizer, since the cutoff
can be used to band limit the signal frequencies.

The amplifier bandwidth (BW) is the range of frequencies for which the gain is
within 3 dB (amplitude down to $0.707A_{mid}$) of that at midband. For any amplifier
stage the product of gain and bandwidth is approximately constant. Therefore, for a
given device, a narrowband (tuned) amplifier may be constructed having a much
larger gain than a broadband amplifier. In order to achieve a wide bandwidth, gain
must be sacrificed or additional stages of amplification incorporated. Figure 42
illustrates two possibilities for a multistage broadband amplifier. The first utilizes
several low-gain stages with identical frequency response. The result is an amplifier
with reasonably uniform passband and extremely sharp cutoff frequencies. Another
design for the amplifier incorporates several stagger-tuned stages. The gain of each
stage can be high because of its limited bandwidth. Although the use of additional

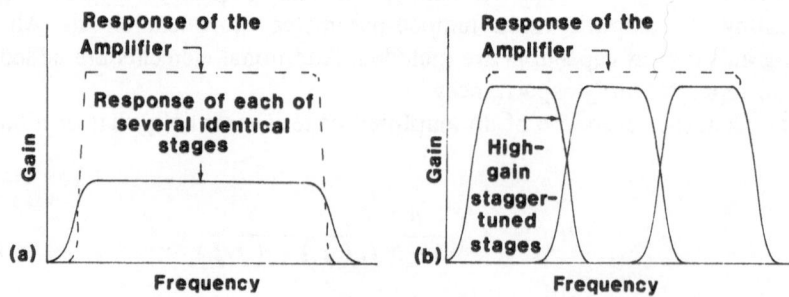

Figure 42. Design possibilities for a wideband amplifier. (a) Design utilizing several identical broadband
stages; (b) design incorporating stagger-tuned high-gain stages.

gain stages can bring the amplification to the desired level, each stage brings with it the attendant problem of increased noise.

In addition to the importance of gain and bandwidth, several other characteristics of the amplifier should be considered. The dynamic range (range of signal amplitudes from the equivalent-noise input to the largest signal at the input which will be amplified without distortion) must be sufficiently wide to encompass all signals expected. The possibility of using a logarithmic stage, for dynamic range compression, should be considered. Additionally, output voltage swing of the amplifier should be within the linear range of operation of other components. Also, linearity of the amplifier should be stringent enough to ensure (for a given frequency) that the output is independent of the amplitude of signal input from the receiving transducer.

Analog Gate

Spectrum analyzers operate on the entire signal input to them. If several discrete waveforms are present, such as echoes from a number of scatterers, the spectrum presented by the analyzer will be that of the complete ensemble. If the spectrum of an echo from an individual scatterer is desired, a method of separating this signal from the others is required. For continuous-wave systems, the separation requires frequency coding. However, if a pulsed system were to be used and the signals were separated in time, then an analog gating circuit could be used to remove all extraneous signals from the waveform, leaving the one desired.

A simple gate could be constructed from a switch (solid state, relay, etc.) opened and closed at appropriate times. This type of device is extremely limited because of the time required to open and close the switch and because of the transients produced during switching.

The use of a double-balanced mixer as a pulse modulator (Figure 43) overcomes the problems of a switch. Switching times on the order of 1 nsec and transients of less than 15 mv can be expected. The ungated signal is applied to the radio frequency (r.f.) port of the mixer and the modulating pulse to the intermediate

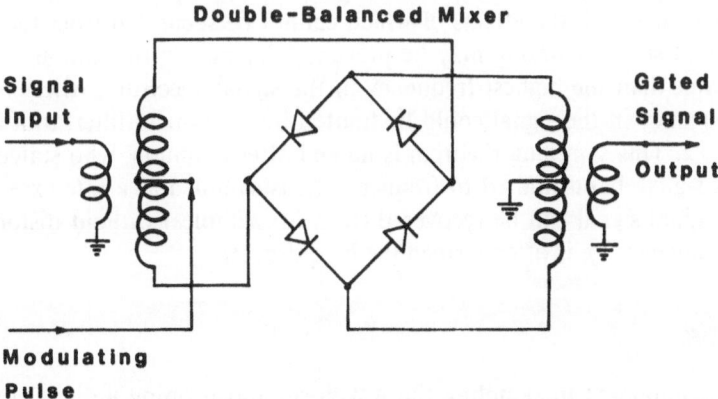

Figure 43. Analog gate utilizing a double-balanced mixer.

frequency (i.f.) port. The gated output is taken from the local oscillator (l.o.) port. The sharp rise in current at the i.f. port due to a sudden pulse input turns the diodes fully on—connecting input to output. When no pulse is applied, only a very highly attenuated portion (around -50 dB) of the input appears at the output. The timing and duration of the modulating pulse may be varied to gate out the desired signals.

Although the switching transients of the double-balanced mixer are small, they may be comparable to the level of the signals to be gated out. For this reason, an amplifier is placed between the receiving transducer and the analog gate. Additionally, in systems utilizing a digitizer, the samples taken during which the switching transients occur should be eliminated.

Sampling and Digitization

The purpose of a digitizer is to create a digital representation of the amplitude of a waveform at particular instants in time. Ideally, the amplitude is sampled instantaneously and the digital representation created is infinitely accurate. More realistically, the sample which is acquired is an average over a short period of time and the digital word length is limited.

Sampling

Sampling of a waveform is illustrated in Figure 44. A succession of sampling pulses investigates the signal at discrete instants of time. Sampling can be thought of as a multiplication of the analog signal by a train of sampling pulses. The result is a waveform whose envelope is that of the original signal, but whose actual values exist at only discrete instants of time. Although in some unusual cases the sampling is not periodic, the analysis which follows proceeds with the assumption of sampling at equispaced intervals.

Sampling affects the spectrum of the signal, causing it to repeat as shown in Figure 45b, every f_s Hz (f_s is the sampling frequency). As is illustrated in Figure 45c, folding about $f_s/2$ Hz can cause problems in interpretation of the spectrum. The frequency components folded into the region of overlap cause distortion (often referred to as aliasing), the effects of which cannot be separated from the spectrum of the original signal. Aliasing may be prevented by raising the sampling rate until $f_s/2$ is greater than the highest frequency in the signal spectrum. Alternatively, the highest frequency in the signal could be limited by a low-pass filter, to a frequency less than $f_s/2$. This sampling theorem is named after Shannon, who stated that if a continuous signal, band limited to frequency f_c, is sampled at a rate exceeding $2f_c$, then the original signal can be recovered (from its samples) without distortion. The sampling frequency $2f_c$ is often termed the Nyquist rate.

Digitization

A key component in sampling the waveform and forming a digital representation is the analog-to-digital converter (ADC). The ADC is basically a quantizer

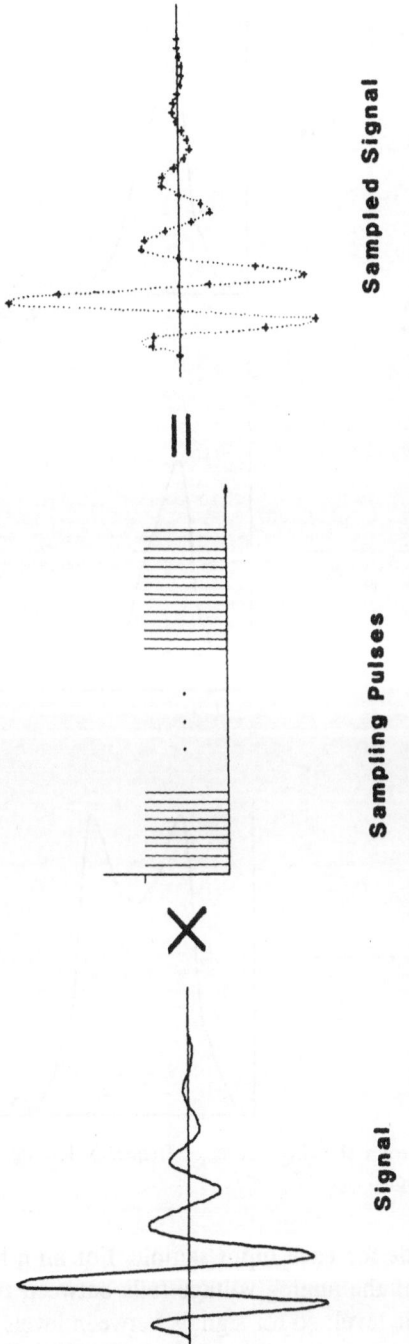

Figure 44. The process of waveform sampling is equivalent to a multiplication of the signal by a train of unit-height sampling pulses.

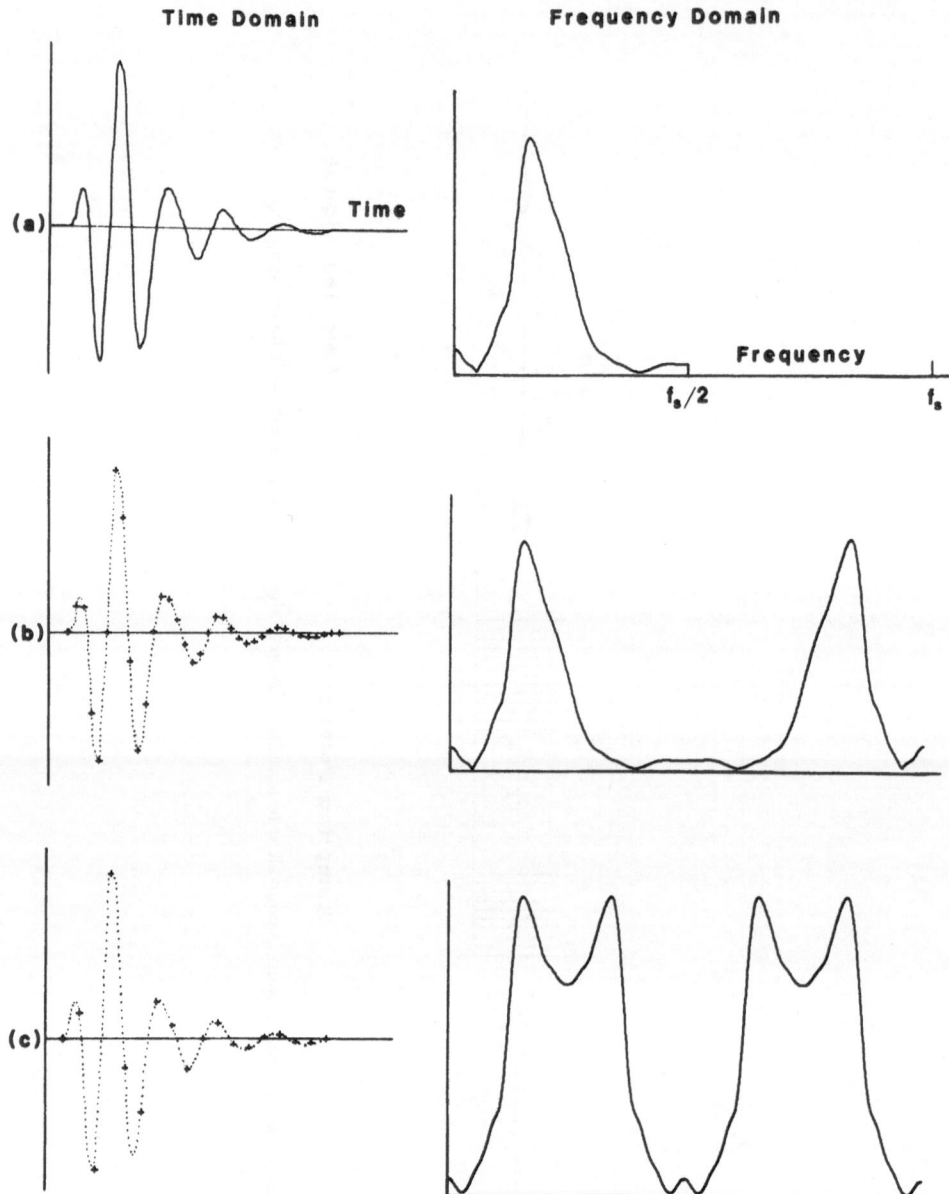

Figure 45. Sampled-signal spectrum showing effects of frequency folding. (a) Continuous signal, (b) sampled signal $f_s/2 > f_c$, (c) sampled signal $f_s/2 < f_c$.

which creates a binary code for each input sample. For an n-bit converter there are 2^n discrete output codes. If the analog voltage falls between two quantizer levels, it can be assigned to only one level; so for signals between levels the peak error will be $\pm 1/2$ level (i.e., $2^n/2$). The rms error is given by $(2^n/2)3^{1/2}$. The error is determined by the number of quantizer levels. For an 8-bit converter the error is $1/2$ in 256 (0.2%). The error reduces to $1/2$ in 4096 (0.01%) for a converter with 2^{12} levels.

During the time an ADC is performing a conversion (t_c) it is sensitive to the voltage level being measured. If the voltage varies, then so will the output code. The error can be reduced to one quantizer level if the conversion time is

$$t_c = \frac{1}{2^n 2\pi f} \tag{81}$$

where f is the frequency of a sine-wave input. For an 8-bit converter to faithfully reproduce a 1-MHz sine wave, a conversion time of 0.6 nsec would be needed. Typical successive-approximation ADCs have conversion times in the 1-μsec range, while very fast "flash" converters require approximately 30 nsec. Thus the converter alone is not capable of digitizing voltages varying at high rates. Fortunately, a simple solution to the problem is provided by the use of a sample and hold circuit (S/H) preceding the ADC.

The purpose of a sample and hold is to rapidly sample the input signal and then hold this voltage constant until the analog-to-digital conversion is complete. The S/H consists of a voltage-holding capacitor connected to the input signal through a switch. The switch is initially open, then on command the switch closes and the voltage on the capacitor begins to track the input. When the switch opens, the voltage is held on the capacitor. There is jitter between the time the command to the switch is given and the time it actually opens. This uncertainty in the actual time of sampling (aperture-time uncertainty$=t_a$) causes an output voltage error which is dependent on the rate of change of the input voltage. The maximum percent error introduced by the S/H should be less than 1/2 quantizer level for the ADC it is used with. That is,

$$t_a = (1/2)t_c \tag{82}$$

where t_c is given in Eq. (81). For an 8-bit ADC digitizing a 10-MHz sine wave (full-scale amplitude), the S/H should have a maximum aperture time uncertainty of 31 psec. If the input amplitude is less than full scale, the acceptable t_a will be larger (or, alternatively, for a given t_a the frequency maximum will be higher).

Adding the sample and hold to the analog-to-digital converter increases the frequency range appreciably. Sample and holds with aperture time uncertainties on the order of 10–30 psec are readily available. If the high frequencies in the input signal are not too large, these S/Hs are probably adequate for use with 8-bit converters at frequencies to approximately 20–40 MHz. Sample and hold modules with shorter aperture uncertainty times are available, but at high cost.

The digitizing system, as outlined in the preceding paragraphs, is diagrammed in Figure 46. The low-pass filter band limits the input signal to prevent aliasing. An oscillator provides the strobe pulses to (1) open and close the switch in the S/H, (2) initiate analog-to-digital conversion, and (3) store the digital code in sequential random access memory (RAM) locations. The rate at which data can be stored in digital memory is limited by the write cycle time. For MOS, bipolar TTL, and bipolar ECL devices average cycle times are, respectively, 200, 60, and 20 nsec. If the

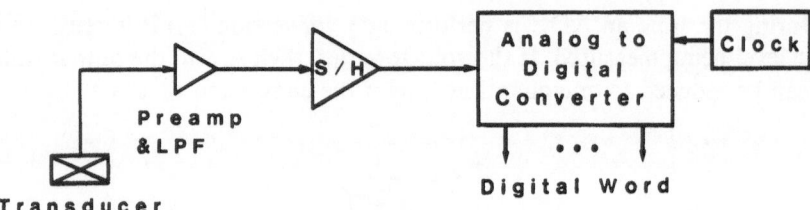

Figure 46. Sampling and digitizing system utilizing a sample-and-hold and fast analog-to-digital converter.

memory were to be the limiting component in the system, these cycle times imply sampling frequency limits of 5, 16.7, and 50 MHz.

Transient Recorders

Perhaps the most widely used instrument for digitizing ultrasonic signals is the Biomation 8100 transient recorder. This device utilizes a patented 8-bit ADC design to provide sample rates up to 100 MHz. Figure 47 presents the essence of the 8100. The folding amplifier is an absolute-value device—it acts as an inverting amplifier for negative input; the polarity of positive input is unchanged. The 6-bit ADC is a parallel converter incorporating 64 comparators. A 2048-word digital memory holds the digitized transient. The memory is available for readout through a computer-compatible interface.

The advantages and limitations of this transient recorder are perhaps best summarized by Elsley (300). He notes that improper use of the recorder can lead to a reduction in the amplitude resolution from 8 bits to 4 bits (1 level in 16). One should also be aware of the possibility of "jitter" which is sometimes encountered and must use care when performing signal averaging.

Another transient recorder, a 9-bit device, is manufactured by Tektronix (model 7912AD). A double-ended scan-converter tube is used for storage. The transient is

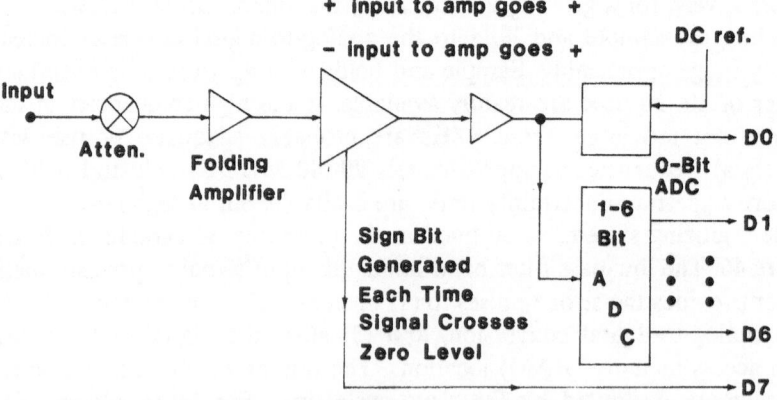

Figure 47. Block diagram of the analog-to-digital converter used in the Biomation 8100.

written onto a small two-dimensional silicon diode array. The data can be nonde-structively read out to a computer interface, or for display, to a television monitor. Up to 50 waveforms can be digitized each second. It is said that signals with frequencies up to 500 MHz can be digitized at 10 mV per division or up to 1 GHz at 4 V/div. Although the 7912AD is expensive, it may provide useful as an ultrasonic waveform digitizer for very high frequencies. At least one ultrasonic laboratory is investigating use of this device; however, no report has appeared on the applicability of the device to ultrasonic work.

Charge-Transfer Devices

Two types of charge-transfer devices (CTD) hold possibilities for sampling ultrasonic signals. Although at present the bandwidths of these devices are limited, they are inexpensive, very small (16-pin DIP) and it is anticipated that bandwidths may increase in upcoming versions.

Single-transfer devices (STD)* are essentially a miniaturized set of sample and holds. The S/H switches are sequentially closed and opened, the sampling rate being determined by an external clock (12 MHz maximum). Up to 64 samples are held internally until read out through an ADC. The advantage of using the STD instead of a single S/H is that a number of samples are taken at high speed and held. Readout of the sample can proceed at a more leisurely pace, with the result that a less expensive ADC could be used and the memory to which the digital data are written could be much slower.

Another type of CTD is the charge coupled device (CCD) shift register.[†] The input voltage is sampled and transformed to a charge packet. As each new sample is acquired, the packet from the previous sample is shifted to the next location in the register. The clock rate which controls the sampling rate can be as high as 40 MHz. Once the register is full, readout can commence. The clock rate is slowed consider-ably, so that the voltage at the output is a low-frequency replica of the input. An ADC converts the output samples to a stream of digital words. As with the STD device, the CCD shift register is used both as a fast sampler and as an analog memory.

Equivalent-Time Sampling

If the ultrasonic waveform is repetitive, a plethora of systems is available for digitizing the signal. These techniques, termed *equivalent-time sampling*, involve obtaining one sample of the waveform for each repetition.

The time interval between the synchronizing pulse (which is time locked to the signal) and the sampling is increased for each repetition of the signal (Figure 48). Note that the time separation of the sync pulse and the signal must remain constant. Also, the waveform must remain unchanged for a number of repetitions equal to the number of samples to be acquired.

---

*Reticon Corp., 910 Benicia Ave., Sunnyvale, California 94086, SAM-64, SAM-128LR.
[†]Reticon Corp., R5102 and R5103 Video Delay. Fairchild Camera and Instrument Corp., 464 Ellis St., Mountain View, California 94042, CCD311 and CCD321A.

The effect of equivalent-time sampling is to create a facsimile of the original waveform which is identical to a sampled version of the original, except that accumulation of the points on the waveform requires many repetitions of the signal rather than a single one. A simple equivalent-time sampling system can be constructed from a S/H, an ADC, and a variable delay generator (Figure 49). Seydel (033) used a digital delay generator to provide the necessary delay between sync pulse and sample command. The counter shown in Figure 49 determines the delay, and is incremented once for each sync generated.

Sampling oscilloscopes are a type of equivalent-time sampling device which can be used for digitizing an ultrasonic signal. Basically, this unit contains a high-quality, low-aperture-time-uncertainty S/H and a variable delay generator. The S/H output determines the vertical deflection of the beam in a CRT, while the delay generator drives the beam horizontally. A dot is displayed on the screen for each sample. An ADC connected to the vertical output, and with conversion synchronized to the scope's time base, will provide accurate sampling within the restrictions imposed by time-base errors. Some sampling oscilloscopes have an external scan input. The voltage applied to this input determines the delay between sync pulse and sampling. A digital-to-analog converter (DAC) is used to provide a staircase voltage. Thus samples are obtained at equispaced intervals if the digital code applied to the DAC is incremented in equal amounts (Figure 50).

A boxcar integrator, commonly used as a signal averager, has been used by several investigators (Simpson, 298; Kennedy and Woodmansee, 086, 134) for equivalent-time sampling of ultrasonic signals.

> This instrument synchronously samples an input signal with a variable width, variable delay gate, which can be fixed at any point on, or slowly scanned across the input signal. That signal passed by the gate is averaged by variable time constant integrations, the output of which is the average of some number of repetitions of the input signal over the gate width interval. If the mean value of noise is zero, an improvement in signal-to-noise ratio occurs upon averaging a large number of repetitions. If the gate is fixed on a single point of the input signal, the boxcar integrator output rises asymptotically toward the amplitude of the synchronous portion of the input signal at the sampled point. If the gate is scanned across the input signal, a low-frequency replica of the synchronous waveform is reproduced at the output.[‡]

As with the sampling oscilloscope, a DAC-generated voltage can be used to position the gate at discrete locations across the signal. However, because the minimum gate width is rather large (100 nsec), a somewhat different mode of operation is used. Kennedy and Woodmansee and Simpson operated the boxcar integrator in its automatic scanning mode, where the gate is repeatedly scanned through the time interval containing the ultrasonic signal. The output is not discrete; rather, a continuous replica of the waveform is produced, but slowed considerably. The frequency content of the output is low enough so that a conventional tape recorder may be used to store the waveform. A slow ADC is adequate for digitizing either the direct output or the recorded version. For good fidelity of the replica, the scan times are relatively long (5–10 sec). Additionally, if an accurate representation

[‡]From the manual describing operation of the Princeton Applied Research Corporation boxcar integrator.

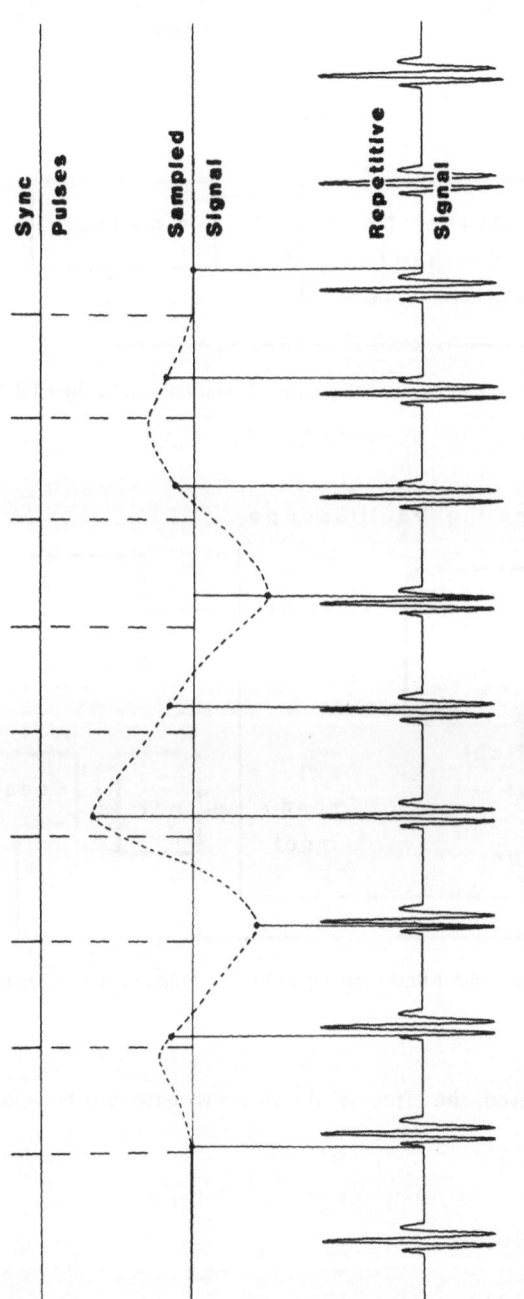

Figure 48. Sampling a waveform in equivalent time (the interval between the sync pulse and the command to sample the waveform is increased for each repetition).

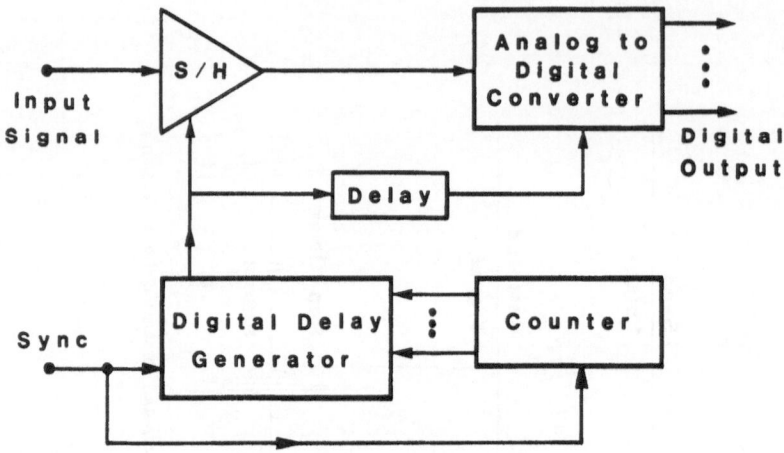

Figure 49. Equivalent-time sampling utilizing a sample-and-hold and a digital delay generator.

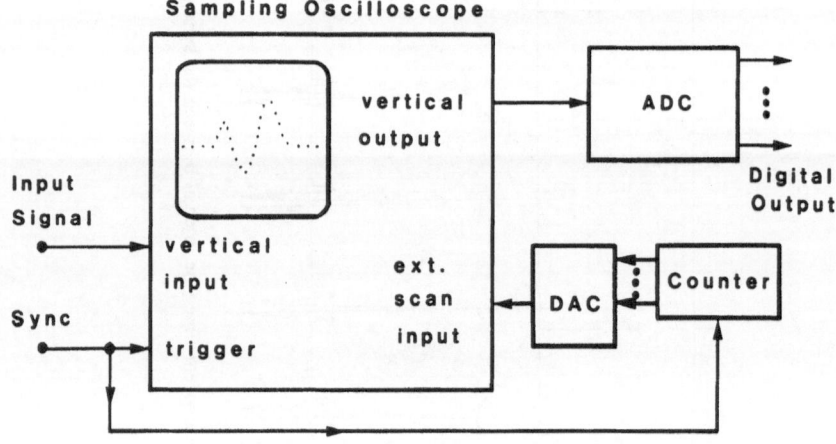

Figure 50. Equivalent-time sampling utilizing a sampling oscilloscope.

of the waveform is desired, the effect of the nonzero gate width should be removed by deconvolution.

Receiving Subsystems

In general, one would like the frequency response of the receiving subsystem to be uniform over a wide range of frequencies. Because of the variations in spectral response of the receiving transducer, some type of compensation must be included elsewhere in the subsystem.

Response Equalization Networks

As was mentioned in the section dealing with the transmitting subsystem, the response of a piezoelectric transducer may be compensated to produce broadband response. The equalization network may be realized acoustically, $C_2(f)$, or electrically, $B_2(f)$, as the reciprocal of the transducer frequency response, $X_2(f)$.

Equalization Network Incorporating a Transmission Line. Lakestani, Baboux, and Fleischmann (223) showed that a length of asymetrically termined transmission line has the required response, for the special case of a transducer backed and loaded such that $R_A = R_B$. Their network (Figure 51) was adjusted for optimum correction by changing the length of the variable-length transmission line and an appropriate choice of Z_L and Z_S. They also mention that when the media on opposite faces of the transducer are different (asymmetrical loading, i.e., $R_A \neq R_B$) correction may be accomplished by cascading two transmission-line correction networks.

The second stage of the two-stage equalization network of Kazys and Lukosevicius (355) is not required when the receiving transducer is operated under open-circuit conditions. The first stage is identical to that described by Lakestani *et al.*

SAW Filter. White (370) has developed a surface-acoustic-wave (SAW) deconvolution filter to provide increased range resolution which also increases the bandwidth of the receiver. The design of the filter is based on a linear, time-invariant model for the receiving subsystem. It closely follows the filter development outlined in the transmitting subsystem section of this report.

The output of the subsystem $v_2'(t)$ is the convolution of the responses of its component [$c_2(t)$, mechanical coupling; $x_2(t)$, receiving transducer; $b_2(t)$, electrical network; and $a(t)$, amplifier]. In the frequency domain

$$C_2(f)X_2(f)B_2(f)A(f) = V_2'(f) \tag{83a}$$

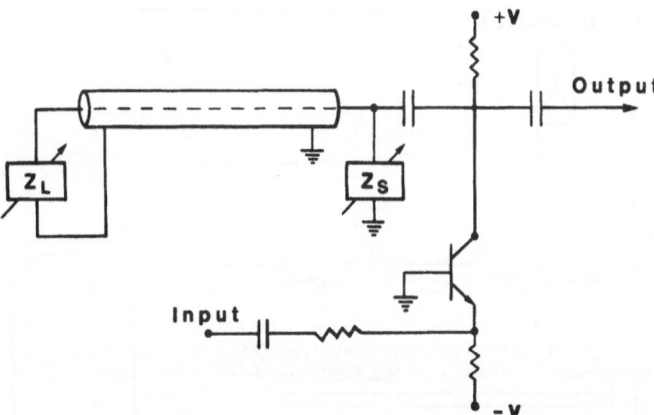

Figure 51. Response equalization network incorporating a variable length of asymmetrically terminated transmission line (after Lakestani *et al.*, 223).

and

$$B_2(f) = \frac{V_2''(f)}{C_2(f)X_2(f)A(f)} \tag{83b}$$

The following steps outline development of a SAW inverse filter:

1. Measure or estimate the frequency response of the mechanical coupling, the receiving transducer, and the amplifier.
2. Choose a $V_2''(f)$: One possible choice would be an approximately flat frequency response over a range of frequencies, with response tapering off at the ends of the range (to minimize ringing in the time domain).
3. Calculate $B_2(f)$ from Eq. (83b).
4. Truncate $B_2(f)$ to a SAW realizable filter.
5. Fabricate the filter.

Figure 52 demonstrates how the SAW filter may be incorporated into the receiving section of an ultrasonic spectroscopic system.

Although the SAW filters worked very well, White noted their limitations:

1. The response is fixed once the filter is fabricated.
2. Since transducers vary, one filter is required for each transducer (or combination of transducers for "pitch-catch" systems).
3. Programmable SAW filters were not commercially available.
4. Limited gain-bandwidth product of interdigital SAWs forces compromise (more "fingers" in the transducer reduce the loss, but also reduce bandwidth.
5. Operation at low frequencies is limited by the large substrate thickness (although heterodyning to higher frequencies is a possibility—the system becomes more expensive).

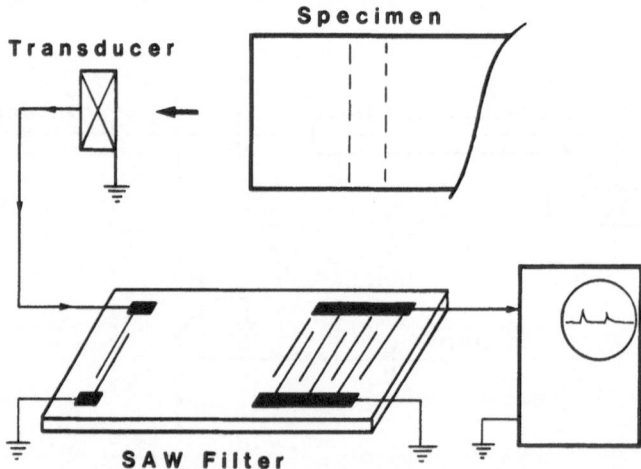

Figure 52. Schematic illustration of ultrasonic receiving subsystem incorporating a SAW filter.

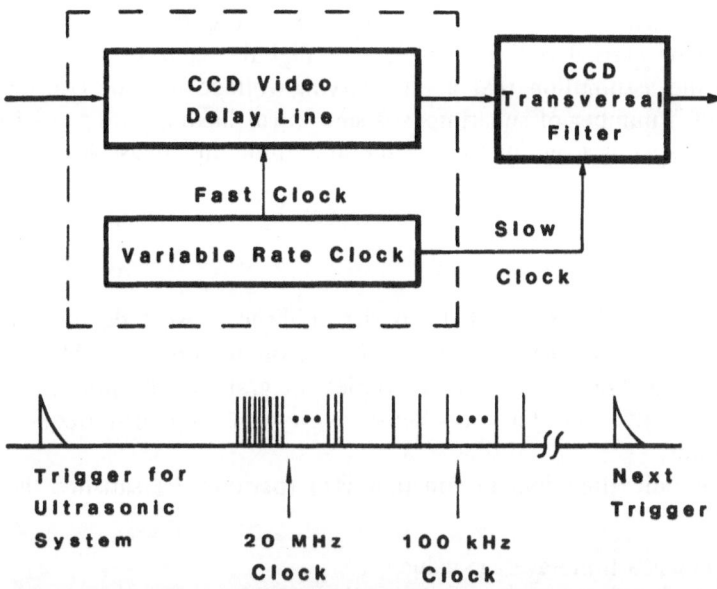

Figure 53. Charge-coupled-device transversal filter incorporating a high speed burst processor (after White, 462).

Programmable CCD Filter. Because of the limitations inherent in SAW filters, White (420) began exploring the use of charge-coupled-device (CCD) transversal filters. The CCD device samples the input voltage and forms discrete charge packets which are shifted along through the device at a rate determined by an external clock. Transversal filtering is carried out by summing a series of weighted samples of the input signal. The individual samples are available at separate taps on the CCD circuit. These programmable filters use a tapped delay line with variable tap weights.

Since the clock frequency may be varied over a wide range, ultrasonic signals with frequency content in the kilohertz to megahertz range may be filtered. In addition to excellent frequency response, CCD filters have negligible insertion loss. Thus the gain-bandwidth product limits of the SAW are overcome.

Because no commercial programmable CCD filters were available, White (462) developed a system with variable (but not as of this writing, programmable) frequency response. Figure 53 shows a block diagram of White's CCD transversal filter system. The video delay line samples the ultrasonic signal (at a rate determined by the fast clock) and temporarily stores them. A slower clock feeds the samples to the transversal filter. The burst processor (video delay line and fast clock) is required because presently available CCD filters are limited to clock frequencies below 5 MHz (ultrasonic frequencies less than 2.5 MHz).

Spectrum Analysis

One may determine the magnitude and phase spectrum at any point in the spectroscopic system where there is an electrical signal. Most commonly, the spectra are determined from the final voltage, $v_2(t)$ (refer to Figure 2). Determination of the spectra may be carried out using analog or digital analyzers. Inherent in all the analysis is the assumption that a time-varying voltage may be considered to be composed of a number of superimposed sine waves having a range of frequencies. The problem becomes one of finding the amplitude and phase of each frequency component.

Fourier Analysis

Fourier Series. In 1822 J. B. J. Fourier published a work detailing a technique for analyzing heat conduction in terms of a trigonometric series. This series forms the basis for the Fourier series and Fourier integral, which allow transformation between physically realizable time-domain waveforms and their frequency-domain representations.

If a periodic time-domain function $y(t)$ (period $= T$) satisfies the Dirichlet conditions

1. $y(t)$ has a finite average value,
2. there is a finite number of finite discontinuities,
3. $y(t)$ has a finite number of turning points (positive and negative maxima),

then $y(t)$ may be expanded into the Fourier series

$$y(t) = \frac{a_0}{T} + \frac{2}{T} \sum_{n=1}^{\infty} (a_n \cos 2\pi f_n t + b_n \sin 2\pi f_n t) \tag{84a}$$

$$= \frac{a_0}{T} + \frac{2}{T} \sum_{n=1}^{\infty} A_n \cos(2\pi f_n t - \theta_n) \tag{84b}$$

where

$$f_n = n f_1 \tag{85a}$$

$$\theta_n = \tan^{-1} \frac{b_n}{a_n} \tag{85b}$$

and

$$A_n = (a_n^2 + b_n^2)^{1/2} \tag{85c}$$

The Fourier coefficients a_0, a_n, and b_n are given by

$$a_0 = \int_0^T y(t) \, dt \tag{86a}$$

$$a_n = \int_0^T [y(t) \cos 2\pi f_n t] \, dt \tag{86b}$$

and

$$b_n = \int_0^T \left[y(t) \sin 2\pi f_n t \right] dt \tag{86c}$$

or equivalently

$$y(t) = \frac{1}{T} \sum_{n=-\infty}^{\infty} C_n \exp(j2\pi f_n t) \tag{87}$$

where

$$C_n = \int_0^T y(t) \exp(-j2\pi f_n t) \, dt \tag{88}$$

C_n is a complex number of the form

$$C_n = a_n - jb_n = |C_n| \phi_n \tag{89}$$

and

$$C_0 = a_0 \tag{90a}$$

$$|C_n| = A_n \tag{90b}$$

and

$$\phi_n = \theta_n \tag{90c}$$

The Fourier series (just presented in several mathematical forms) states that a periodic signal $y(t)$ may be considered to be the sum of a dc level and a series of superimposed sine waves. The lowest frequency component (f_1) (sine wave), called the fundamental, has a period equal to that for $y(t)$ (i.e., T). The remaining frequency components (harmonics) are integer multiples of the fundamental (i.e., $2f_1, 3f_1, \ldots$). The dc component is given by C_0/T, and the magnitude of the nth harmonic is $2|C_n|/T$.

Several points should be noted concerning the Fourier series. An infinite number of frequencies may be required to describe an arbitrary signal $y(t)$. In order to determine the magnitude of the harmonics, one must evaluate an integral. This limits analysis to waveforms described by simple analytical expressions. Finally, one is limited to periodic waveforms. Although the process may appear somewhat complex, the phase (θ_n) as well as the magnitude of the harmonics may be computed.

Fourier Integral. One may approach the problem of analysis of transient signals by considering the Fourier series and letting the period (T) approach infinity. The series reduces to

$$y(t)=\int_{-\infty}^{\infty} Y(f)e^{j2\pi ft}\,df \tag{91}$$

The Fourier coefficients have become a continuous function of frequency, and are given by

$$Y(f)=\int_{-\infty}^{\infty} y(t)e^{-j2\pi ft}\,dt \tag{92}$$

Equation (92) is called the direct Fourier transform, and Eq. (91) is termed the inverse Fourier transform. Together they form a Fourier transform pair.

Discrete Fourier Transform. Thus far we have expressions, in the Fourier series and integral, for determining the frequency content of periodic or transient time domain signals. However, as has been noted, evaluation of the required integrals limits frequency analysis to rather simple waveforms. Fortunately, an expression denoted the discrete Fourier transform (DFT) is available which is applicable to a wider class of waveforms. The DFT operates on a set of samples of the time-domain signal (taken at equally spaced intervals, t_s). It gives a set of frequency-domain coefficients (at equally spaced frequencies, f_s).

The DFT pair is

$$Y(kf_s)=t_s \sum_{n=0}^{N-1} y(nt_s)\exp(-j2\pi kf_s nt_s) \tag{93}$$

$$y(nt_s)=f_s \sum_{k=0}^{N-1} Y(kf_s)\exp(j2\pi kf_s nt_s) \tag{94}$$

where N is the number of samples, k and n are the frequency- and time-domain indices (they take on the values $0,1,2,\ldots,N-1$), $y(nt_s)$ is the set of discrete time samples, and $Y(kf_s)$ is the set of Fourier coefficients. Additionally,

$$\text{time-record length (window length)}=Nt_s \tag{95a}$$

$$\text{frequency sampling interval}=f_s=1/(Nt_s) \tag{95b}$$

$$\text{frequency range}=Nf_s=1/t_s \tag{95c}$$

Expressions (93) and (94) allow one to calculate a set of frequency-domain samples given a set of time-domain samples and vice versa. Since no integral expressions are involved, any sampled signal satisfying the Dirichlet conditions may be transformed to give its magnitude and phase spectra. The discrete Fourier transform should not be thought of as an approximation to the Fourier integral. Rather, it is an exact transform which operates on discrete-time data.

Fast Fourier Transform. Although the class of waveforms which are easily transformable has been increased by changing from the Fourier integral or series to the DFT, the calculation time is still relatively long. For each Fourier coefficient, N

Table 2
Number of Mathematical Operations Required to Evaluate the Fourier Coefficients

Number of samples	Number of operations	
	Discrete Fourier transform N^2	Fast Fourier transform $N\log_2 N$
16	256	64
32	1024	160
64	4096	384
128	16384	896
256	65536	2048
512	262141	4608
1024	1048576	10240
2048	4194304	22528
4096	16777216	49152

complex multiplications and additions are required. Since there are N Fourier coefficients, a total of N^2 operations must be performed. Even with a computer implementation of the DFT, processing times are quite long for a moderate number of samples. If N is 1024, then $(1024)^2$ or 1,048,576 mathematical operations (on complex numbers) are required.

An algorithm for reducing the number of operations required to evaluate the DFT was advanced by Danielson and Lanczos in 1942. In the early 1960s Cooley and Tukey independently "rediscovered" this algorithm. Evaluation of the DFT for $N=2^m$ (m an integer) by this technique requires only $N\log_2 N$ operations. Table 2 demonstrates the dramatic decrease in computation time made possible by the fast Fourier transform (FFT).

The term *fast Fourier transform* refers to any algorithm which reduces the number of operations to less than N^2. Most commonly the sample size is restricted to a power of two (radix 2); however, mixed-radix algorithms have been developed. The radix 2 algorithms remain the fastest.

Implementation of the FFT in hardware as well as software is possible. Dedicated FFT analyzers and array processors are capable of performing a 1024-point transform in as little as 77 μsec. Digital-processing oscilloscopes capable of a wide variety of signal processing (including FFT) at the touch of a button provide another· analysis alternative. As for software, the FFT has been programmed in almost any computer language imaginable. A FORTRAN implementation of the radix-2 FFT is included as Figure 54. An ALGOL version of the FFT may be found in an article by Singleton (349). Singleton (350) has also published a FORTRAN sub-routine for a mixed-radix FFT.

Analog Spectrum Analyzers

Bandpass Filters. Probably the simplest analog device for extracting the amplitude of a frequency component is a bandpass filter tuned to that frequency. Ideally, only the center frequency (f_0) would pass through the filter. The peak-to-peak voltage at its output is the magnitude of that component. In actuality, the ideal

```
      SUBROUTINE FFT(A,M)
      COMPLEX A(1024),U,W,T
      N= 2**M
      NV2 = N/2
      NM1 = N-1
      J=1
      DO 7 1=1,NM1
      IF(1.GE.J) GO TO 5
      T = A(J)
      A(J) = A(1)
      A(1) = T
    5 K=NV2
    6 IF(K.GE.J) GO TO 7
      J = J-K
      K=K/2
      GO TO 6
    7 J= J+K
      P1 = 3.14159265358979
      DO 20 L=1,M
      LE = 2**L
      LE1 = LE/2
      U = (1.0,0.)
      W=CMPLX(COS(P1/LE1),SIN(P1/LE1))
      DO 20 J=1,LE1
      DO 10 1=J,N,LE
      1P = 1+LE1
      T=A(1P)*U
      A(1P)=A(1)-T
   10 A(1)=A(1)+T
   20 U=U+W
      RETURN
      END
```

Figure 54. FORTRAN subroutine for calculation of the radix-2 FFT (Cooley, Lewis, and Welch, 348).

bandpass filter is not realizable. Real devices constructed of passive or active components have a passband spreading over a range of frequencies. If the signal being measured contains other frequencies adjacent to the center frequency (within the passband), the filter output will include contributions of these nearby frequencies. Although not ideal, filters are available which are quite selective.

If there is interest in more than a single frequency, a bank of bandpass filters could be used, with each tuned to a different frequency. One implementation of the multifilter approach and its frequency characteristics are shown in Figure 55.

An alternative to utilizing many bandpass filters is to employ a single filter whose center frequency may be swept over a range of frequencies. Figure 56 illustrates the relationship between the frequency and display for such a device. During the time the center frequency of the filter sweeps through the range f_1-f_3, the display (strip chart or CRT) changes position from 0 to 10 divisions. As with any of the bandpass filters, the frequency domain resolution may be increased by narrowing the filter's passband.

Superheterodyne Analyzers. The present state-of-the-art in electronically variable bandpass filters places a limit on the sweep range. To circumvent this problem, a sweeping superheterodyne system is used. The swept bandpass filter relies on moving the passband of the device relative to the frequencies in the signal. The same

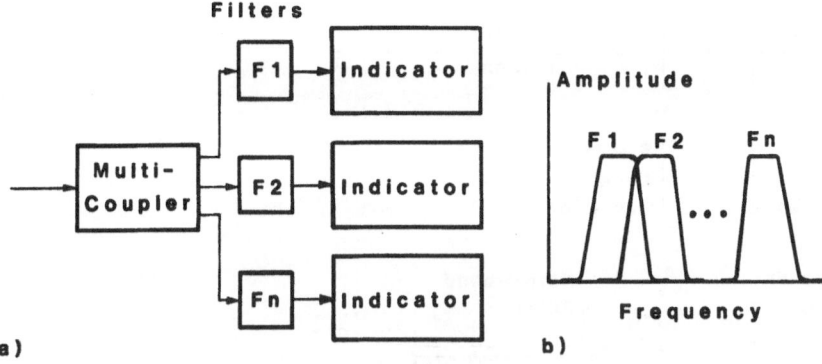

Figure 55. (a) Filter bank system for spectrum analysis and (b) its frequency characteristics (after Engelson, 154).

effect can be obtained by sweeping the frequencies of the signal past a bandpass filter with a fixed center frequency. The signal sweep is performed by a mixer. This device has two inputs (an rf frequency and one from a local oscillator, l.o.) and one output. The signal leaving an ideal mixer will be intermediate frequencies (i.f.) with amplitudes equal to those of the signal input, but whose frequencies consist of the algebraic sum and difference of the frequencies of the l.o. and rf inputs.

Consider the superheterodyne system shown in Figure 57. Suppose that the output at the mixer is filtered such that only the frequencies given by

$$f_{\text{i.f.}} = f_{\text{l.o.}} - f_{\text{rf}}$$

remain. Let us follow signal rf frequencies f_1 and f_2 through the system. The local oscillator sweeps from f_3 to f_4. As it does so, the mixer output, with f_1 as input, ranges from $f_3 - f_1$ to $f_4 - f_1$. Similarly, for input f_2, the output ranges from $f_3 - f_2$ to $f_4 - f_2$. The output of the mixer is connected to a bandpass filter (narrowband i.f. amplifier) having a fixed center frequency. When the signals $f_{\text{l.o.}} - f_1$ and $f_{\text{l.o.}} - f_2$ are within the passband of the filter, the indicator rises above baseline to a height which is directly related to the amplitude of the frequency components f_1 and f_2.

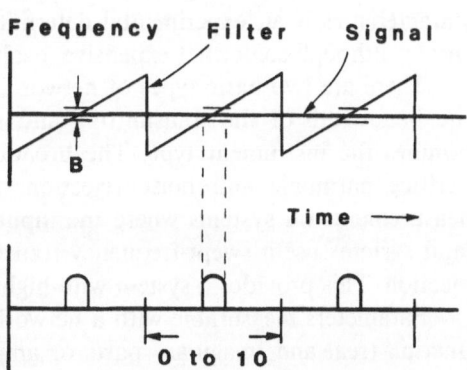

Figure 56. Frequency–time position relationships for a swept filter system. B is the filter bandwidth (after Engelson, 154).

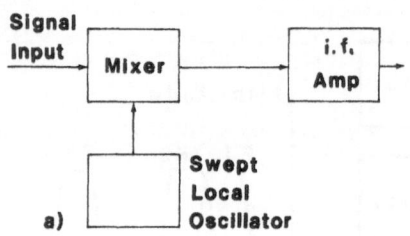

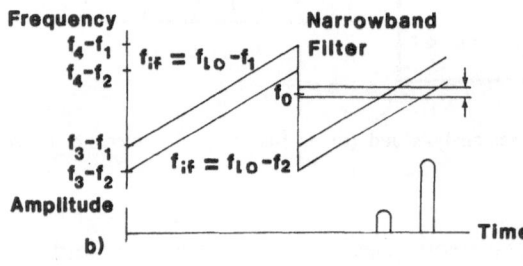

Figure 57. Superheterodyne system for spectrum analysis (after Engelson, 154). (a) Swept signal system, (b) frequency–time relationships.

Most modern spectrum analyzers incorporate several superheterodyne and bandpass stages to increase frequency resolution (Figure 58). Analyzers covering the subhertz to gigahertz range are commercially available and perhaps offer the easiest path to spectrum analysis. Connecting the signal output from a stepless gate to the analyzer produces a display of the magnitude spectrum.

Other Analog Analyzers. In addition to spectrum analyzers, several other analog devices should be mentioned as possibilities for performing signal analysis. These include wave analyzers (a manually tunable bandpass filter; see Frederick and Seydel, 001 and 009), vector voltmeters (Ting and Sachse, 253), and network analyzers. Since network analyzers have received little attention in ultrasonic work and yet may be of some use, a brief description is in order.

Network Analyzer. The purpose of a network analyzer is to characterize the behavior of linear electrical networks. For frequencies in the kilohertz range, systems can be evaluated by analyzing the network consisting of discrete components. As the frequency is raised higher, parasitic capacitances and inductances begin to become important. Since the circuit layout and variation in individual components usually cannot be accounted for, the best approach to the problem of defining transfer characteristics is an experimental determination. The network analyzer provides a simple, although somewhat expensive, evaluation of network characteristics.

There are two basic types of network analyzers—narrowband and broadband. The bandwidth of the transmitting and receiving sections of the instrument determines the instrument type. The broadband instruments are lower in cost, but sacrifice harmonic and noise rejection. Broadband instruments can also make measurements on systems where the input and output frequencies differ. Narrowband systems use a swept-frequency transmitter and tuned receiver to attain noise rejection. This provides a system with high accuracy and wide dynamic range.

Parameters measurable with a network analyzer (NA) include: system transfer function (real and imaginary parts or amplitude and phase), complex impedance,

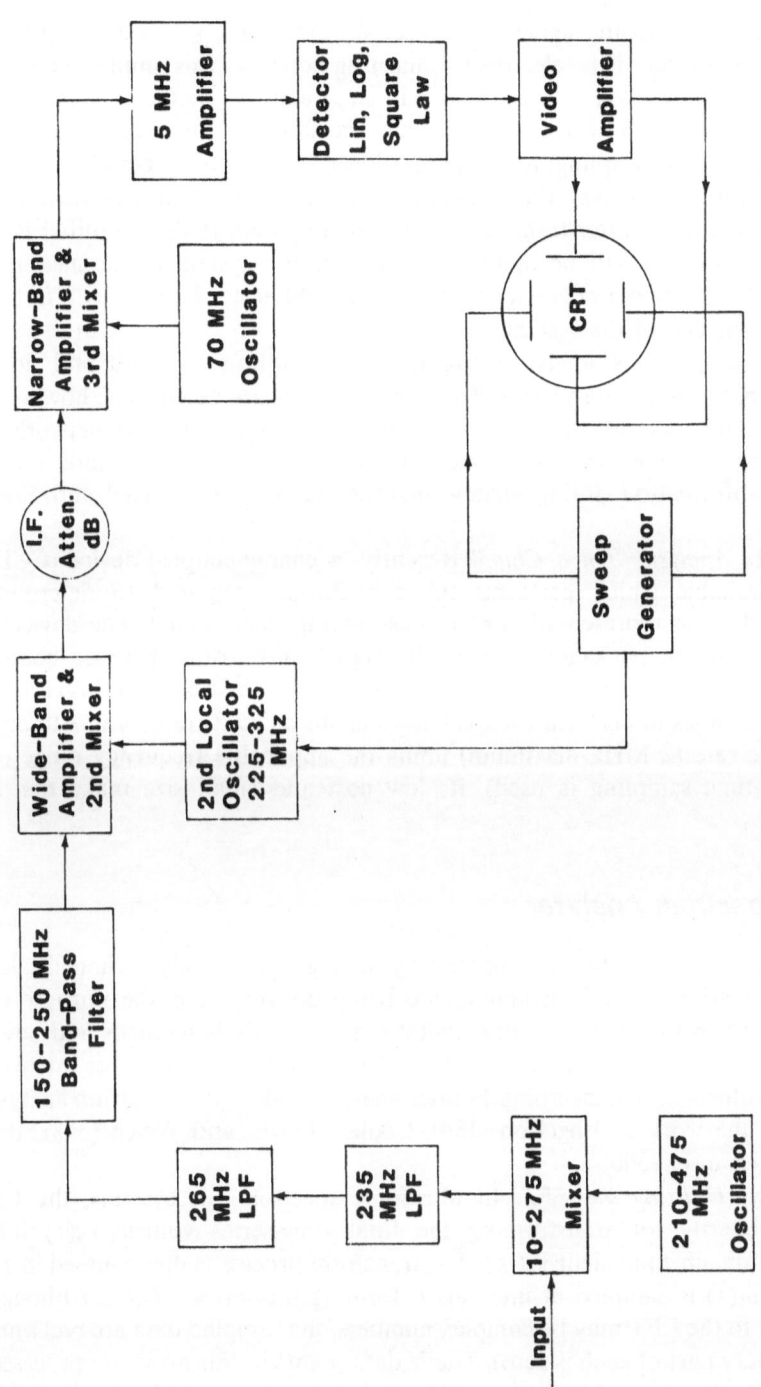

Figure 58. Block diagram of a wideband spectrum analyzer (Tektronix, model IL20).

transmission and reflection characteristics, and group delay (first derivative of phase with respect to frequency).

The amplitude of the signal source used for excitation is very low (1 μV to 1 V). For characterization of the electrical connecting networks this amplitude is satisfactory. However, if the NA were to be used in an attempt to characterize the system of electrical coupling network, transmitting transducer, ultrasonic coupling, test medium, ultrasonic coupling, receiver transducer, and electrical coupling network, a voltage amplifier between the network analyzer source and electrical coupling network to the transmitter transducer would be required. If this amplifier is added, its transfer function will be included in that of the system under measurement. However, if the transfer characteristics of the amplifier are known, its effect can be removed from that of the system.

In general, analog analyzers provide only amplitude spectra and no phase information. Network analyzers and vector voltmeters are exceptions; however, both devices require a swept reference oscillator as an input to the network under investigation. A somewhat complicated, though straightforward, method for retaining phase information during analog analysis was used by Seydel and Frederick (001, 009).

Fourier Analysis "On a Chip." Recently, a charge-coupled-device (CCD) has become available which performs the convolutions required to determine the frequency domain representation of a time-varying input signal. The device* (in a 22-pin dual-in-line package) samples the signal and stores these as charges on internal capacitors. Mathematical operations (digital transversal filtering) are carried out on the charges to perform what is known as the chirp Z transform. Although the low sample rate (2 MHz maximum) limits the applicable frequency range (unless equivalent-time sampling is used), its low cost and small size make this device attractive.

Digital Spectrum Analyzer

Because of the limitations imposed by analog signal analysis, namely, lack of phase information, digital techniques are being utilized. It is the purpose of the following section to briefly describe digital spectral analysis methods and how they may be used to advantage in ultrasonic spectroscopic systems. Readers interested in additional information concerning Fourier analysis and digital spectrum analysis are referred to the work of Engelson (154), Cooley, Lewis, and Welch (348), Burgess (351), and Ramirez (500).

Analysis Utilizing the FFT. In ultrasonic spectroscopic systems, the FFT is utilized primarily for transforming the final time-series voltage $v_2(t)$ into its frequency-domain equivalent $V_2(f)$. The transform process is diagrammed in Figure 59. Signal $v_2(t)$ is sampled at intervals t_s forming the series $v_2(nt_s)$. Although the input array to the FFT may be complex numbers, the sampled data are real numbers (the imaginary part of each is zero). The N data points in this array are processed by a FFT algorithm to yield a set of N complex numbers which are the real and

*Quad chirped transversal filter, R5601, Reticon Corp., 910 Benicia Avenue, Sunnyvale, California 94086.

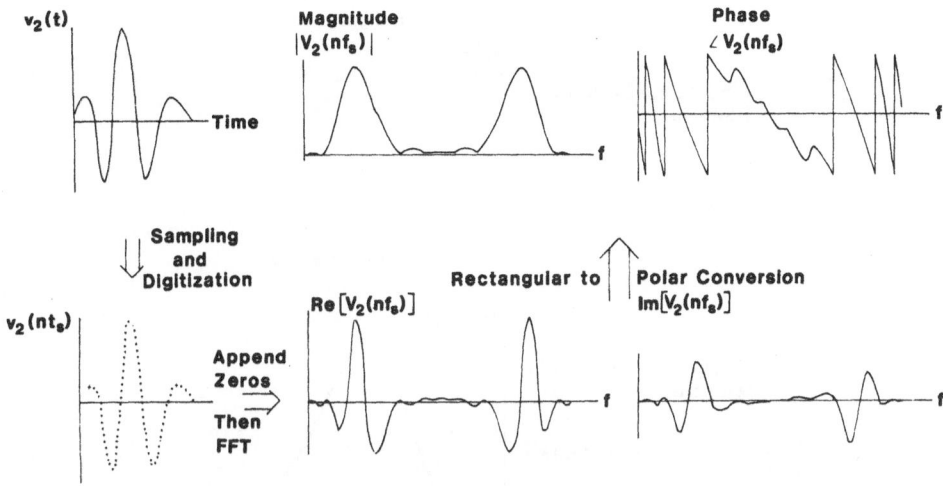

Figure 59. Determination of the frequency-domain equivalent of a time-domain signal utilizing the FFT.

imaginary parts of the frequency components $V_2(nf_s)$. The magnitude and phase spectra are then calculated from these complex numbers.

Notice that the spectrum from $(N/2)f_s$ to $(N-1)f_s$ is a mirror image of the spectrum in the range 0 to $(N/2-1)t_s$. That is, the spectrum is folded about the point $[(N/2)-1/2]f_s$. Spectral folding arises from sampling of $v_2(t)$.

The aspects of sampling a waveform were covered in a previous section. Recall that no information is lost from a band-limited signal when sampled at a rate greater than twice the highest frequency in the signal. With this limitation of minimum sample rate in mind (to eliminate aliasing), the choice of sampling frequency $(1/t_s)$ must be based on considerations of the desired frequency range (95c) and resolution (95b). Ideally the time interval between samples should be long and the number of samples large.

For those new to Fourier analysis the fact that coarse time-domain sampling leads to fine resolution in the frequency domain may seem a bit paradoxical. Figure 60 demonstrates the spectra obtainable with large and small intersample periods, but both having sample rates above the minimum Nyquist rate. The same sample size, N, was used for each.

A useful technique for increasing frequency-domain resolution which does not involve changing the sample rate is to increase the sample size (N). The sampled data occupy part of the array while the remaining positions in the array are filled with zeros. It is perhaps easiest to let the actual data occupy the first part of the array (Figure 61a); however, placing the data as shown in Figure 61b is preferred because it results in the minimum variation of the phase spectrum. Shifting the data points in the array (because of the assumed periodicity of the time-series signal) leaves the magnitude spectrum unchanged.

In addition to providing the frequency-domain representation of a time series, the FFT permits one to estimate the frequency response of linear systems, and to easily perform convolution, deconvolution, and correlation. At the beginning of this

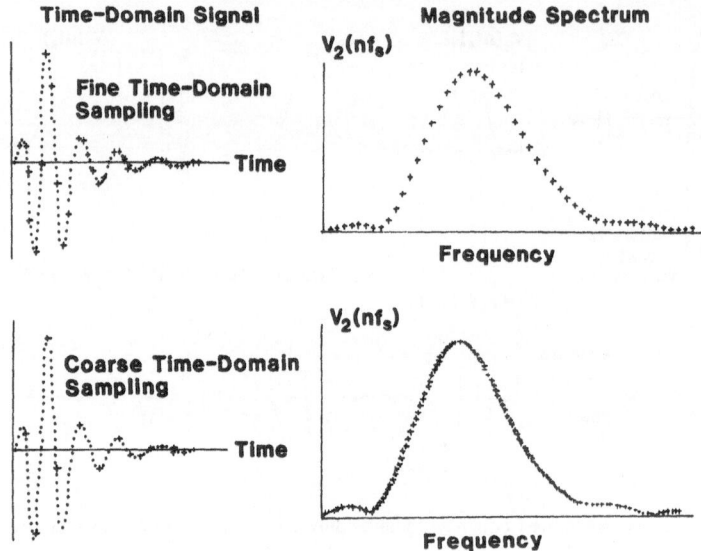

Figure 60. For the same sample size, coarse time-domain sampling (above the Nyquist rate) results in a more closely spaced frequency sampling interval.

chapter of the report the ease was demonstrated with which time-domain convolution and deconvolution could be performed in the frequency domain. The FFT is used to transform the measurable time series into frequency components where convolution becomes a multiplication and deconvolution a division.

Now that we have established that FFT analysis is valuable, we would like to offer some hints for improving results.

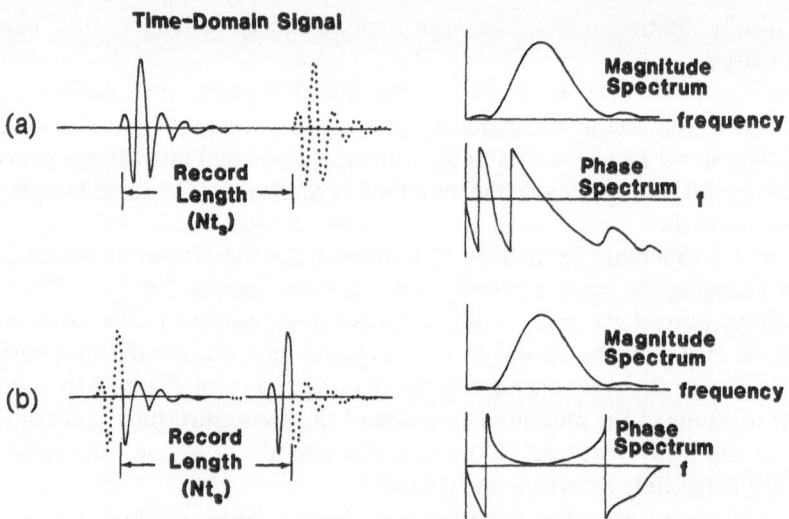

Figure 61. Shifting the time-series data points within the window affects only the phase spectrum. The assumed periodicity in the time series is shown dotted outside the data record. (a) Signal at the beginning of the record, (b) signal split between the beginning and the end of the record.

1. Perform signal averaging on the time-series data. Noise and jitter in the electronics probably have a zero mean value, and so averaging tends to improve signal to noise (S/N). The improvement in this ratio for M averages is $M^{1/2}$. Thus, if $M=2^m$ signals are averaged, the S/N will improve by

$$m \cdot 3 \quad \text{dB}$$

A minimum of 32 averages (Ramierez, 300) is recommended for relatively noise-free signals, while for moderately noisy signals 128–512 averages may be required.

2. Determine the highest frequency component (f_{max}) of the signal, or alternatively, use a low-pass filter to limit the highest frequency to f_{max}. Choose a sample rate $(1/t_s)$ which is at least twice (and preferably three times) f_{max}. This choice sets the frequency range for the FFT to a value greater than f_{max}. Next, decide on the desired spacing of points in the frequency domain (f_s'). The smallest sample size N for a radix-2 FFT is the minimum value which satisfies the inequality

$$N \geq 1/f_s' t_s$$

subject to $N=2^m$ ($m=$ an integer). The time-series array should contain the samples of the signal with zeros filling the remainder of the array to bring the total number of data points to N.

3. Remove the mean value. If the acquired time-series samples have a dc bias, subtract it from all samples before performing the transform.

4. Be aware of the assumed periodicity in the sampled data. Arrange the signal within the window so that there are no discontinuities.

5. If a stepless gate is utilized, remove the samples containing the switching transients created at the beginning and end of the gated interval.

6. When dealing with data arrays of real numbers, two data arrays can be simultaneously transformed in one transform. Let the two arrays be denoted y_1 and y_2. Form the complex data array z as

$$z(n)=y_1(n)+jy_2(n) \tag{96}$$

Perform the Fourier transform, which gives

$$Z(k)=Z_1(k)+jZ_2(k) \tag{97}$$

Then

$$\mathrm{Re}[Y_1(k)]=\frac{Z_1(k)+Z_1(N-k)}{2} \tag{98a}$$

$$\mathrm{Im}[Y_1(k)]=\frac{Z_2(k)-Z_2(N-k)}{2} \tag{98b}$$

$$\mathrm{Re}[Y_2(k)]=\frac{Z_2(k)+Z_2(N-k)}{2} \tag{98c}$$

$$\mathrm{Im}[Y_2(k)]=\frac{Z_1(N-k)-Z_1(k)}{2} \tag{98d}$$

Additional Analysis Techniques

Multiple Signals. Before closing the discussion on signal analysis, several additional techniques which have been found to be useful in analysis of NDT-produced spectra will be presented. One of these methods involves analysis of the case when two separate signals (two ultrasonic echoes) are Fourier transformed as an ensemble (Simpson, 002, 013). The magnitude spectrum of two identical signals separated in time by $2t_0$ is shown to be

$$|2\cos 2\pi ft_0||Y(2\pi f)| \tag{99}$$

where $Y(2\pi f)$ is the Fourier transform of *one* echo. Thus if the two echoes are gated into a spectrum analyzer, the envelope of the spectrum will be identical to that for a single echo. However, the spectrum will now be modulated. The spacing of the frequency minima (Δf) may be used to determine the time separation of the ultrasonic echoes (Δt):

$$\Delta f = 1/2t_0 = 1/\Delta t \tag{100}$$

Additionally, if the ultrasonic velocity is known, the thickness of material between the echo-producing structures may be found. Where Δt is small (the signals overlap), a wide-bandwidth spectrum analyzer is required because Δf will be large.

For two ultrasonic echoes, one modified by frequency-dependent attenuation $k(2\pi f)$, the spectrum is given by

$$\left\{[1-k(2\pi f)]^2 + 4k(2\pi f)\cos^2 2\pi ft_0\right\}^{1/2}|Y(2\pi f)| \tag{101}$$

The spectrum is similar to that for the two identical echoes except the minima do not reach zero. Equating the modulation in a measured spectrum to that predicted from Eq. (101) allows the attenuation to be determined. In addition to measurement of $k(2\pi f)$, Simpson (13) also suggests the possibility of measuring acoustic impedance through analysis of the phase shift produced at boundaries.

The Cepstrum. Returning to the spectral modulation produced as the result of placing two signals within the data window, we note that the variation is periodic. The modulation frequency can be determined by performing a Fourier transform on the spectrum. Since the Fourier transform is taken twice during this type processing, the units of the ordinate will be time. A peak will be found which corresponds to the time separation of the echoes. This technique, called cepstrum processing, was first outlined by Bogert, Healey, and Tukey (347). By definition the cepstrum is the inverse Fourier transform of the logarithm of the Fourier transform of the time-domain signal. In some cases deleting the logarithm step results in improved performance in the presence of noise (Loew, Shankar, and Mucciardi, 222). As noted by Morgan (085) cepstral processing can be carried out by analog processing. An analog implementation requires two spectrum analyzers—the first forms the log spectrum, the second (fed the vertical display signal from the first) displays the cepstrum.

When more than two ultrasonic echoes having unequal time separation are included in the data window, the spectrum is modulated by the time interval between each pair of echoes. Although the spectrum may look quite complex, analysis of the cepstrum is straightforward. There will be peaks in the cepstrum corresponding to the interval between each pair of echoes. Figure 62 demonstrates the cepstral technique applied to three ultrasonic echoes. The power cepstrum, outlined above, is but one of several cepstral techniques discussed in a review article by Childers, Skinner, and Kemerait (206).

When a reference spectrum is available (system response interrogating a lossless media), a modified version of the cepstrum (Morgan, 085) may be calculated as follows:

1. Divide the experimentally determined spectrum by the reference spectrum (a "deconvolution") leaving the modulating part of the spectrum introduced by the experiment.

2. Remove the dc component.

3. Take the logarithm of the spectrum derived in steps 1 and 2.

4. Fourier transform the log spectrum, giving the desired cepstrum. This technique gives superior results when some of the echoes are weak and the spectral modulation is small.

Other Transformations. Throughout this report, and most of the references in the Bibliography, the Fourier transform has been used as the link between the time and frequency domain representations of a signal.

The Fourier transform "decomposes" a time series into a weighted set of orthogonal functions, namely, sinusoids. However, analysis of time-domain waveforms may also be performed utilizing transformations to other sets of orthogonal

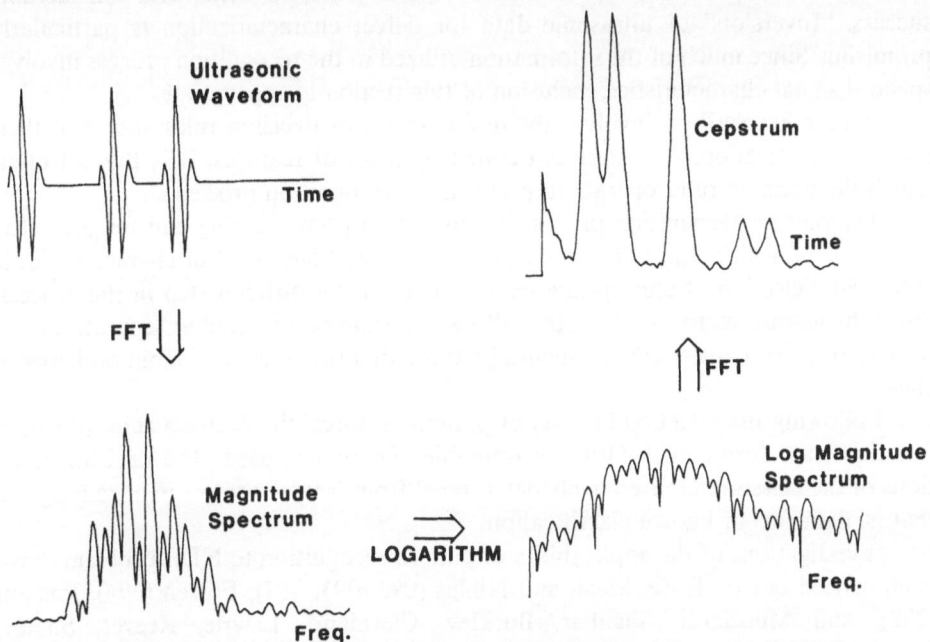

Figure 62. Cepstral analysis for determining the separation of time-domain signals.

functions (e.g., square waves, or polynomials). These alternate transformations may prove useful for extracting information carried by the ultrasonic waves (Mast and Rose, 092).

Maximum-Entropy and Maximum-Likelihood Spectrum Analysis. When performing analysis utilizing the fast Fourier transform, the frequency domain resolution may be improved by appending zeros to the data record before transformation. However, processing time increases as the record is lengthened. Recently, several alternate methods have been used for spectral estimation to achieve improved spectral resolution (499). These techniques, the maximum-entropy method (MEM) and maximum-likelihood method (MLM) are attractive when the available data record is short and the signal is contaminated by noise. The methods are data adaptive in that when an estimate is being made at one frequency, the methods adjust themselves to be least affected by all other frequencies.

For power-spectral-density determinations, using the FFT, one normally calculates an estimate of the aurotcorrelation function, windows it, appends zeros, and performs a Fourier transform. The maximum-entropy method utilizes a technique for estimating the autocorrelation function beyond the range limited by the finite number of data samples. The estimate is that which is the most random (maximum entropy) of any power spectrum consistent with the limited data record.

The maximum-likelihood method, like the MEM, provides increased spectral resolution. The MLM performs a filtering operation which passes the power in a narrowband about the frequency of interest while minimizing the effects of other spectral components (e.g., noise). The MLM gives a minimum-variance, unbiased estimate of the spectrum.

Pattern Recognition. Mathematical pattern recognition techniques, although applied to NDE problems only in the last several years, have met with considerable success. "Inversion" of ultrasonic data for defect characterization is particularly promising. Since much of the information utilized in the recognition process involves spectral signal characteristics, inclusion of this section is appropriate.

Pattern recognition involves the development of decision rules and then their use for classification. A pattern is defined by a set of features. It is this set upon which the decision rules operate to perform the recognition process.

The pattern recognition process begins with a preprocessing and feature selection step, in which the ultrasonic signal is acquired and a set of characteristics is extracted. Selection of appropriate features is the most difficult step in the process. Since the feature vector must contain all the information required to separate classes of patterns, the features chosen should be those that provide discrimination between classes.

Following the selection of a set of pattern features, the decision rules (decision functions) are formulated. Often, a trainable classifier is used. The decision functions of the classifier are iteratively determined from feature vectors in a training set, that is, data sets of known classification.

Investigations of the applicability of pattern recognition to NDE problems have been carried out by Rose, Mast, and Niklas (092, 093, 187), Forsen (394), Preston (288), and Mucciardi, Shankar, Buckley, Cleveland, Lawrie, Reeves, Shaley, Johnson, and Whaley (094, 126, 141, 380, 393, 426, 447, 454, 481).

Complete Ultrasonic Spectroscopic Systems

In previous sections, the components of ultrasonic spectroscopic systems were reviewed. Incorporation of these parts into complete systems is discussed in the paragraphs which follow. The advantages and shortcomings of each type of system are noted.

Continuous-Wave System—Single-Frequency (Narrowband) (Figure 63). The simplest spectral analysis system is a continuous-wave (cw) device operated at a number of discrete frequencies. An oscillator providing a single-frequency sinusoidally varying voltage drives a transducer at (or near) its fundamental resonant frequency (or its odd harmonics). A tuned amplifier boosts the received signal. The amplitude of the received signal is its peak-to-peak voltage, while the phase may be calculated from a measurement of the time shift of the received waveform relative to the transmitted signal. Although the cw technique is most often associated with the measurement of attenuation and velocity, it could be used to determine the ultrasonic energy scattered from a defect or interacting with a surface.

Examples of systems of this type may be found in references 082 and 237.

Advantages of the system include the following:

1. It is simple.
2. It is relatively inexpensive (unless the "optional" items in Figure 63 are included).
3. Very accurate measurements are possible by use of resonance techniques.

Disadvantages of the system include the following:

1. It is not able to pinpoint the region of investigation (the region of interaction is the entire section of the sample intercepted by the ultrasound beam).
2. Only a limited number of frequencies (fundamental and several odd

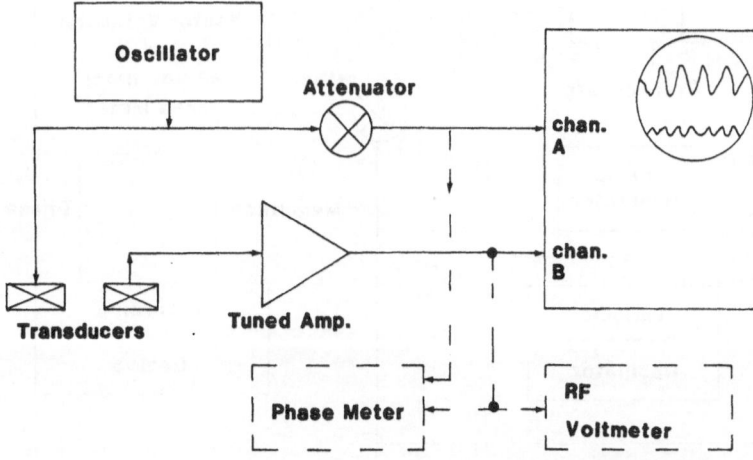

Figure 63. Narrowband continuous-wave ultrasonic spectroscopic system.

harmonics) may be investigated with a single transducer. The spectrum is limited to a small number of points (poor frequency resolution).

3. If several transducers are utilized, the changing coupling conditions must be accounted for.
4. Absorption of the ultrasonic waves can cause appreciable heating.

Further characteristics of the system are as follows.

Transmitter: Continuous-wave radiofrequency oscillator; $V_1(f)$ should have a very narrow bandwidth.

Transducers: Usually quartz disks, air backed to produce narrowband $X_1(f)$ and $X_2(f)$ (alternatively, broadband transducers could be utilized with some degradation in spectral purity).

Amplifier: Tuned amplifier; small bandwidth $A(f)$, so high gain and signal-to-noise ratio are possible (variable tuning or wide bandwidth required if a broadband transducer is used).

Analysis: Amplitude and phase measurements from amplitude and time shift of waveform displayed on an oscilloscope (a vector voltmeter may be used for greater accuracy).

Continuous-Wave System—Swept Frequency (Figure 64). The frequency resolution of a continuous-wave system may be improved by slowly sweeping the oscillator over a range of frequencies. For highest signal-to-noise ratio a variable tuned amplifier should be used which sweeps in unison with the oscillator. The output voltage of the oscillator must be constant or known over the band of frequencies. Unless the frequency sweep is slow enough to permit oscilloscope readings, a rf voltmeter and phase meter are required, with their outputs connected to a recording device.

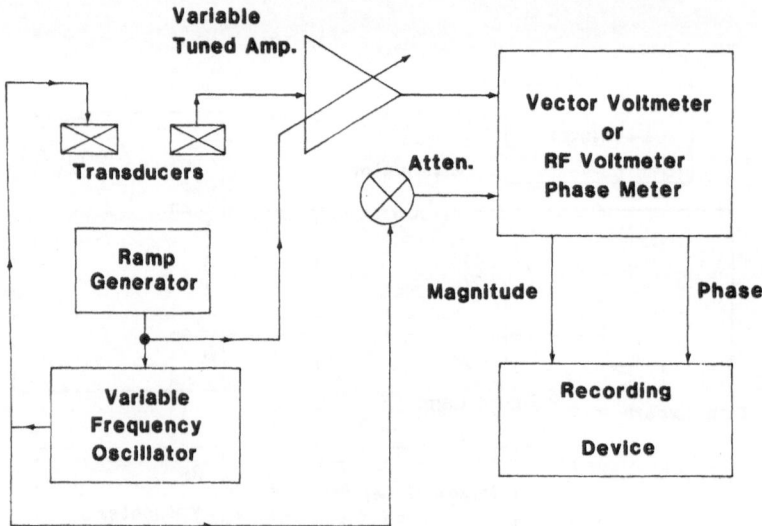

Figure 64. Swept-frequency cw ultrasonic spectroscopic system.

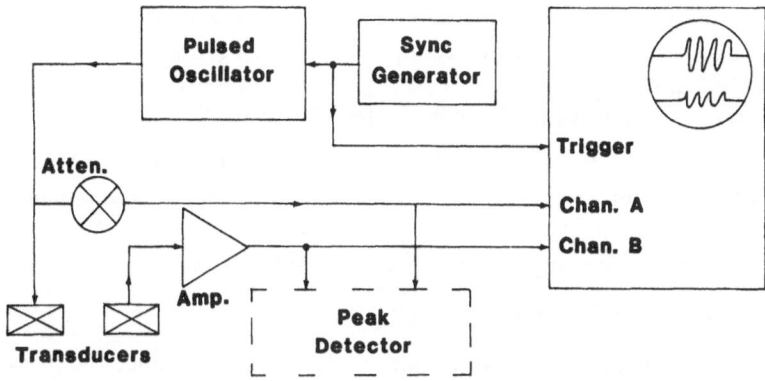

Figure 65. Single-frequency pulse-burst ultrasonic spectroscopic system.

Examples of systems of this type may be found in references 064, 082, 229, 253, 360, and 434.

Advantages of the system include the following:

1. It provides increased frequency resolution.
2. "Automated" measurements are possible over a range of frequencies.

Disadvantages include the following:

1. One is not able to precisely locate the region of investigation in the sample.
2. Absorption can cause undesirable heating.
3. Broadband transducers are required.
4. The system is more expensive because of incorporation of the vector voltmeter and recording device.

Further characteristics of the system are as follows.

Transmitter: rf oscillator having a narrowband $V_1(f)$ at any instant of time, but which is capable of swept frequency operation over a wide range of frequencies. The output amplitude over the frequency sweep range should be constant or known.

Transducers: $X_1(f)$ and $X_2(f)$ should be broadband devices, preferably with as uniform a response as possible covering the frequency sweep range.

Amplifier: Variable-tuned or broadband amplifier with known $A(f)$.

Analysis: Outputs from the vector voltmeter (or rf voltmeter and phase meter) recorded over the range of frequencies (calibration of the sweep is required to correlate the recorded output with the frequency of measurement).

Pulsed System—Single-Frequency rf Burst (Figure 65). Several limitations of the continuous-wave systems may be eliminated by modulating the cw output to produce pulses. Because the wave train is limited in time, one may associate different intervals of the received waveform with interactions of the ultrasonic pulse at different depths in the sample. An additional benefit of pulsed operation is that heating effects in the sample may be minimized by making the duty cycle low.

A sine-wave burst (often thought of as a cw carrier signal modulated by a rectangular pulse) contains frequencies centered about the frequency of the carrier.

The greater the duration of the modulating pulse, the narrower will be the range of frequencies contained in the burst. Examples of systems of this type may be found in references 082 and 098.

Advantages of the system include the following:

1. Because the system is pulsed rather than cw, a more precise location of the region of interaction of the ultrasonic waves is possible (smaller regions may be investigated).
2. There is less heating because of the low duty cycle.

Disadvantages include the following:

1. One sacrifices accuracy in frequency determination, since the measurement made is the system response, $H(f)$, averaged over a frequency range determined by the pulse width. As the pulse burst is made longer, the frequency band over which the $H(f)$ is averaged narrows; however, the region of interaction is known less precisely (there may be overlap from interactions in adjacent regions).
2. There is poor frequency resolution because of operation at only the fundamental and harmonics of the transducer.

Further characteristics of the system are as follows.

Transmitter: rf-pulsed oscillator, $V_1(f)$, depends on the width of the rectangular modulating pulse.

Transducers: Undamped quartz or ceramic disks, for relatively narrowband $X_1(f)$ and $X_2(f)$.

Amplifier: Tuned amplifier with the passband encompassing at least the main lobe of the receiving transducer's output.

Analysis: The peak-to-peak voltage at the output of the receiving subsystem is related to the amplitude of the system response. Phase is calculated from the time shift of the received waveform referenced to the transmitted signal.

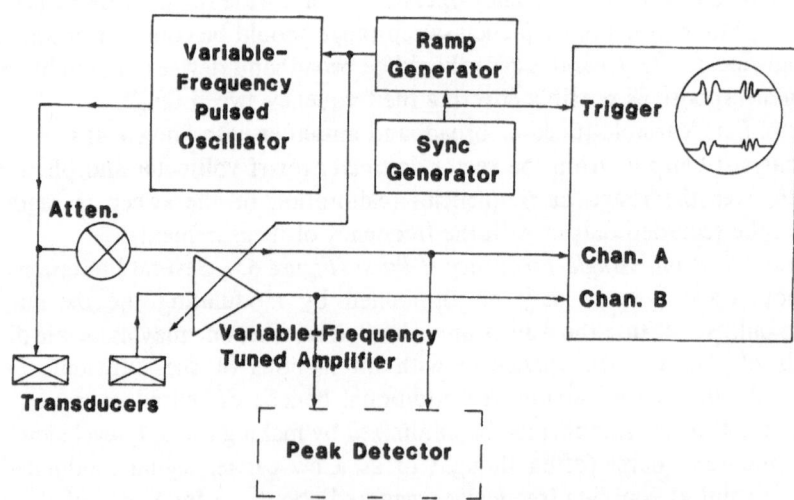

Figure 66. Frequency-modulated pulse-burst ultrasonic spectroscopic system.

Pulsed System—Swept Frequency (Figure 66). By slowly sweeping the carrier frequency of the pulsed oscillator, the frequency resolution of a pulse-burst system improves. The frequency sweep should be slow enough to keep the frequency practically constant during each burst. Broadband transducers are utilized, so the frequency sweep range may be wide.

Examples of systems of this type may be found in references 050, 064, 082, and 268.

Advantages of the system include the following:

1. It has a wide frequency coverage.
2. Benefits derived from using a pulsed system include better location of interaction region and less sample heating than cw.

Disadvantages include the following:

1. Acquisition of spectra takes time; a slower sweep to improve frequency fidelity increases the acquisition time.
2. Broadband transducers are required.

Further characteristics of the system are as follows.

Transmitter: Pulsed rf oscillator with the capability of sweeping the carrier frequency. The modulating pulse should start and stop at the points where the carrier sine wave goes through zero.

Transducers: Damped ceramic disks, having broadband $X_1(f)$ and $X_2(f)$.

Amplifier: Tuned amplifier, whose passband encompasses the frequencies contained in the received waveform and which is swept in synchrony with the carrier frequency of the pulsed oscillator.

Analysis: Peak-to-peak voltage and time shift of received waveform are related to a weighted average of the sample transfer function, $H(f)$, over the range of frequencies contained in each ultrasonic pulse burst.

Pulsed System—Broadband Analog (Figure 67). In a broadband pulsed system, a short ultrasonic signal containing a very wide range of frequencies interrogates the specimen. Elements of the transmitting subsystem are matched to produce the widest bandwidth and most uniform response. The signal from the broadband receiving subsystem is connected to an analog spectrum analyzer. Ordinarily, the analysis is performed by a superheterodyne-type instrument. The frequency resolution of the system is determined by the sweep rates of the local oscillators in the analyzer. Since the analysis is performed in "real time," dynamic processes may be monitored (provided the process does not change appreciably during one sweep of the analyzer). Phase information is not available unless the received signal is mixed with a reference.

Examples of systems of this type may be found in references 001, 005, 008, 041, 042, 064, 082, 114, 122, 167, 174, 180, 268, and 372.

Advantages of the system include the following:

1. It has a wide frequency coverage.
2. There is "real-time" spectral display.
3. Precise definition of the interaction region in the sample is possible.
4. There are no heating effects.
5. There is good frequency resolution.

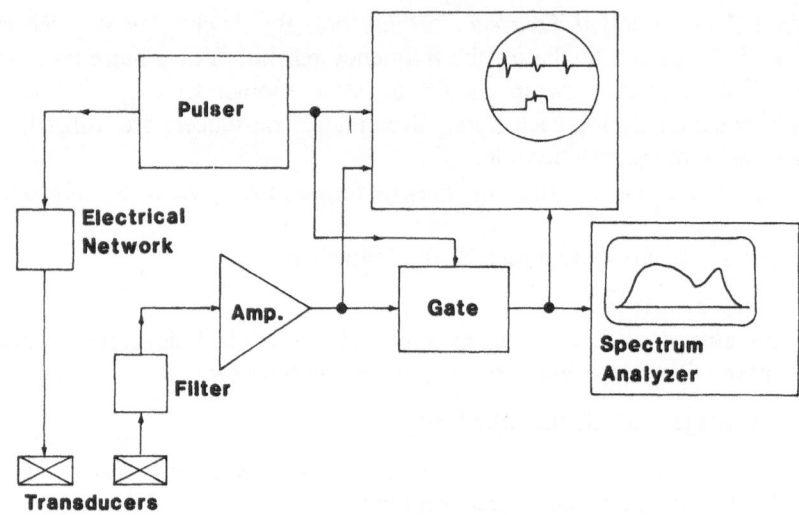

Figure 67. Broadband pulsed ultrasonic spectroscopic system utilizing analog signal analysis.

Disadvantages include the following:

1. Phase information is difficult to extract.
2. The spectrum analyzer is expensive.
3. Frequency accuracy is usually not as precise as with cw methods.
4. Broadband transducers are required.

Further characteristics of the system are as follows.

Transmitter: $v_1(t)$ is usually a very short pulse; $V_1(f)$ is matched to the transmitting transducer response, $X_1(f)$.

Transducers: Damped ceramic or electrostatic devices with broadband $X_1(f)$ and $X_2(f)$.

Amplifier: Broadband $A(f)$, with flat response over the frequency range of interest.

Gate: Range and width selectable analog gate, broad passband $G(f)$.

Analysis: Bandpass filters; superheterodyne spectrum analyzer; dispersive surface-acoustic-wave filter.

Pulsed System—Broadband Digital (Figure 68). Incorporation of digital analysis components into a broadband pulsed system allows the phase spectrum to be easily extracted from the received waveform. The advantages of a broadband pulsed system are retained.

Examples of systems of this type are contained in references 001, 110, 123, 146, 219, 298, 299, 300, 371, 384, 442, 448, 465, and 475.

Advantages of the system include the following:

1. It is wideband.
2. Phase spectra are easily computed.
3. It provides good frequency resolution.
4. Precise definition of the interaction region in the sample is possible.

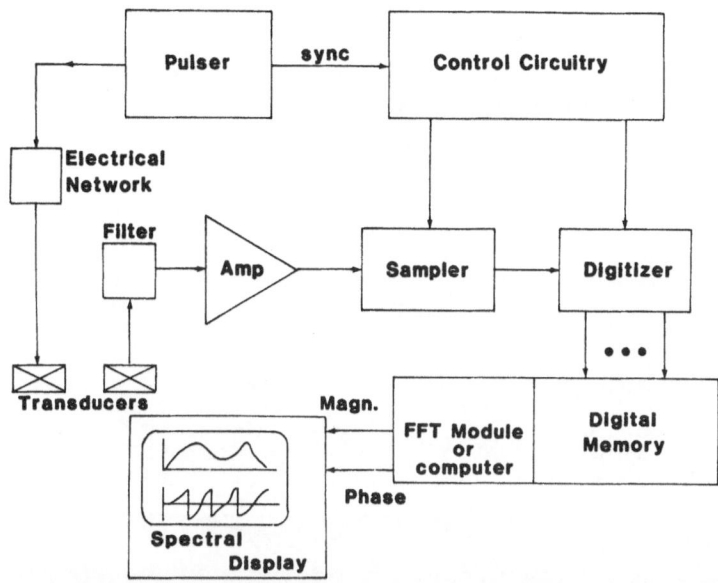

Figure 68. Broadband pulsed ultrasonic spectroscopic system, utilizing digital signal analysis.

5. There are no heating effects.
6. It is possible to study fast dynamic processes or scan a wide area, finding $H(f)$ at each point, *if* real-time sampling is used.
7. It provides for ease of additional processing.

Disadvantages include the following:

1. The system is expensive.
2. It is not usually a real-time system.
3. Broadband components are required with transfer functions known over a wide range of frequencies.

Further characteristics are as follows.

Transmitter: Broadband pulser, with $V_1(f)$ matched to the response of the transmitting transducer, $X_1(f)$.

Transducers: Damped ceramic or electrostatic transducers having wide bandwidth and uniform $X_1(f)$ and $X_2(f)$.

Amplifier: Broadband, with flat $A(f)$ over the frequency range of interest; upper cutoff frequency corresponding to the Nyquist sampling rate.

Sampler: Wideband device, having a very small aperture-time uncertainty; samples taken at precisely spaced time intervals.

Digitizer: High resolution to minimize quantizing error.

Analysis: Hardwired or software version of the fast Fourier transform operating on the sampled and digitized signal from the receiver subsystem; amplitude and phase spectra are calculated from the real and imaginary parts of the Fourier components.

Applications of Ultrasonic Spectroscopy to Materials Evaluation

Defect Characterization

The determination of flaw size and shape is becoming increasingly important in nondestructive evaluation of materials. No longer is it sufficient merely to locate flaws, but a more definitive description of flaws is required for input into sophisticated design analysis which will give a better assessment of the probability of failure and hence reduce the number of costly components.

In conventional ultrasonic inspection of a material the size of the defect is estimated by the use of standard reference blocks with flat-bottomed holes, artificial flaws machined into the test material itself, or to a lesser extent, the Krautkramer AVG diagram. Each of these methods is based on the amplitude of the pulse reflected from the defect and gives accurate results if the natural flaw is oriented parallel with the surface of the material. Because the reflected amplitude depends strongly on the orientation, shape, and attenuation, flaw sizes are often under- or overestimated by an order of magnitude.

A new approach to flaw characterization was suggested by Gericke (006) in 1963. He used a broadband ultrasonic pulse and found that the spectrum of the echo from a void is altered by the geometry of the void. Whaley and Cook (005) and Whaley and Adler (036) have adapted the spectral analysis technique to an immersed system, introduced a multitransducer system, and developed a model to measure size and orientation of circular reflectors in water (simulating planar flaws in solids) from the scattered spectral distribution. In this model it was assumed that the edges of the reflectors behave as Huyghens sources. To explain some of the anomalies observed in the received spectrum, Simpson (002, 013) formulated a model based on Fourier analysis. Several recent developments have provided further insight into the understanding of the relationship between the ultrasonic spectra and flaw geometry. These are the work of Sachse and Pao; Ying, Chang, and Couchman; Rose, Gilmore, Tittmann, and Elsley; Nabel and Mundry; and Klein, Brown, and Lloyd.

An impetus to the development of ultrasonic spectroscopy for defect characterization was the program on quantitative nondestructive testing organized by Rockwell Science Center for the Advanced Research Projects Agency and the Air Force Materials Laboratory. Through this program it was realized that meaningful flaw characterization is possible only when the basic physical mechanism of scattering (diffraction) by discontinuities is understood. This requires strong theoretical foundations in elastic wave scattering by flaws and the confirmation of these theories with appropriate experiments. Once reasonable agreement is achieved between experiment and theory, then the inverse problem can be attacked, namely, how the measured scattered spectrum can be related to the geometry of the scattering center (flaws).

General Considerations

When an ultrasonic wave is scattered from a defect the information is contained in a pair of scattered displacement fields, which are in the far field, of the form

$$\frac{\mathbf{A}}{k_l r} \exp(jk_l r) + \frac{\mathbf{B}}{k_t r} \exp(jk_t r)$$

where k_l and k_t are the longitudinal and transverse wave vectors and r is the distance from the defect. The scattering amplitudes **A** and **B** can be measured as functions of angle, polarization, and frequency. It is from these data that the characterization of the defect is performed.

Scatterers are usually classified according to the parameter ka, where a is a characteristic length of the defect. In the following section, theoretical predictions are discussed for the behavior of scattering amplitudes in various ranges of ka. Experimental approaches to the measurement of the scattering amplitudes and their analysis based on various theories will follow.

Theoretical Considerations

This section of the report concerns itself with investigating the presence and characteristics of flaws in materials and structures by the methods of ultrasonic spectroscopy. The fundamental idea is that ultrasonic (elastic) waves with known properties are introduced into a medium, whose properties when it is flawless are also assumed to be known. The waves interact with the flaws, are modified by them, and finally are reexamined by the observer; the modifications to the waves give information about the flaws. More explicitly, an elastic wave is introduced into the material by some transducer, is then scattered or diffracted by the flaw or flaws, and then either the backscattered wave is observed by the original transducer or obliquely scattered waves are observed by other transducers (of course both types of scattered waves may be observed). The scattering depends on the dimensions of the flaw, so a change in length scale (i.e., a change in wavelength) alters the result; for this reason the scattering is frequency dependent.

The experimental situation is thus the examination of the scattered waves for a particular incident wave. If this is to give information about the flaw it is clear that the relationship between flaw properties and wave scattering must be understood. This means that one must be able to predict theoretically the scattering from a flaw of known properties. This first part of our report reviews this problem.

The scattering and diffraction of waves of any type is a notoriously difficult problem in mathematical physics. Virtually all the theoretical work that has been done with elastic waves has simplified the problem as much as possible by considering the elastic medium in question to be linear and isotropic. That it is linear means that no elastic constants of higher than second order are needed to describe it; that it is isotropic means that only two (combinations of) second-order constants are needed (the well-known Lamé constants, λ and μ, in the usual notation). It is also assumed that this medium is homogeneous except for the flaws; that is, λ, μ, and the density ρ do not depend upon position.

Even with these simplifications, problems that have exact solutions are relatively few. Nevertheless, exactly soluble problems are at the heart of all general investigations, since they provide benchmarks by which all approximate methods of solution may be judged.

Elastic wave scattering problems may be described as partial differential equations with prescribed boundary conditions. The method of obtaining exact solutions of these problems is well known. First, a system of coordinates is chosen such that the boundary conditions become simple—the surface of the flaw is described by having a constant value of one of the coordinates. Next, the partial differential equations are separated. With this done the problem may be completed by separation of variables; or a Green's function may be found, and Green's theorem may be applied; or integral transforms may be found to eliminate derivatives. However, if the equation does not separate, then in general none of the three methods (separation of variables, Green's functions, integral transforms) may be applied (although in a few *special* cases it may be possible to make one of them work, e.g., finding a Green's function). From this it is seen that the types of flaw become crucial in determining the possibility of an exact solution, since the shape of the flaw will determine the coordinate system required, which will then determine whether or not the variables separate.

It is thus not surprising to learn that the problem of the scattering from a single spherical flaw has been solved exactly, since the partial differential equation separates in a spherical coordinate system. This problem has been treated in detail by Truell and co-workers, and appears in the literature (301, 302, 303). Unfortunately few other surfaces, even spheroids, lead to separable problems. In fact the only other available exact solution for the elastic wave scattering problem is for another very simple boundary surface, the infinitely long cylinder (305).

Exact results are available also for a two-dimensional flaw ("crack") which occupies a half-plane. These solutions exist for both the weak crack (the crack is empty so that the stress vanishes on the surface) and the rigid crack (the crack is filled with a rigid substance so that the displacement vanishes on the surface). The first solution for the rigid half-plane crack was given by Fridman (306) and the weak crack solution was given by Maue (307). The reason the half-plane problem is so tractable is that the displacements parallel to the edge of the half-plane decouple, so that the remaining problem is two dimensional. This fortunate situation does not occur for other crack problems, so that no other exact solutions for two-dimensional flaws exist.

We should also mention that the description of cracks as rigid or weak is quite idealized. Experimentally a crack may show different characteristics depending on its loading. Clearly more theoretical work is needed to describe realistic boundary conditions. It should also be pointed out that the exact solutions, and most of the approximate solutions available, are for cracks in an infinite elastic medium, again not the true experimental situation.

Beyond the analysis of particular cases, general scattering theory provides some useful results. These are usually analogs of familiar theorems in other areas of wave theory. One of them is the familiar "optical theorem" which relates the total power scattered and absorbed by the obstacle to the amplitude of the scattered wave in the forward (parallel to the incident wave) direction in the wave zone. Treatments of this have been given recently by Tan (312) and Varadan (313). See also Gubernatis *et al.* (314). Another valuable result is the set of "reciprocity relations," which relate results obtained by interchanging the transducers used for creating the incident and observing the scattered wave; again Tan (315) and Varadan (313) have given treatments recently. Finally we mention that results on power spectra, based on normal-mode theory, have been presented by Pao and Mow (316).

When exact analytic methods fail to give a solution (as will invariably be true in situations of actual experimental interest) recourse must be had to approximation methods. There are several useful ones.

One which is quite useful is the application to elastic wave scattering of the quantum mechanical Born approximation. This follows quite naturally from a description of the scattering problem in terms of integral equations. (As is well known a differential equation plus boundary conditions is equivalent to an integral equation.) With an integral equation one can generate an iterative scheme. Stopping with the first iteration is called in quantum mechanics the (first) Born approximation; this nomenclature has been extended to elastic wave theory by Gubernatis and co-workers, who have studied this method extensively (318).

As in quantum theory the Born approximation is better for long wavelengths. It is easy to apply it to scatterers of complicated shapes. Gubernatis *et al.* have tested the Born approximation by using it to compute the scattering from a spherically shaped flaw, for which as mentioned above exact solutions exist. They find that it works well for backscattering and for situations where the incident wavelength is about an order of magnitude or more greater than the length scale of the scatterer. The principal failure of the method is its inadequate description of forward scattering for short wavelengths. They also find some interesting features that may prove helpful in nondestructive evaluation (NDE) applications; for example, the backscattered power appears to increase with frequency for a cylindrical scatterer but not for a spherical scatterer. Because of the success of the Born approximation in describing backscattering they suggest that it would be most useful in experimental applications in the reflection mode. It might be added that they found that for elastic inclusions, where properties of medium and flaw differed by 20–40%, the Born approximation did well for all angles and even at short wavelengths.

Another method which is strongest where the Born approximation is weakest, namely, at short wavelengths, is the geometrical theory of diffraction of Keller. This originated in the attempt to extend the methods of geometrical optics (i.e., the construct of rays) to the diffraction of light. The fundamental idea is that laws are postulated for the direction of diffracted rays. The variation of amplitude along these rays is determined by the conservation of energy. The initial value of the amplitude of a diffracted ray is proportional to the amplitude of the incident ray at that point. Since this is a point relationship it does not depend on the large-scale nature of the scattering surface, and therefore a problem with simpler geometry, which is more easily (perhaps exactly) soluble, can be used to give the ratio between scattered and incident amplitudes. In Keller's nomenclature the solvable problem is a "canonical problem" and it yields a "diffraction coefficient." Thus the method depends on the existence of solutions to canonical problems which yield diffraction coefficients, which in turn yield the initial amplitude along a diffracted ray; these amplitudes (and phases) then vary with distance in a prescribed manner, and the field at the observation point is obtained by superposition. The exact solutions play a doubly important role here—besides serving as a benchmark as in all other approximation theories, they serve as canonical solutions. The concept of rays is of course not meaningful when the wavelength is long, so that this is a short-wavelength approximation.

The extension of geometrical diffraction theory to elastic waves was first made by Karal and Keller (319).

The first application of Keller's theory to the flaw characterization problem was carried out by Adler and his group (14, 145, 322). They use as a canonical solution the work by Maue (307) previously noted. Recently, considerable work using the Keller theory has been done by Achenbach and co-workers (323, 324). They present a systematic procedure of approximations of increasing accuracy. The first approximation corresponds to geometrical optics; the second, to the geometrical theory of diffraction, and is good to order $k^{-1/2}$, where k is the wave number. This result blows up in certain places, which is corrected by the third approximation, which is equivalent to the uniform asymptotic theory known in acoustics and electromagnetism. This approximation is good to order k^{-1}. Achenbach *et al.* have tested the approximation on the case of diffraction by a plane crack of a normally incident plane longitudinal wave, for which an exact solution can be computed. They find that the approximation method produces surprisingly good results for this case even at relatively long wavelengths and even relatively close to the crack tips.

Another long-wave method is Datta's method of matched asymptotic expansions (325, 333). He has applied this to the case of ellipsoidal-shaped inclusions. He shows that the approximation depends partly on the strain field inside the inclusion, which may be quite different from the strain field in the medium in the absence of the inclusion. Thus the Born approximation, which approximates the strain in the inclusion by that in the medium, may not give a good description of the scattered field.

Datta's theoretical calculations show that the scattering is distinctively influenced by the shape of the scatterers, which is of course what is desired for application to NDE. Comparison of his results with experiment shows that they give correct qualitative predictions; for quantitative predictions, however, it is necessary to extend the validity of the theory to shorter wavelengths.

A different approach to scattering problems is the scattering matrix method, originally developed by Waterman for acoustic and electromagnetic scattering. It was shown by Bolomey and Wirgin that for acoustic waves this technique is especially useful for numerical computation. It has recently been extended to elastic waves by its original developer (326) and by Varadan and Pao (327). It is applicable to obstacles of arbitrary shape. The matrices appearing are symmetric and unitary, which is extremely important since it allows a check on the accuracy of the numerical calculations. Another highly desirable property of this method is that when mode conversion effects are arbitrarily suppressed (by setting certain matrix elements equal to zero) the equations reduce to a simultaneous description of acoustic and electromagnetic scattering. In this way the entire body of acoustic and electromagnetic results becomes available to serve as a guideline for elastic investigations.

Varadan has applied this method to study the scattering from elliptic cylinders (329). She finds that at long wavelengths the cylinder behaves as a point scatterer, so long-wavelength results are not useful for NDE applications. However, for shorter wavelengths she finds the matrix approach to be a powerful and economical way of calculating the scattered field for a wide range of frequencies, even relatively near the obstacle. (These are numerical results, not comparisons with experiment.) Finally she states that "the task of interpreting in a practical manner the enormous amount of data generated by these calculations is still far from complete."

Finally we mention that there has recently been considerable progress, mostly motivated by seismic applications, in the development of finite difference and finite element methods (331, 332), which are purely numerical techniques.

Experimental Work

Spectra of Waves Scattered from Immersed Reflectors. In order to understand the relationship between the parameters of a defect (size, shape, orientation) and its interaction with ultrasonic waves, an experimental study was undertaken (Whaley and Adler, 012, 036, 037). They immersed variously shaped metal reflectors in water to simulate flaws in solids. Such an approach seemed to be reasonable since the large impedance mismatch between water and metal reflectors is analogous to that which occurs between solids and their internal discontinuities. In these experiments, broadband ultrasonic pulses illuminated the reflector. Scattered pulses were spectrum analyzed after reflecting from the ends of the shaped brass rods (diameters ranged from 0.78 to 7.1 mm). Figure 69 shows the relative positioning of the transducer and the reflecting surface.

Major variations of the reflected spectrum as a function of incident angle were observed (Figure 70). The frequency spectrum of the reflected signal was also sensitive to the size of the reflector, as shown in Figure 71, where the frequency dependence of the scattered waves from a 3.2-mm reflector is shown for various angles. The variation of the frequency spectra with angle and reflector size can be explained by the frequency and angular dependence of scalar wave diffraction by circular obstacles. Adler and Lewis (014) calculated the spectra using J. B. Keller's geometrical theory of diffraction and made comparisons to their experimental data (shown in Figure 72).

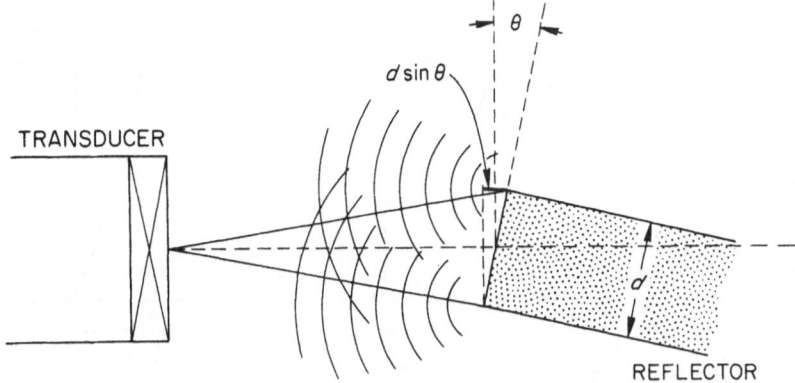

Figure 69. Geometrical relationship of ultrasonic transducer and reflection for measurement of scattered signals. The wavelets arising from the edges of the reflector (inclined at angle θ) are shown.

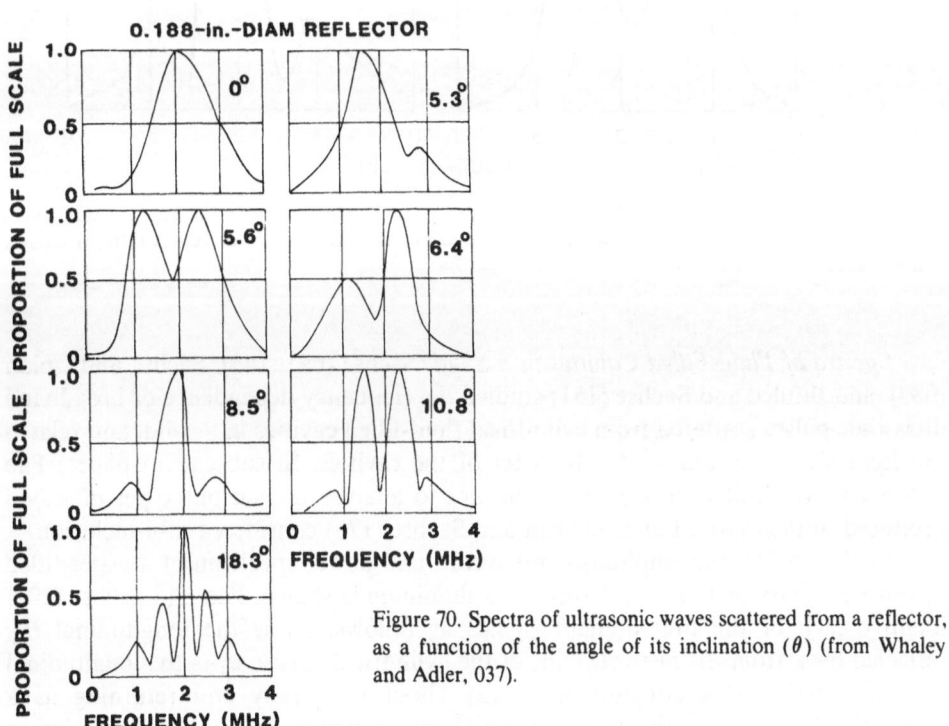

Figure 70. Spectra of ultrasonic waves scattered from a reflector, as a function of the angle of its inclination (θ) (from Whaley and Adler, 037).

It was pointed out by Whaley and Adler (038, 039, 040, 041) that the periodicity in the frequency spectrum of waves scattered from a discontinuity can be used to measure its size. They used a model based on the premise of interference of wavelets arising from the edge of the discontinuity. Other interpretations of the observed frequency spectrum have been given by Simpson (002, 013), who used a Fourier model, and Gilmore and Czerw (203), who used a radiation field theory.

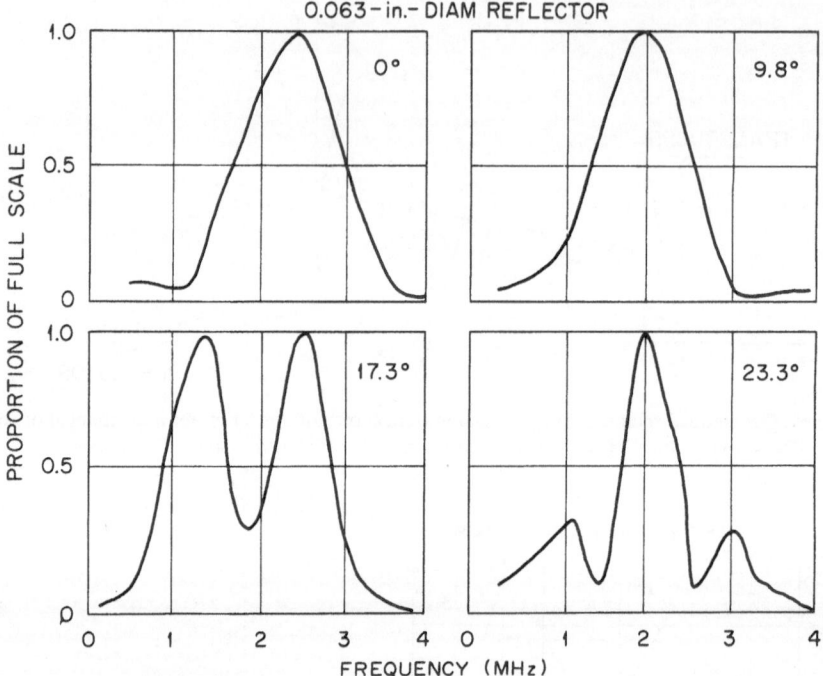

Figure 71. Spectra of ultrasonic waves scattered from a small reflector. Note the spectral differences between these plots for a 0.063-in. reflector and those in Figure 70 for a 0.188-in. reflector (from Whaley and Adler, 037).

Spectra of Fluid-Filled Cavities in a Solid. Sachse (091, 149), Sachse and Chian (087), and Bifulco and Sachse (151) studied the frequency dependence of broadband ultrasonic pulses scattered from cylindrical fluid-filled cavities in a solid, and related the frequency spectrum to the diameter of the cavities. In subsequent papers Pao and Sachse (207, 388) applied ray acoustics to analyze the various types of waves produced in the cavity. Later, Sancan and Sachse (176) considered a bi-inclusion.

In Figure 73a, the amplitude-time record and power spectrum of a water-filled cylindrical cavity of 1/16-in. diameter in aluminum is shown. Pao and Sachse (207) identify each of the time-domain signals as follows: A_1 is the longitudinal ray reflected back from the nearest point of the cylindrical cavity; a_1 is the longitudinal ray converted into a circumferential ray about the cavity and returning as a longitudinal ray; a_2 is the circumferential ray refracted into the fluid inclusion (following Snell's law) and returning as a longitudinal ray; A_2 is the longitudinal ray reflected back from the inner back surface of the cavity; A_3 is the second multiple of A_2; and B_1 is the reflection from the back of the sample. Figure 73b is the spectrum of pulses A_1, a_1, a_2, and A_2, and Figure 73c is the spectrum of pulses A_1 and A_2.

Simulated Defects in Solids. In the ARPA/AFML (Rockwell) program on quantitative NDE, diffusion-bonding techniques were developed to produce samples with simulated defects. Sets of titanium samples were prepared having internal defects of well-characterized shapes and a variety of sizes. The simulated defects

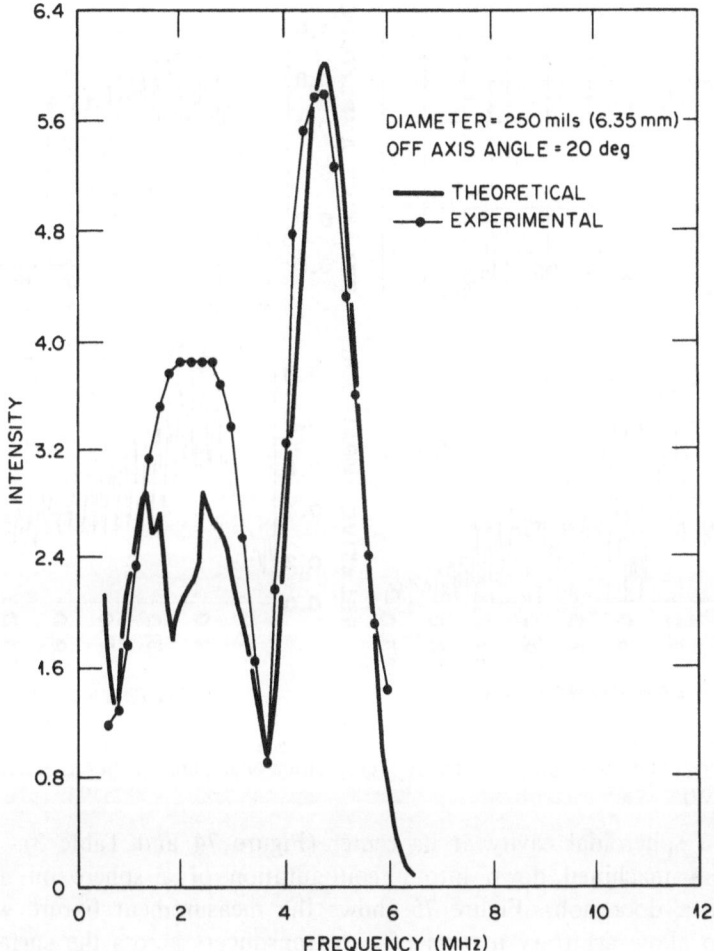

Figure 72. Experimentally (●) and theoretically (——) determined scattered wave spectra from voids in an immersed disk (from Adler, 391).

(Figure 74 and Table 3) are cavities with the shape of spheres, oblate and prolate spheroids, flat-bottomed holes, as well as circular and elliptical cracks. The diffusion-bonded samples were machined into two different shapes: disks with flat surfaces for studies in an immersed system and spherically shaped (doorknob) samples for use with contact transducers.

Defect Characterization Utilizing Contact Measurements

A unique experimental system was used by Tittmann (374, 390, 412, 443), Cohen and Tittmann (133, 379), Tittmann, Elsley, Nadler, and Cohen (423), and Tittmann and Elsley (451) for measurements of ultrasonic wave scattering from simulated defects in solids. Two cylinders of titanium alloy were machined and their bases diffusion bonded together in such a way that the final longer cylinder

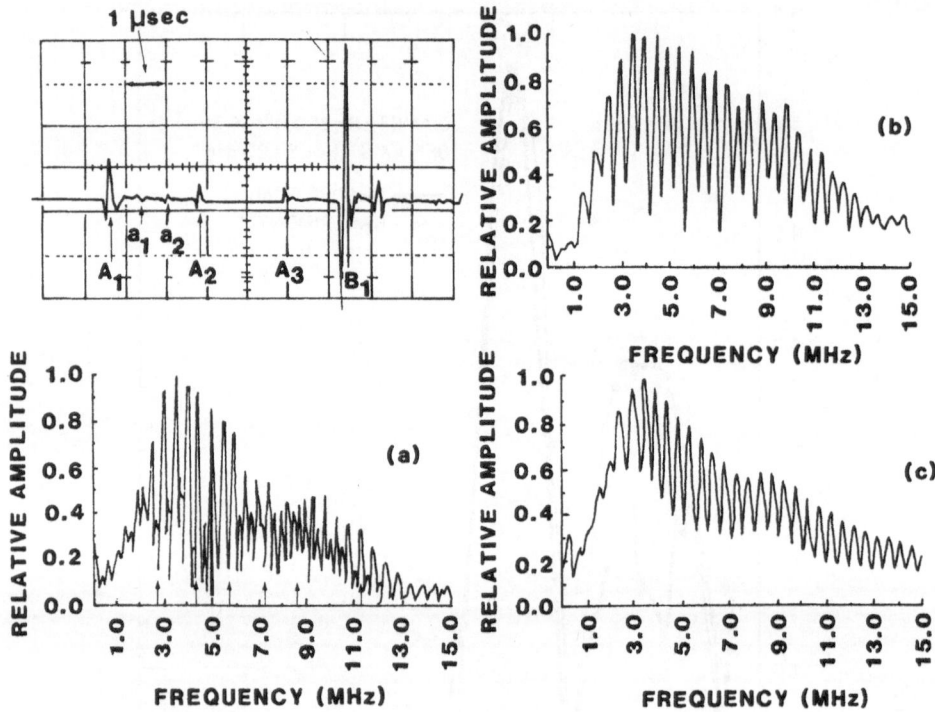

Figure 73. (a) Amplitude-time record and power spectrum of a water-filled cavity (1/16-in. diam.); (b) spectrum of amplitude pulses A_1, a_1, a_2, and A_2; (c) spectrum of amplitude pulses A_1 and A_2 (from Pao and Sachse, 207).

contained a spheroidal cavity at its center (Figure 74 and Table 3). Then this cylinder was machined down into a configuration of a sphere on a pedestal (resembling a doorknob). Figure 75 shows the measurement fixture which was designed to allow arbitrary motion of two transducers across the surface of the sphere.

Both angular and frequency dependence of the scattered energy were analyzed from these simulated defects. In Figure 76 the angular dependence of pulse–echo power is plotted for variously shaped defects. The plot demonstrates the dramatic differences in the angular dependence for various shapes when the sizes of the defects are approximately comparable. In Figure 77 representative experimental data are given and calculations shown for the frequency dependence of the backscattered power from an oblate spheroid, with 2:1 aspect ratio, at polar angles 10° and 60°.

Characterization of Defects in Flat Immersed Samples

The experimental technique used by Adler and Lewis (145, 446, 452) to study broadband pulse spectra from simulated defects in titanium samples is shown in Figure 78. Both longitudinal and shear wave scattered amplitude spectra are studied. The L and S waves can be separated by time gating owing to their differing velocities. Corrections are made to the spectrum for the transducer transfer characteristics and losses at the liquid–solid interface.

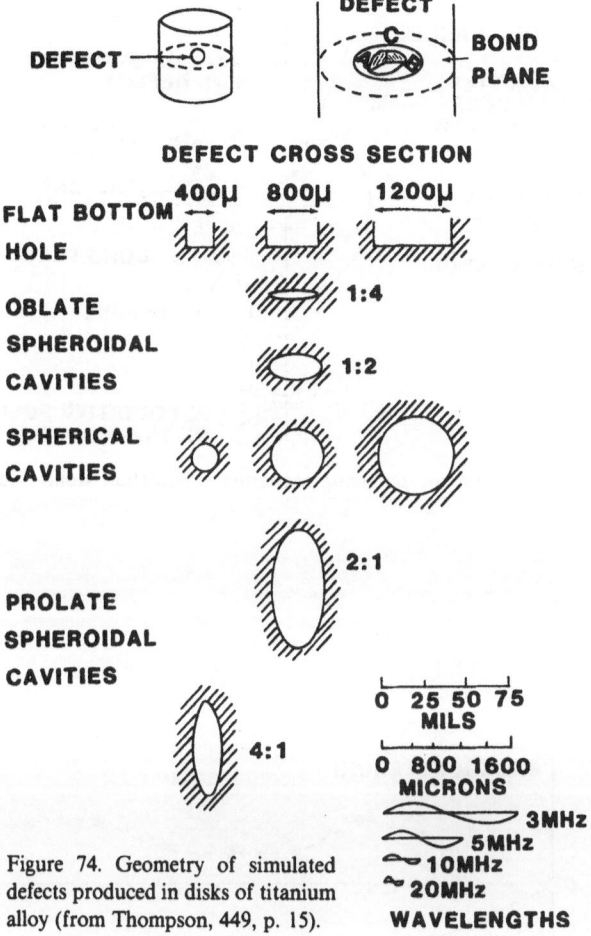

Figure 74. Geometry of simulated defects produced in disks of titanium alloy (from Thompson, 449, p. 15).

Table 3[a]

Dimensions of the Simulated Defects Shown in Figure 74

Description	a, μm	b, μm	c, μm
Prolate spheroid	200	200	800
Prolate spheroid	400	400	800
Sphere	200	200	200
Sphere	400	400	400
Sphere	600	600	600
Oblate spheroid	400	400	200
Oblate spheroid	400	400	100
Circular disk	600	600	100
Elliptical disk	2500	600	250
Simulated crack	600	600	—

[a]Source: Tittmannn and Elsley, 451.

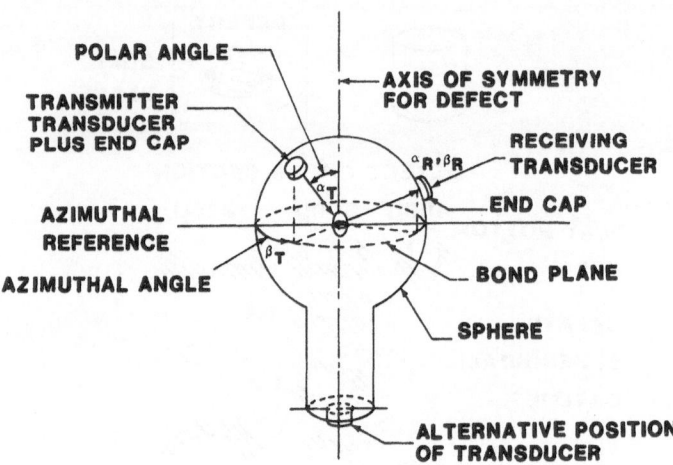

Figure 75. Measurement fixture for obtaining scattering information from voids in a solid (from Tittmann, 445).

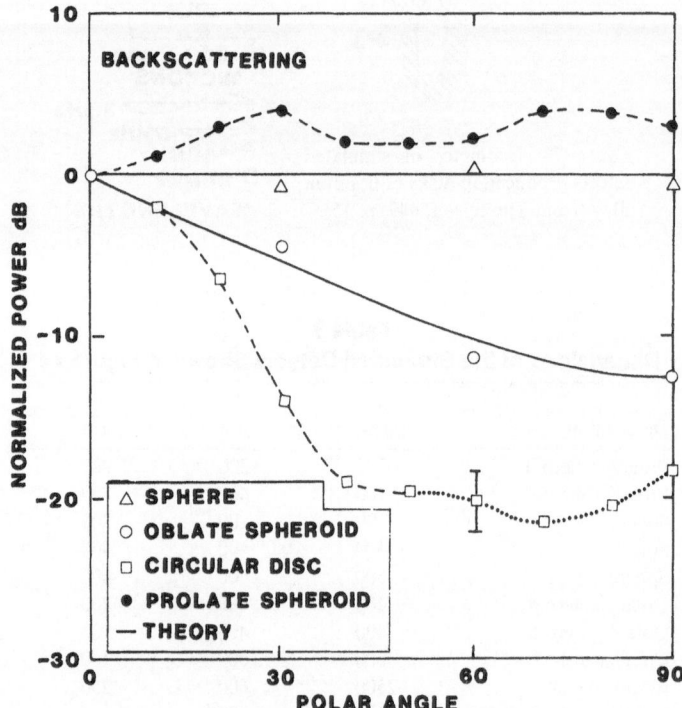

Figure 76. Angular dependence of pulse–echo power (data normalized at zero polar angle) (from Tittmann, 445).

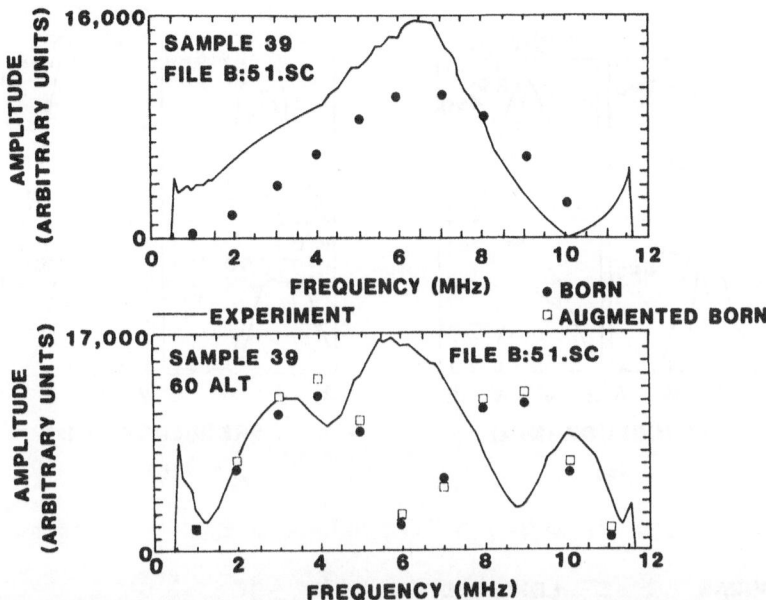

Figure 77. Comparison of approximate theories and data for backscattering from an oblate spheroid for 10° (upper figure) and 60° (lower figure) off the axis of symmetry (from Tittmann and Elsley, 451).

In Figure 79 experimental and theoretical scattered L and S wave amplitude spectra are shown for spherical, oblate, and prolate spheroidal cavities. Information concerning size and shape is apparent in both frequency and angular dependence of the scattered waves.

Additional experiments compared amplitude spectra for waves diffracted from circular cracks to the results of a theoretical analysis using elastodynamic ray theory (Achenbach, 323, 324). For a 2500-μm circular crack, the experimental amplitude

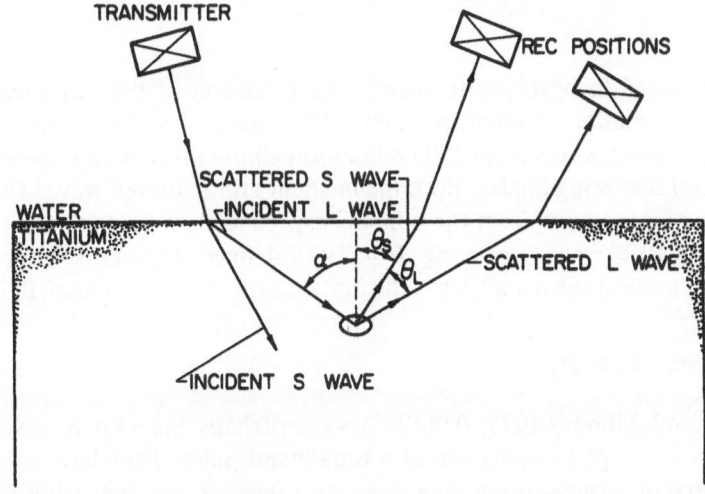

Figure 78. Experimental technique for the study of scattering from defects in immersed specimens.

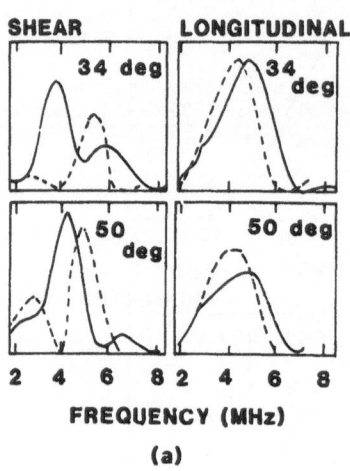

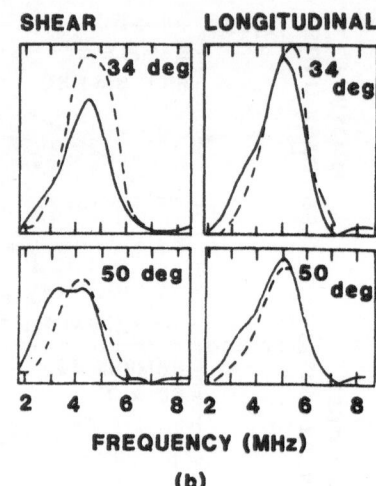

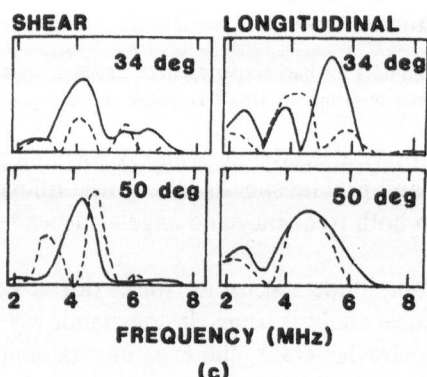

Figure 79. Comparison of experimental data (——) to the Born approximation (– – –) for the intensity of scattered wave spectra [for cavities in titanium: (a) 800-μm sphere, (b) 200×800-μm oblate spheroid, and (c) 1600×800-μm prolate spheroid) (from Adler, 446).

spectra compared favorably with theory. An extension of this ray theory may be used for consideration of elliptical cracks. Experimental and theoretical amplitude spectra of scattered waves from 2500×1250-μm elliptical cracks are shown in Figure 80 for several scattering angles. Both the incident and scattered waves are longitudinal. The favorable results from these model experiments indicate that the amplitude spectra of elastic wave scattering from well-defined geometrical flaws can be predicted from theoretical analysis—the first criterion of defect characterization.

Phase Spectroscopy

Nabel and Mundry (073, 079, 232) were perhaps the first to investigate the information in the phase spectrum of a broadband pulse. They have calculated the phase spectra of various input functions, and have shown the value of retaining phase information during deconvolution.

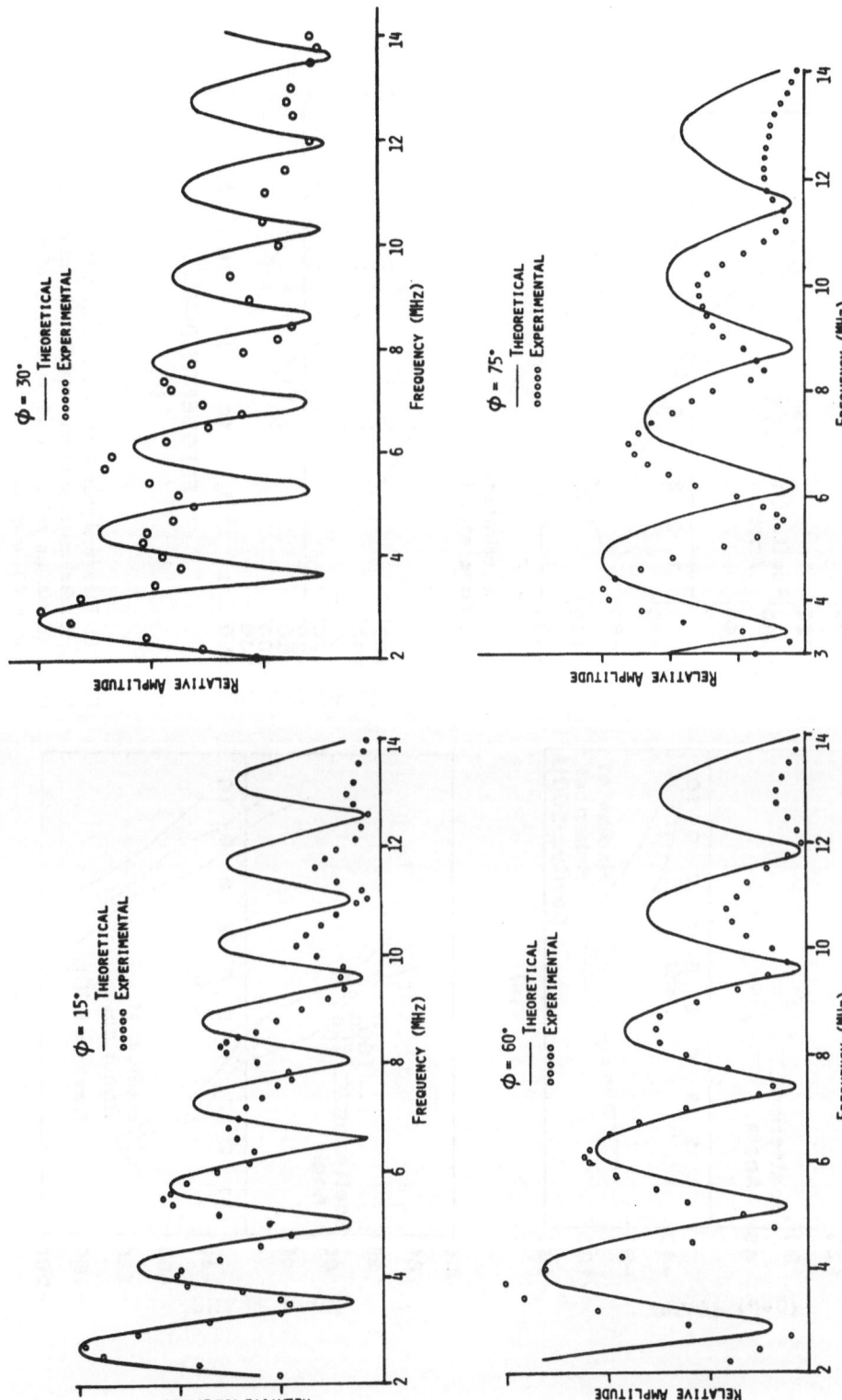

Figure 80. Amplitude spectra of scattered L waves from a 2500×1250-μm elliptical crack in titanium along different azimuthal directions; polar angle$=60°$ (from Adler, 505).

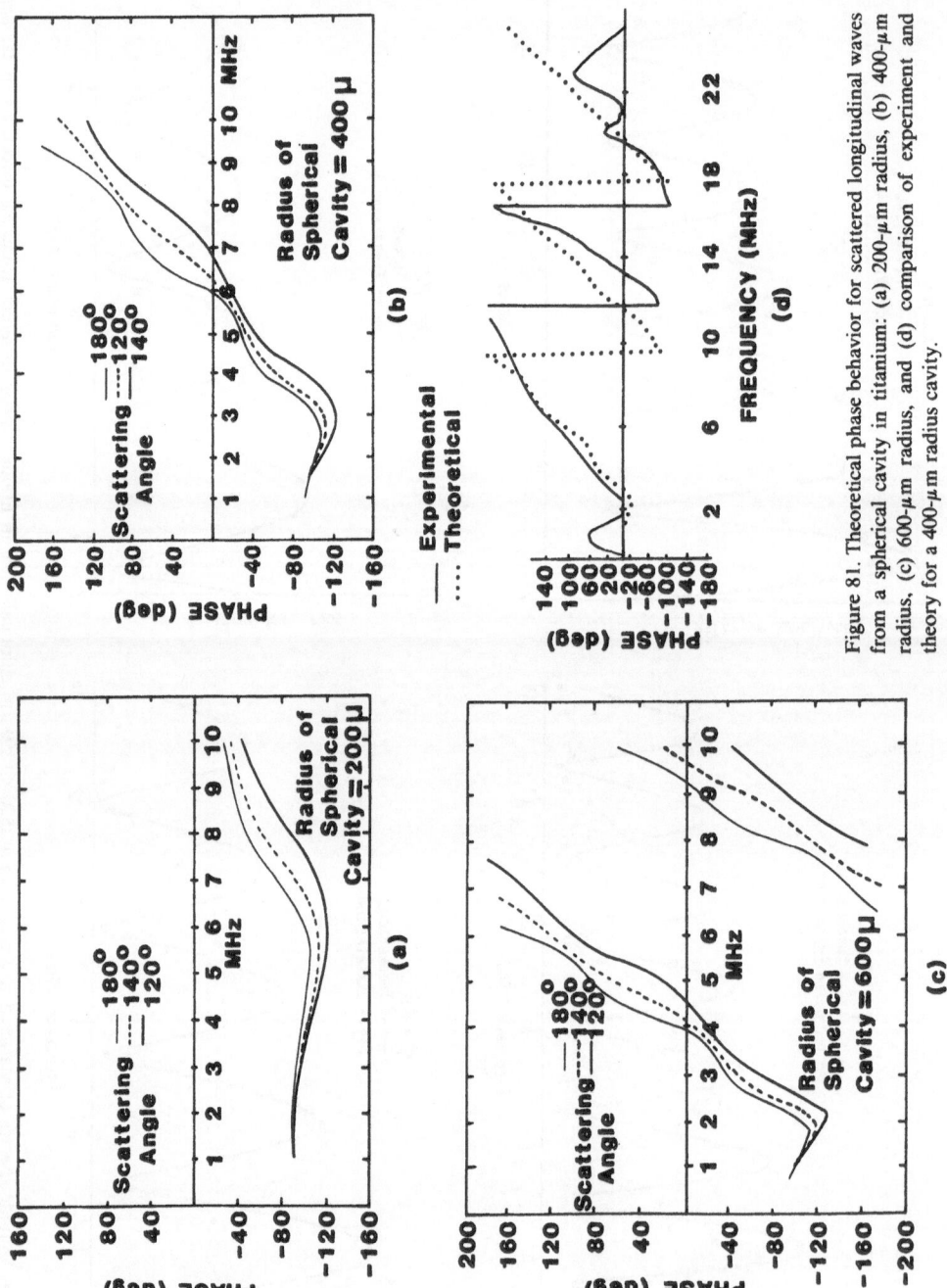

Figure 81. Theoretical phase behavior for scattered longitudinal waves from a spherical cavity in titanium: (a) 200-μm radius, (b) 400-μm radius, (c) 600-μm radius, and (d) comparison of experiment and theory for a 400-μm radius cavity.

Adler and Lewis calculated the scattered phase spectra from spherical cavities in metals using Ying and Truell's analysis. The phase spectra for scattered L waves from spherical cavities shows size dependence. Figure 81 (a, b, c, and d) presents the calculated phase spectra for 200-, 400-, and 600-μm spherical cavities. For a normally incident L wave, the 180° (back reflection), 140° and 120° scattered angles are shown. The experimentally obtained phase spectrum for a 180° backscattered L wave from a 400-μm spherical cavity compares favorably with the theoretical calculation (Figure 81d) in the range 2–10 MHz (where most of the energy in the ultrasonic pulse is concentrated). Tittmann and Elsley (451) have also studied the phase spectrum of waves scattered from oblate spheroidal cavities. It appears that phase spectroscopy may complement amplitude spectroscopy.

Inversion Techniques

In order to estimate the size and orientation of hidden flaws from ultrasonic scattering data, a number of inversion techniques are being developed. Mucciardi and co-workers (094, 119, 126, 447, 454) use trainable classifiers [adaptive learning networks (ALN)]. The ALN is trained on theoretical models. Key parameters of the scattered rf waveform and its spectrum, which are characteristic of a particular defect, are identified during a feature extraction step. ALN networks tested on experimental data have given favorable results.

Other promising techniques of inversion rely on analytical techniques. Representative of such approaches is the work of the following: Bleistein and Cohen (455), who have shown that high-frequency scattering information may be processed to yield flaw shape (physical optics far-field inverse scattering); Rose and Krumhansl (502), who developed an inversion algorithm based on extended quasistatic approximation of Gubernatis; Richardson's (503) treatment of the inversion problem in the Rayleigh (long-wavelength limit); and Achenbach's (504) application of an inversion integral from elastodynamics.

A simple approach to determination of size and orientation of cracks is based on the recognition that the amplitude spectra contains modulation (504). This inversion technique has been used with success to determine the size of 2500-μm circular cracks in titanium (as shown in Table 4).

Table 4
Crack Size Calculated from Spectral Modulation
(2500-μm Circular Crack in Titanium)

Diffraction angle	Frequency Spacing of maxima (avg.)	Computed radius of the crack, μm
35°	2.18	2530
40°	1.87	2630
45°	1.83	2450
50°	1.68	2460
55°	1.60	2410
60°	1.47	2500
65°	1.39	2510

Adhesive Bonds

Laminates of metal and polymer (adhesive bonds) have been studied for a number of years by utilizing ultrasonic spectroscopic techniques. The strength of the bond is of primary importance and so many of the investigations have been aimed at deducing this property from nondestructive ultrasonic measurements. Overall strength of an adhesive bond is determined by both the cohesive strength of the polymer and the adhesive strength of the metal–polymer bond.

A number of models for the propagation of an ultrasonic wave in an n-layer laminate have been developed. A good deal of experimental work has been done to test the validity of the models. Most of the investigations demonstrate good agreement between theory and experiment. However, these tests are usually performed on highly idealized bonds. More detailed models are required to describe wave propagation in the complex "real" adhesive bonds. The fracture mechanics–acoustic model of Crane and Nayfeh (506) provides an example of this new direction for investigation.

Models and Theoretical Developments

A thorough treatment of wave propagation in layers may be found in the book by Brekhovskikh (296). Expressions for the amplitude transmission coefficient (T_{13}) and the amplitude reflection coefficient (R_{13}), derived in this reference for the three-layer problem, are reproduced here. For a plane wave incident from layer 3

$$R_{13} = \frac{R_{23} + R_{12} \exp(j2k_2 d)}{1 + R_{12} R_{23} \exp(j2k_2 d)} \tag{102}$$

$$T_{13} = \frac{4Z_1 Z_2}{(Z_1 + Z_2)(Z_2 + Z_3)} \frac{1}{\exp(-jk_2 d) + R_{12} R_{23} \exp(jk_2 d)} \tag{103}$$

where the Z_i are *normal* acoustic impedances and the reflection coefficients R_{12} and R_{23} are given in Figure 82. Note that

$$Z_i = \frac{\rho_i c_i}{\cos \theta_i} \tag{104}$$

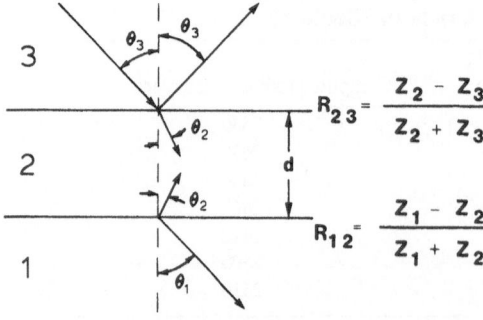

Figure 82. Propagation of ultrasonic waves in a system of three layers.

reduces to the characteristic acoustic impedance

$$Z_i = \rho_i c_i$$

at normal incidence. Attenuation in any of the layers may be accounted for by making the k_i complex.

Lloyd (026, 054) developed a lumped-parameter, mass compliance model for a three-layer adhesive bond (Figure 83). The masses of the metal layers are represented by m_1, m_3, and the compliances of k_1 and k_3. Compliance k_2 is that of the adhesive. Investigation of the response of this model to a range of driving frequencies indicates equispaced resonances and antiresonances.

Chang, Couchman, and Yee (256) formulated a transmitted energy flux coefficient for a six-layer laminate, then used it for theoretical investigations of materials with one, two, and three layers. As was the case with Lloyd's model, peaks appeared in the transmission spectrum corresponding to resonances of individual layers and to resonances of combinations of layers. They (150) also found that the depth of a delamination (which acts as a reflector) could be inferred from the spacing of the antiresonance dips, if the ultrasonic velocity were known.

With an analogy from transmission-line theory, Highmore (237) noted that the acoustic impedance of a multilayer structure is a complex number. From the complex impedance, he was able to calculate the magnitude and phase of the reflection coefficient as a function of frequency. He studied the resonance behavior for nonbonds (air gaps) at various levels in the laminate. He noted [as did Lopilato and Carter (051)] that the phase change on reflection from unbonds was 180°.

Scott (048) and Scott and Gordon (153) solved for the steady state solution of wave propagation in a single layer. The Fourier transform was used to find the solution for transient excitation. An iterative procedure was developed for tabulating the transmission and reflection coefficient in a medium having an arbitrary number of layers of varying thickness and acoustic impedance. In the derivation, the incident wave was assumed to be normal to the surface, and the attenuation in the adhesive was assumed to be negligible.

After a review of the adhesive bonding process, Rose and Meyer (034) developed several analytical models for the bond. A "reference bond" model was considered in which the adhesive is a homogeneous and isotropic layer. In their "material property gradient" model the adhesive has a variation in mechanical properties through its thickness. This model may be constructed by considering the adhesive to be comprised of several sublayers, each with different mechanical properties. The adhesive was considered to have many points of contact separated by microscopic voids in the "surface preparation" model. Their "combined property

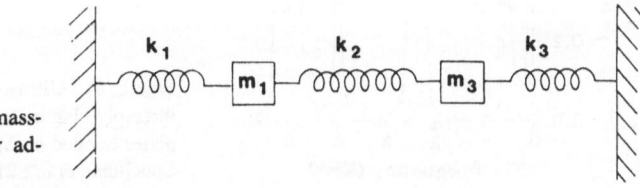

Figure 83. Lumped-parameter mass-compliance model of three-layer adhesive bond.

gradient and surface preparation" model included the effects from two of the simpler models.

Development of the models by Rose and Meyer was along the lines outlined by Brekhovskikh (296). First the steady state (cw) case was considered, then the Fourier transform was used to find the solution for transient excitation. Computer programs, listed in an appendix, were used to generate the reflected signal as a function of input pulse shape and bond line characteristics. Multiple reflections (though only a few) were accounted for. Parametric studies of time and frequency domain reflected signals were generated for all models.

The "reference bond" model was expanded (Rose, 118) to include the effects of mode conversion and oblique incidence. A solution to the problem of energy partitioning at a solid–solid interface into longitudinal and shear wave components was incorporated to account for mode conversion. Reflections were ignored, since a pulse–echo system was assumed to be used with the beam striking the surface of the laminate at an angle.

The effects of an attenuating adhesive layer (Rose, 378; Chang, Flynn, Gordon, and Bell, 177) are to broaden resonance peaks and decrease the depth of minima. Apparently, α in the metal is still considered negligible in the models now used.

The anomalous low-frequency peak and resonance "splitting" observed in spectra of waves transmitted through laminated structures (Figure 84) was investigated by Couchman, Chang, Yee, and Bell (218). They found, through parametric theoretical studies, that the resonance splitting is really a resonance shifting. That is, the peaks represent resonances of combinations of layers, as well as individual layers.

Alers, Flynn, and Buckley (200) found that where a jump in displacement at the metal–adhesive bond is allowed (to account for the effects of improper adhesion) the lowest antiresonant dip shifted its position by a greater percentage than other minima, for a given change in interface properties.

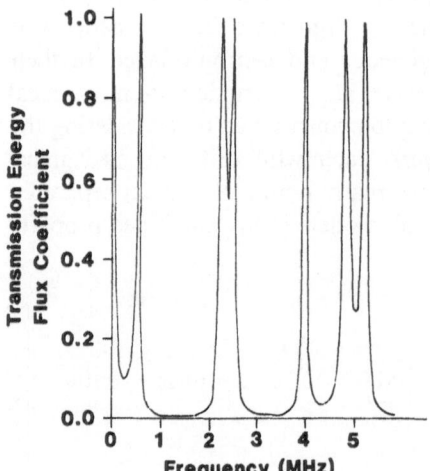

Figure 84. Ultrasonic transmission spectrum of a three-ply laminate (two 0.05-in.-thick aluminum plates bonded by a 0.012-in.-layer of adhesive) (after Couchman *et al.*, 218).

Experimental Investigations

Perhaps the first "spectroscopic" testing of laminates was along the lines suggested by Lopilato and Carter (051), that disbonds (air gaps) caused a 180° phase change in the signal reflected from a layer. The depth of the nonbond could be inferred from the velocities and thickness of substrate materials and the transit time until a phase-reversed pulse was noticed. This concept was carried a bit further by Highmore (237), who designed a system for monitoring the shifts in resonance frequency caused by unbonds. He also noted that the input of multilayer structures is complex, but becomes real at resonance.

Chang, *et al.* (256) measured transmission spectra for several single layers of various thicknesses and three-layer sandwiches of aluminum, adhesive, and air. Resonance behavior was noted in the spectra. For the single layers the spectral structure was simple, with the peaks equally spaced. Transmission spectra of the laminates were considerably more complex. Qualitative agreement with theory was achieved.

Pulse–echo measurements on a step-lap joint were performed by Rose and Meyer (020) in an attempt to deduce bond strength. They found that the smaller the ratio of backwall echo to frontwall echo the greater was the strength of the bond. Two spectra were presented to show that bond line thickness could be found from the spacing of the resonance dips. They also made some general suggestions about the possibility of relating shifts of frequency peaks or magnitude changes to bond strength.

Lloyd (026, 054) compared experimental reflected spectra of strong and weak step-lap bonds. He found no simple correlation of the response peaks in the spectra with bond strength. He noted several possible problems with the measurements: the fillets formed by excess adhesive changed bond strength appreciably; the failure mechanism may not be simply related to interfacial or bulk properties of the adhesive; and the model was too simplistic. A more complex model based on complex reflectivities at all interfaces was suggested.

An equivalent-time sampling system was used by Chang *et al.* (150) to observe spectral changes caused by drilling flat-bottomed holes into a graphite–epoxy laminate. The hole, penetrating the layers, decreases the thickness of the overlying material and so changes the thickness resonance. The depth of delaminations could be found from the spacing of the antiresonance minima.

Seydel (033) developed an equivalent-time sampling system built around a digital delay generator. In his investigations of the reflection spectra from strong and weak adhesive bonds he found "little if any" correlation of spectral features in the frequency range 4–22 MHz with surface preparation (although some variations were noted outside this range). He suggested that phase measurements may prove fruitful, since when the bond thickness is much less than a wavelength, the "adhesive contributes an acoustic reactance term which primarily affects the phase portion of the reflectivity function."

Scott (048) and Scott and Gordon (153) studied the transmission spectra of laminates. They observed equally spaced transmission resonance peaks for a single glass plate. When additional plates were added (with thin layers of water separating them) the periodicity became much longer and there was splitting of resonance peaks, making the spectra much more complex. They attributed the splitting to thickness resonances of the entire array of plates. They also noted that the "broader resonances remain constant in number, but rapidly deepen as layers are added... producing frequency bands having increasingly high acoustic attenuation." For an infinitely periodic medium there will be "forbidden" frequency bands where there is no transmission.

In addition, Scott and Gordon found, for wavelengths shorter than the period of the laminate, a periodic laminate behaves acoustically much as a dispersive monolithic material with its transmission frequency spectrum exhibiting thickness resonances. If the wavelength is approximately equal to an integer times the period of the laminate there will be anomalous dispersion and high attenuation. They also found the 4th and 18th resonance peaks were missing in the frequency spectra of 4- and 18-ply graphite–epoxy panels. For testing laminates, they used the output of a bandpass filter, centered at a resonance peak, to modulate the intensity of a C-scan. Peak shifts, caused by changes in elastic modulus, density, or thickness, appeared in the image as dark areas.

Scott and Gordon suggested that a generalized model for interaction of waves with laminates be developed. It would include the effects of scattering, shear wave propagation, and wave propagation at oblique incidence to the laminate interfaces.

Chang, Flynn, Gordon, and Bell (177) studied the effect of control and process variations on the strength of step-lap bond specimens. They measured the ratio of reflected amplitudes (R = frontwall/backwall) and the bandwidth (B) of the antiresonances. Since their parametric studies had shown that increased attenuation or acoustic impedance of the adhesive broadens resonance dips and decreases the depth of minima, R and B were expected to change depending on the properties of the adhesive. Upon examination of their experimental results, they concluded the following:

1. R indicates bond strength (Figure 85a).
2. Bond-line thickness may be found from the spacing of the antiresonance dips.
3. Bond strength correlates with $1/B$ (Figure 85b).
4. R and $1/B$ are related to Z and α of the adhesive.
5. R and $1/B$ are linearly related (Figure 85c).
6. R and the mechanical Q are related to bond strength through Z of the adhesive.

Gericke and Monagle (028) made pulse–echo spectral measurements on laminated panels. Stacks of various thicknesses of plastic, aluminum, and stainless steel, separated by layers of glycerin, were studied. Numerous theoretical and experimental spectra are presented for various three-layer combinations. Spectra for simulated disbonds, where the ultrasonic penetrates only one or two layers, are also given.

Correlations of adhesive bond strength with the ultrasonic velocity and attenuation in the adhesive were discovered by Alers, Flynn and Buckley (200). Additionally, they performed low-frequency spectral measurements (using a 0.5-MHz

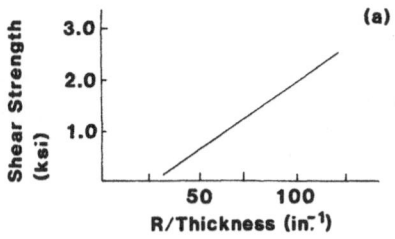

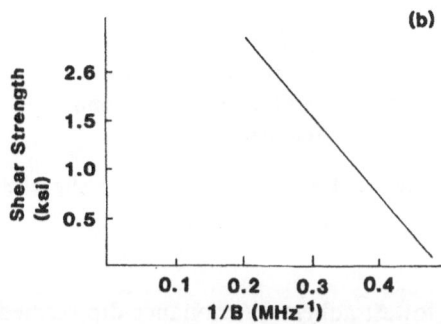

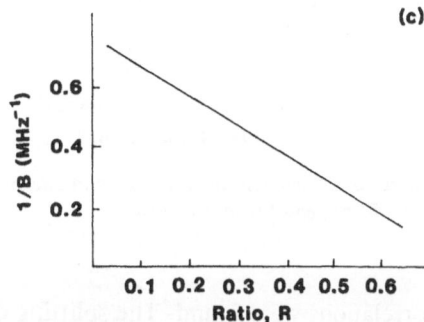

Figure. 85. Experimentally determined relationships between adhesive bond strength and acoustic properties of the adhesive. (R is the ratio of the front- and backwall echo amplitudes and B the bandwidth of the antiresonances) (after Chang *et al.*, 177).

transducer) on the bond and found a correlation between shear strength and the lowest resonant frequency in the spectrum. This resonance corresponds to a mode of vibration where the aluminum plates act as rigid masses with the adhesive providing a restoring (spring) force.

Couchman, Chang, Yee, and Bell (218) investigated resonance splitting in transmission and reflection spectra. Qualitative agreement of experimental measurements and theoretical predictions was achieved on an idealized laminate (two aluminum plates separated by a variable water gap).

Wave modes propagating parallel to the metal–adhesive bond were studied by Alers (432). Since this is a waveguide effect, a high sensitivity to boundary conditions was expected. At frequencies where the dispersion was large, the boundary conditions changed the velocity to a greater extent. However, special transducers imbedded in the bond were required.

The effects of fillers and scrim cloths on adhesive bond strength (compression-shear) and ultrasonic parameters of the adhesive were studied by Flynn and Henslee (430). Some correlation of shear strength and ultrasonic velocity was noted, but only for those samples with the adhesive "close to optimally cured." They found poor correlation of shear strength with attenuation because of small air voids (trapped in the scrim cloth) which increased attenuation, but apparently did not decrease strength. Studies were also directed toward determining the effects of moisture on the adhesive. They noticed a large attenuation of the high frequencies in moisture-induced damaged panels.

Alers, Elsley, and Flynn (459, 466) made improvements in their ultrasonic and mechanical tests with the aim of verifying the correlation of strength and lowest resonant frequency observed in earlier studies (200). Although not much correlation of bond strength and lowest frequency was noted in these careful studies, other

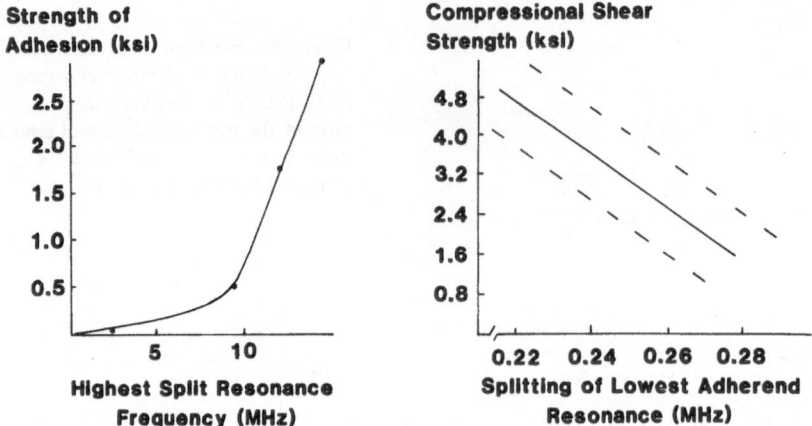

Figure 86. Correlations of adhesive bond strength with features in the ultrasonic reflection spectra (after Alers, Elsley, and Flynn, 459, 466).

correlations were found. The splitting of the lowest adherend resonance dip seemed to correlate with bond strength (Figure 86). A method of correcting measured parameters for bond-line thickness improved the correlation. Additionally, the frequency of the highest split resonance was found to be an indicator of strength (Figure 86).

Zuckerwar's work (132) provides a good example of the detailed examination of test data which may, in some cases, be required to form a signature of a laminated structure.

Surface Properties

Surface Roughness

Rough surfaces may be broadly categorized as being randomly rough or having a texture which is periodic. Ultrasonic spectroscopy has been demonstrated to be useful in assessing parameters characteristic of such surfaces (de Billy, Doucet, and Quentin, 137; Quentin et al., 138; de Billy et al., 139; Jungman et al., 140).

A rough surface is described by its rms roughness, "h," and correlation length, "L." Ordinarily, these two parameters are obtained from profilometer measurements of the variation in height (y_i) of the surface in a given direction. However, for fragile surfaces or for roughness in the low-to-submicrometer range, mechanical profilometer measurements are impossible.

Ultrasonic measurement of surface roughness is carried out in the following manner. An ultrasonic transducer, operated in the pulse–echo mode, is used to record the intensity of backscattered ultrasonic compressional waves as the surface is rotated about an axis perpendicular to the axis of the ultrasound beam (see Figure 87). Plots of relative intensity, at a given frequency, versus angle demonstrate that

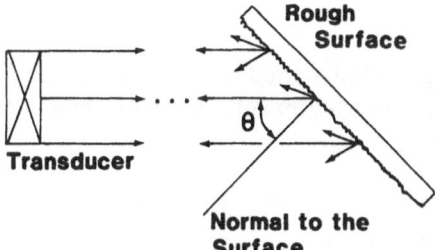

Figure 87. Geometry for measurement of back-scattered ultrasound from rough surfaces (after de Billy *et al.*, 139).

the locus of data points for a surface with a certain roughness are separate from other loci for surfaces with different h. Alternatively, one may plot relative intensity against roughness known from profilometer measurements (Figure 88). These plots for known roughness are required as a standard, since theoretical prediction of the curves has not yet been made.

The determination of the roughness of a surface with unknown h may be summarized as follows:

1. Measure the backscattered intensity, $I(\theta)$, several times and find the average, $\bar{I}(\theta)$.
2. Use a least-squares method to fit an analytic function $J(\theta)$ to $\bar{I}(\theta)$.
3. Plot relative $J(\theta)$ on the standard curves (Figure 88) for the single frequency used in the measurement.
4. The mean abscissa, $\hat{h}$, gives the "best" estimate of h; Δh gives the error.

Although ultrasonic waves of a single frequency are required to estimate h, an optimal estimate is obtained when the ultrasonic wavelength bears a certain relationship to h. No exact relationship has been found; but generally, higher frequencies are preferred for small h and low frequencies for large h. Thus a wideband spectroscopic system, although not necessary for roughness measurement, permits a single run of data acquisition with unknown h. Optimized frequencies may be selected after the experiment.

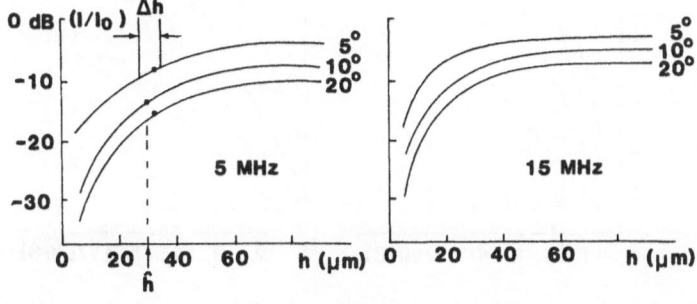

Figure 88. Intensity of backscattered ultrasound versus profilometer-measured rms roughness (after de Billy *et al.*, 137).

Periodic Structure

The characteristics of surfaces exhibiting spatial periodicity may also be assessed with ultrasonic spectroscopic methods. The experimental measurements are identical to those just described for determining roughness; however, the analysis differs.

The periodic surface is assumed to act as a diffraction grating. Peaks in the intensity of the backscattered ultrasound are found when

$$2\sin\theta = m\lambda/d_1$$

where m is the order of diffraction and d_1 is the grating spacing constant. Thus diffraction lines in the spectrum of the echo will be observed at frequencies

$$f_m = m\frac{c}{2d_1}\frac{1}{\sin\theta} \tag{105}$$

The velocity, c, is that of ultrasonic waves in the liquid coupling the transducer to the surface.

To calculate d_1 from ultrasonic measurements, the frequency at a diffraction peak (f_m) is plotted versus the reciprocal of $\sin\theta$ (Figure 89). For a given diffraction order, m, the experimental data will form a set of points scattered about a straight line. The slope of the line fit to the data is

$$m\frac{c}{2d_1} \tag{106}$$

Since for a perfect grating m and c are known, the grating spacing, d_1, may be found.

The spectral peaks observed when backscattering measurements of a periodically rough surface are made are the result of gating all of the signals returned from the surface into the analyzer. As was mentioned in the analysis section, the spectrum of two echo signals is a modulated version of the spectrum of one signal. Similarly,

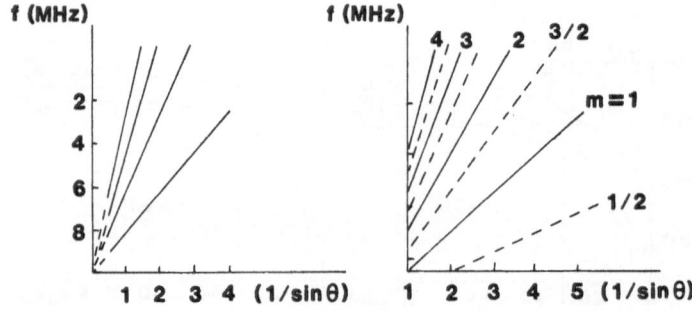

Figure 89. Plots of the frequency of diffraction peaks versus the reciprocal of the sine of the angle at which the backscattered signal was recorded. (a) A perfectly periodic surface, (b) an imperfectly periodic surface (flawed) (after Quentin *et al.*, 138).

many echo signals, all equally separated (which would arise from equally spaced scatterers on a surface), will again give a modulated spectrum. The spectral maxima and minima will occur at the same frequencies as in the spectrum of two signals; however, the peaks and valleys will be much sharper. For maximum sharpening of features to occur, the entire signal should be analyzed as a whole. That is, a wide enough time gate should be used to encompass all echo signals from the surface.

In the case of imperfect gratings, an example of a flawed periodic surface, the scatterers are *not* all equally spaced. Again, because a number of signals arise from the surface, analysis will give a modulated spectrum. For this case, spectral peaks will not be equispaced. Because each pair of scatterers creates a characteristic modulation, the spectrum may be quite complex. Discerning all of the grating spacings is probably best carried out using cepstral analysis. The cepstrum would give the time separations of signals, from which the spacings can be calculated.

Quentin *et al.* (138), de Billy *et al.* (139), and Jungman *et al.* (140) have investigated backscattering of ultrasonic waves from both one-dimensionally and two-dimensionally periodic surfaces. Scattering from perfect and flawed gratings gave the expected modulated spectra. The modulation was quite uniform for the perfect grating, but increased in complexity for gratings with an increasing number of imperfections. The grating spacings were deduced quite accurately from the backscattered ultrasonic measurements. A grid with square meshes and a two-dimensionally engraved surface were also studied. The spatial periodicity of the grid and surface, determined by the ultrasonic technique, agreed quite closely with the true spacing.

In addition to the applications presented in his published work, Dr. Quentin has suggested that the inside surfaces of vessels also might be investigated. Monitoring corrosion inside nuclear reactors or chemical containers appears promising.

Corrosion, Deformation, and Fatigue

Pouliquen and Defebvre (179) utilized a Rayleigh (surface) wave technique to monitor changes occurring when a surface was subjected to corrosive substances, deformed or damaged by fatigue. Their method is extremely sensitive to changes in surface wave velocity ($\Delta v/v = 10^{-6}$) and attenuation (as low as 0.01 dB).

The surface under study is incorporated as a delay line into the feedback loop of an amplifier (Figure 90). Interdigital transducers on quartz substrates produce

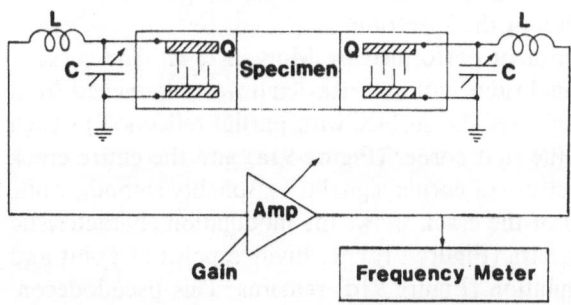

Figure 90. Instrument for the study of microdeformation, fatigue, and corrosion. Q is the quartz crystal with inter digital comb, L and C the inductors and capacitors for impedance matching (after Pouliquen and Defebvre, 179).

Rayleigh waves on the sample surface and then convert the waves back to electrical signals. The amplifier gain at which the oscillation just cuts off is determined by the surface wave attenuation. Changes in surface-wave velocity in the sample alter the frequency at which the device oscillates.

Since the depth of penetration of Rayleigh waves is wavelength dependent, the depths to which the subsurface properties are probed is determined by the frequency of the waves. For the instrument described above, the frequency is determined by the spacing of fingers in the interdigital combs. In order to change ultrasonic frequency the transducers must be interchanged.

Fatigue tests, at a fixed frequency, demonstrated that attenuation (α) increased with the degree of fatigue damage and increased abruptly upon crack initiation. Measurements of attenuation versus crack length demonstrated an increase in α for increasing length. Corrosion tests, also at fixed frequency, showed an unexpected behavior. The attenuation (and velocity) decreased as the corrosion layer was made thicker. The authors postulated that the Rayleigh waves travel beneath the corrosion layer. Because cold working (residual stress) decreases with depth, the attenuation decreases with increasing depth of corrosion as the waves are forced deeper into the less-attenuating material.

Although Pouliquen and Defebvre did not report measurements at a number of frequencies, their methods could easily be extended to several fixed frequencies. Each test frequency would require separate transmitting and receiving transducers.

Surface-Breaking Cracks

The interactions of ultrasonic waves and surface-breaking cracks have been studied by a number of investigators. The varied techniques employed to date have been reviewed by Doyle and Scala (205). In addition to several time-of-flight and mode conversion methods, ultrasonic spectroscopy has been applied to the problem of determining crack depth.

Morgan (085) and Hudgell, Morgan, and Lamb (083) directed surface waves normal to the plane of the crack. The surface waves followed the surface of the crack; however, each feature (corner, etc.) returned some of the surface wave. The time interval between reflected waves can be used to determine the crack dimensions if the Rayleigh wave velocity is accurately known. Gating all of the returned echoes into a spectrum analyzer produced a spectrum with modulation corresponding to each pair of features. If the reflected echoes from the top and bottom of the crack are large compared to those from other features then crack length may be determined from the major modulation in the spectrum.

Figure 91 illustrates the experiment performed by Morgan *et al.* The "crack" was a milled slot having an additional ridge within it. Rayleigh waves launched from a surface-mounted transducer travel over the surface with partial reflection at each ridge or corner. The signals from the first corner (Figure 91a) and the entire crack (Figure 91b) are recorded. The spectrum of corner signal is reasonably smooth, while the spectrum of the signal from all of the crack shows the modulation characteristic of multiple signals. If these two spectra (Figure 91c) are divided point of point and unity is subtracted, only the modulation (Figure 91d) remains. This pseudodecon-

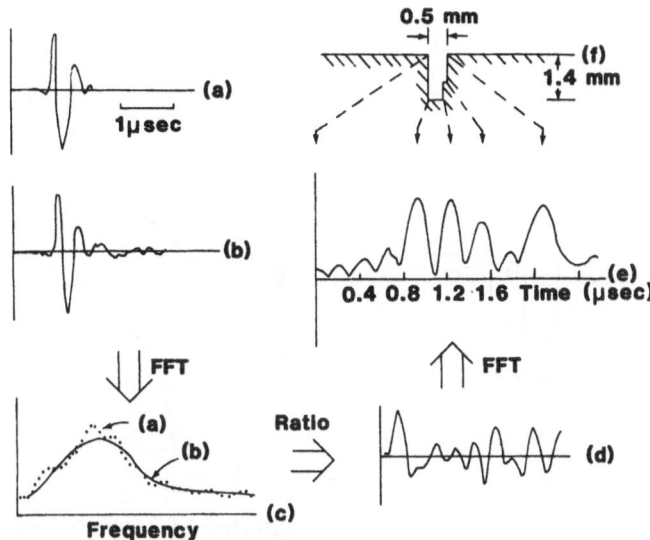

Figure 91. Inference of crack depth from pulse-echo surface wave data. (a) Echo from the first crack corner; (b) entire ensemble of echoes; (c) spectra of signals in (a) and (b); (d) ratio of the spectra in (c) minus unity; (e) Fourier transform of modulation in (d); (f) simulated crack (after Hudgell *et al.*, 083).

volution presents the modulation in a manner amenable to measurement. The Fourier transform of the modulation gives a set of points (Figure 91e) where peaks represent the time separation of echoes returned from crack features. This method of analysis is essentially that of cepstral analysis; however, a deconvolution step is added and the logarithm-taking step has been deleted. All major features of the crack can be identified and their separation distances calculated.

In addition to the modified cepstral analysis described above, Morgan (085) used a "time reconstitution" method to determine the separation of scattering features. Although the effects of dispersion and multiple reflections are neglected in his analysis, the results are comparable to that obtained using the modified cepstral technique.

Silk (196), noting that the interaction of surface waves and surface-breaking defects is wavelength dependent, suggested that spectroscopic methods could be useful for the study of surface cracks. When the wavelength of the ultrasound is much greater than the defect depth, little energy is reflected or delayed by the crack. Conversely, for the case of very small wavelengths the ultrasonic wave follows the surface of the crack. Reflections can occur at each corner or along the rough surface of the crack. If transit time measurements are made for surface waves passing across a defect, the time can be expected to change with the frequency of the wave.

No detailed theoretical treatment of the interaction of a surface wave and crack was available to Silk. He proposed a simple model, the predictions of which were in reasonable agreement with experiment. Three simplifying assumptions are required:

1. The ultrasonic penetration (p) is equal to the depth at which the displacement amplitude is 10% of that at the surface.

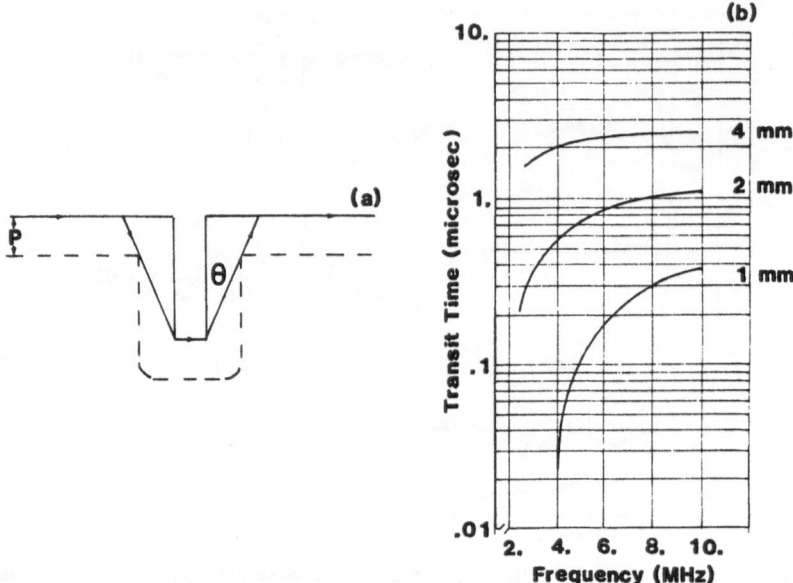

Figure 92. Surface wave propagation about a crack. (a) One possible path for a surface wave subject to the three conditions of Silk; (b) the transit time variation with frequency for the path indicated in (a) (for crack depths of 4, 2, and 1 mm).

 2.   The surface wave is limited to depths of zero to p.
 3.   The wave will take the shortest path (consistent with assumption 2).

One possible path (the path considered by Silk is more complicated) of the wave is shown in Figure 92. The change in transit time (relative to the transit time across an uncracked surface) is plotted versus frequency for three crack depths. The depth of the surface-breaking crack may be inferred from the slope of the curve.

Subsurface Gradients

Rayleigh waves are not strictly constrained to the surface; they penetrate beneath it. Figure 93 illustrates the variation in displacement amplitude with depth for a surface acoustic wave (SAW). Although most of the energy in the wave is concentrated within a fraction of a wavelength beneath the surface, the displacement

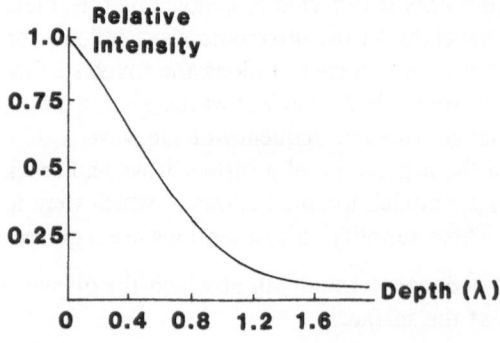

Figure 93. Variation of the maximum intensity of displacement with depth for surface waves (after Silk, 196).

amplitude does not reach 10% until a depth of approximately 1.3λ. Because the wave penetrates the surface, it can interact with subsurface defects or gradients. Since the depth of penetration of the wave depends on wavelength, the probing depth can be altered by changing the ultrasonic frequency.

Auld (501), Tittman and Thompson (cited in references 136, 211, and 321), Richardson and Tittman (321), Richardson (403), and Szabo (136, 211, 402) have considered the case of a Rayleigh wave traveling over a surface of a material for which the velocity of wave propagation varies with depth. Variation of elastic constants or density can give rise to the changes in velocity. Since the penetration depth of the surface waves is (wavelength) frequency dependent, the subsurface gradient will cause dispersion. The formulas are available for calculating the dispersion from a known gradient. However, the practical problem is to infer the subsurface gradient from the dispersion data.

Auld (501) derived an expression which relates a mechanical perturbation to the SAW velocity dispersion. Tittmann and Thompson have shown that this expression may be written in a much simplified form:

$$\frac{\Delta V_R(f)}{f} = \int_0^\infty \left\{ \sum_{i=1}^3 M_i \exp\left[-(q_i f)z\right]\right\} F(z)\, dz \tag{107}$$

where ΔV_R is the change in the SAW velocity, $F(z)$, the subsurface gradient, is a function of depth, z, q_i are constants determined from the Rayleigh wave equation for the material of interest, and M_i are constants resulting from mathematical operations implicit in the original expression of Auld.

Two methods for inversion of Eq. (107) are available. Richardson and Tittman (321) used a technique based on estimation theory to give the most probable gradient for the measured dispersion data. This approach was deemed necessary because the dispersion data are sparse enough that interpolation could cause an improbable estimate of the gradient. The other method of inversion (Szabo, 136, 211) utilizes a Laplace transform technique.

This second method begins by identifying terms in Eq. (107) as scaled Laplace transforms:

$$\mathcal{L}(F(z)) = \bar{F}(q_i f) = \int_0^\infty \exp(-q_i fz) F(z)\, dz \tag{108}$$

Inversion can be carried out by inverse Laplace transformation:

$$F(z) = \mathcal{L}^{-1}(F(s)) \tag{109}$$

$$\mathcal{L}^{-1}\left(\frac{\Delta V_R(f)}{f}\right) = \sum_{i=1}^3 \frac{M_i}{q_i} F(z/q_i) \tag{110}$$

If $\Delta V_R(f)/f$ is in analytical form, then the inverse Laplace transform can be performed and $F(z/q_i)$ found [hence $F(z)$].

Two types of gradients have been studied: those gradually decreasing from the surface, and layers. Szabo parametrically studied typical gradients, the dispersion which they produce, and the density or elastic constant variation required to give the calculated dispersion.

For gradients of the first type and low frequencies (extent of gradient much less than probing depth of SAW) there is little dispersion. At higher frequencies, the velocity asymptotically approaches its value at the surface. At low frequencies, SAWs are insensitive to layered gradients and the dispersion curve is determined by the area of the gradient. Maximum dispersion is achieved when the SAW wavelength is equal to the depth of the layer. For very high frequencies the surface wave may not penetrate to the layer.

The problem of sparse dispersion data, which can have a profound effect on the inferred gradient, may be solved by using a broadband ultrasonic spectroscopic system. Dispersion information can be extracted from the phase spectrum (Pao and Sachse, 221). Ting and Sachse (253) describe a swept-frequency system for measurement of dispersion.

Strength-Related Properties

Ultrasonic spectral analysis techniques have been used in attempts to determine the strength of composite materials. In some cases (Vary, 488; Stone and Clarke, 162) the results have been encouraging; in others (Tamburelli, 233) no correlation was apparent.

Stone and Clarke (162) described a technique for determining the void content of carbon-fiber-reinforced plastics from ultrasonic attenuation measurements. They found a close correlation between attenuation and interlaminar shear strength. They noted attenuation increased with frequency. If their data are replotted as attenuation spectra at various percentage void contents (Figure 94), one notices an increase in $d\alpha/df$ with increasing void content (decreasing strength). It appears this slope could provide an additional estimate of shear strength.

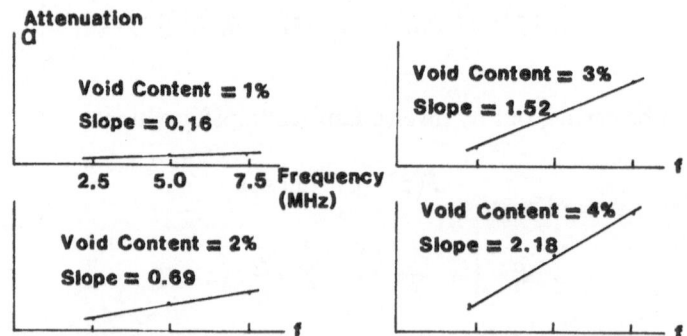

Figure 94. Attenuation spectra for panels with increasing void content (replotted from data given by Stone and Clarke, 162).

Vary (488) defined a "stress wave factor," calculated from ultrasonic parameters, which correlated extremely well with the interlaminar shear strength of graphite–polymide composite panels. This factor was calculated from measurements acquired in the following manner. A broadband transmitting transducer was electrically excited to emit repeating (at rate$=r$) bursts of ultrasonic waves. Each burst produces simulated stress waves having the characteristics of acoustic emission signals. The simulated stress waves are influenced by factors which might alter an actual stress wave arising from the area excited by the transmitter. A waveguide, located some distance from the transmitter, couples longitudinal waves to a wideband acoustic-emission transducer. A counter records the number of oscillations per burst (n) having amplitudes above a threshold. If g represents the interval between counter resets, the

$$\text{stress wave factor} \equiv grn \qquad (111)$$

is a measure of the efficiency of stress wave propagation. The stress wave factor, an indicator of interlaminar shear strength, increases in direct proportion to strength.

Frequency spectra of the received signals showed a strong peak, believed to be representative of a plate resonance. As the percentage of voids in the panel increased the peak height decreased. In addition, some modulation of the low-frequency portion of each spectrum was noted.

Tamburelli (233) used ultrasonics to assess the susceptibility of steel to lamellar tearing. Although several ultrasonic factors correlated with susceptibility to tearing, no correlation was found with the slope of the attenuation versus frequency curve (as measured at two closely spaced frequencies, 14 and 18 MHz).

Fracture Mechanics Parameters

Several authors (Elsley, Richardson, and Thompson, 004; Budiansky and Rice, 234) have described techniques for estimating the fracture mechanics parameter, k_I, from long-wavelength scattering measurements:

$$k_I = (K_I)_{max}/\sigma \qquad (112)$$

where K_I is the mode I stress intensity factor associated with stress, σ, normal to the plane of a crack. The subscript "max" refers to the largest value of K_I on the crack perimeter.

The scatterer is first located in the host medium by short-wavelength techniques such that the distance to the defect from the wave source is known. Low-frequency (long-wavelength) ultrasonic waves (of known amplitude and direction) are then launched toward the flaw. The scattered waveforms $v_2(t)$ are recorded. If possible, calibration waveforms are acquired for a defect-free specimen of identical geometry.

The scattered and reference waveforms are subtracted and a deconvolution performed. The resultant, in the frequency domain, represents the scattering interactions of the ultrasonic waves with the defect as a function of frequency (wavelength).

An expansion of this function in terms of frequency takes the form

$$A = A_2 f^2 + A_3 f^3 + \cdots \tag{113}$$

where A_0 and A_1 are zero for isolated scatterers. The coefficient A_2 is extracted from the spectrum.

The A_2's for a number of scattering configurations are sent to an inversion routine. One possible configuration for inversion is a "feedback" configuration. Tentative parameters (size, shape, and orientation) of the defect are sent to a scattering theory routine which returns an A_2'. This coefficient is compared with the measured A_2, the defect parameters readjusted, and the process repeated until a set of parameters is found which would produce the measured A_2. From the defect parameters, one may estimate fracture mechanics parameters. For example (in the case of a spheroidal void),

$$k_I \simeq (\pi \times \text{longest dimension of the void})^{1/2} \tag{114}$$

Elsley, Richardson, and Thompson (004) describe a probabilistic approach to the inverse scattering problem, as well as their experimental work. They calculated k_I and the standard deviation of k_I from pitch-catch and pulse–echo data of scattering from a spherical void. The pulse–echo measurements gave more consistent results.

The estimation of the k_I from long-wavelength scattering from a flat crack of arbitrary orientation and dimensions was considered by Budiansky and Rice (234). By assuming that the scattering data are accumulated under prescribed conditions, it is possible to find the crack normal and a factor they label P. From P one may estimate the fracture mechanics parameter k_I as

$$(k_I)_{\max} \simeq \left(\frac{8P}{\pi^3} \right)^{1/6} \tag{115}$$

This approximation is said to be good for most elliptic cracks and exact for a circular crack. For elliptic cracks (with axes lengths a and b) with

$$1 \geqslant b/a \geqslant 0.05$$

the estimation is accurate to within 10%.

Fracture Toughness

Fracture toughness is a property of a material. It may be expressed as the critical stress intensity (K_c) at which a crack will propagate abruptly. It is known that the fracture toughness is related to microstructure, and since the attenuation (α) is also dependent on microstructure, it is reasonable to assume that K_c and α are related. Vary (017,334) proposed that fracture toughness is related to the slope of the attenuation versus frequency curve:

$$K_c = \phi \beta_f = \phi \frac{d\alpha}{df} \tag{116}$$

If attenuation measurements are made in the frequency range where Rayleigh and stochastic scattering occur, then α may be represented as

$$\alpha = Cf^m \tag{117}$$

Also,

$$\beta = \frac{d\alpha}{df} = mCf^{(m-1)} \tag{118}$$

When β is evaluated at the point where $\alpha = 1$, $\beta_{\alpha 1} = mC^{1/m}$. Vary measured $\alpha(f)$ over the frequency range 5–40 MHz on two grades of steel having known plane strain fracture toughness (K_{IC}). Plots of β versus K_{IC} (Figure 95) demonstrated that a linear relationship did indeed exist between the two quantities, as proposed in Eq. (116).

Fracture toughness measurement may be outlined as follows:

1. Make repeated measurements of $\alpha(f)$ over a wide frequency range and average them.

2. A plot of $\log \alpha(f)$ versus f will be approximately linear at high frequencies. Using a least-squares linear regression fit to the data, find C and m [Eq. (117)]. These two variables are assumed constant over the entire frequency range. A more accurate description would be to calculate $d\alpha/df$ as a function of frequency.

3. Repeat steps 1 and 2 for materials of known fracture toughness for a range of K_{IC}. This forms the standard curve for this particular grade of material.

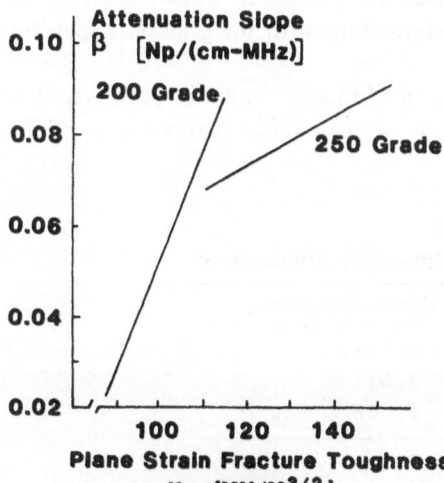

Figure 95. Correlation of attenuation slope with plane strain fracture toughness for two maraging steel specimens (after Vary, 017, 334).

4. To determine fracture toughness for samples of unknown K_{IC} (for which a standard curve is available, as in step 3), repeat steps 1 and 2 and using the standard curve appropriate to the sample material, read off K_{IC}.

Vary (003) later extended his measurements to include specimens of a titanium alloy. He also derived empirical relationships which relate measurable material parameters [$\alpha(f)$, longitudinal wave ultrasonic velocity (c_L), and yield stress (σ_y)] to plane strain fracture toughness. Let β_d be β [Eq. (118)] evaluated at the frequency where the ultrasonic wavelength is equal to the grain dimensions. When $c_L\beta_d$ was plotted versus K_{IC}^2/σ_y, data from the titanium as well as the steel specimens grouped about a straight line. A regression fit to his data has the equation

$$K_{IC}^2/\sigma_y = 8.12 \times 10^6 (c_L\beta_d)^{0.344} \tag{119}$$

where, as noted before, β_d is found at a frequency dependent on grain size.

Microstructure

The microstructure (grain size distribution) of a material is amenable to study by ultrasonic techniques because the interactions of elastic waves and material are wavelength dependent. Ultrasonic attenuation is particularly sensitive to changes in microstructure. The dependence of attenuation on average grain size ($\bar{D}$) and frequency (f) is summarized in Table 5. Entries in this table having coefficients A_i and B_i represent, respectively, scattering and absorption mechanisms. Figure 96 graphs this dependence of attenuation on frequency for the three ranges of $\lambda/\bar{D}$.

As was noted previously, attenuation may be categorized as scattering or absorption. The terms in Table 5 having B_i's for coefficients represent absorption losses (hysteresis and thermoelastic), while the terms having A_i's for coefficients represent scattering losses. The scattering mechanisms are observed to depend on average grain size as well as frequency. Thus an ultrasonic wave traversing an attenuating media will bear the imprint of interactions with the grain size distribution.

A comprehensive review of the scattering of ultrasonic waves in polycrystalline materials has been given by Papadakis (490). Mercier (082) performed a bibliographic review of this same problem.

Table 5
Functional Dependence of Ultrasonic Attenuation[a]

Wavelength range	Functional dependence[b]
$\lambda > 2\pi\bar{D}$	$B_1 f + A_4 \bar{V} f^4$
$\lambda < 2\pi\bar{D}$	$A_2 \bar{D} f^2$
$\lambda \ll D_{min}$	$B_1 f + B_2 f^2 + A_0/\bar{D}$

[a]Source: Papadakis, 490.
[b]Note: $B_1 f$ is elastic hysteresis loss, $B_2 f^2$ is thermoelastic loss, $\bar{D}$ is the average grain diameter, and $\bar{V}$ is the average grain volume.

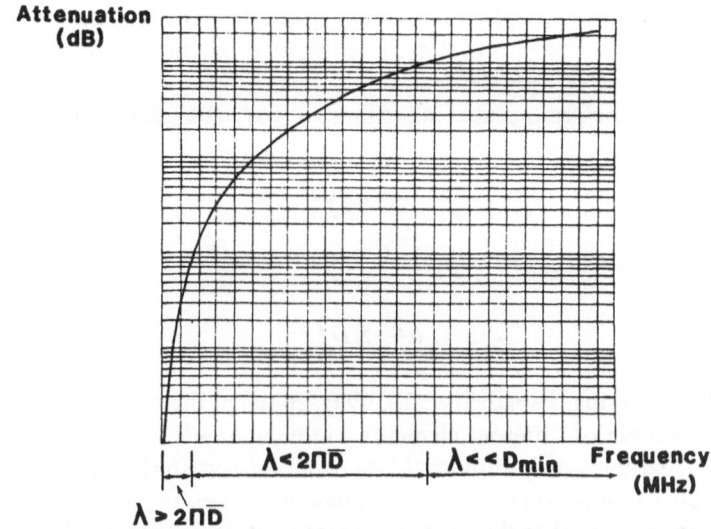

Figure 96. Frequency dependence of ultrasonic attenuation for a hypothetical material. The three ranges of $\lambda/\overline{D}$ define the regions of Rayleigh, stochastic, and diffusion scattering.

Grain Size Distribution

Most polycrystalline materials have a distribution rather than a single grain size. The standard method for determining grain size distribution is from optical measurements on a polished surface of the specimen. Since one obtains only a single section through the distribution, it is necessary to account for the decrease in apparent grain size in a section when the center of the grain lies some distance from that section. Papadakis (186) considered this problem and derived a transformation which relates the number of spheres (of various sizes) per unit volume to the number of circles per unit area smaller than a certain radius seen in a section of the sample.

He also applied this transformation to a polycrystalline metal to determine a grain size distribution consistent with the grain dimensions observed on a photomicrograph. In this same paper Papadakis notes that in the Rayleigh scattering region $(\lambda > 2\pi\overline{D})$ a grain size distribution requires the $\overline{V}$ in Table 5 be determined as $(4/3)\pi\overline{D}^6/\overline{D}^3$ rather than $(4/3)\pi\overline{D}^3$.

Quantitative Scattering Formulae

Mason and McSkimin (055,056) proposed expressions for attenuation due to Rayleigh scattering. Although these formulations were qualitatively correct, the effects of mode conversion were omitted. Later, work by Liftshits and Parkhomovskii and additions by Merkulov and by Bhatia (cited in references 021, 112, and 490) showed that the mode conversion effects accounted for a major part of the scattering loss and could not be neglected. The L.P.M.B. work produced expressions for the linear attenuation coefficients of longitudinal and transverse waves for two ranges of the ratio wavelength to grain dimension $(\lambda > 2\pi\overline{D}$ and $\lambda < 2\pi\overline{D})$. These expressions involve the elastic constants of an individual grain, the density, elastic wave velocity, frequency, and grain size. Formulas for crystallites having cubic, hexagonal, and orthorhombic symmetry were derived.

The forms of the grain scattering expressions are (Papadakis, 112)

$$\alpha = S\bar{V}f^4 \tag{120}$$

for Rayleigh scattering, and

$$\alpha = \sum \bar{D}f^2 \tag{121}$$

for stochastic scattering. Papadakis (112, 490) tabulated the scattering coefficients, S and Σ, for a number of different elements and compounds.

Recently, Serabian and Williams (358) reviewed a scattering theory proposed by Roney in 1950. Comparison of predicted attenuation spectra with their experimental work seems to indicate that the theory is capable of describing the scattering losses through all three ranges of the ratio of wavelength to grain dimension (Rayleigh, stochastic, and diffusion). The attenuation is written as

$$\alpha\bar{D} = B\mu + \frac{S_f}{\mu^2} \sum_{m=0}^{\infty} (2m+1)\sin^2\delta_m \tag{122}$$

where B is a constant, indicating the magnitude of the hysteresis loss, S_f is a scattering factor which is constant for a given material and grain size, μ is a dimensionless parameter equal to $(\pi\bar{D}/\lambda)$, and δ_m is a function of Bessel and Neumann functions.

Serabian and Williams summarize their experimental work and their review of Roney's theory by tabulating the hysteresis (B) and scattering factors (S_f) for several types of steel and for pure nickel.

Experimental Determination of Microstructure

In conventional, single-frequency ultrasonic attenuation studies of polycrystalline materials, attenuation increases as the average grain dimensions become larger. Work, such as that by Meyer (255) using samples of grey cast iron, indicates one may also monitor the size and quantity of scattering inclusions (in his work, graphite flakes).

Roderick and Truell (022) measured attenuation in a number of chrome-molybdenum steel samples over the frequency band 5–50 MHz. The measured loss, using a narrowband pulse burst technique, was quite sensitive to changes in microstructure (as altered by heat treatment).

Investigation of microstructure from information contained in attenuation spectra was performed by Papadakis (021) on several specimens of nickel, stainless steel, brass, and copper. His narrowband measurements spanned the frequency range 0.5–50 MHz. This band included two wavelength ranges—those in which

Rayleigh or stochastic scattering were dominant. He found the attenuation spectra to be in good agreement with L.P.M.B. grain scattering theory.

Gericke (061) was the first to suggest the use of a broadband pulse technique for the study of microstructure. A wideband ultrasonic pulse reflected from the backwall of the sample was analyzed and compared with the spectrum of the backwall echo of a "nonattenuating" (aluminum) sample. His system produced a spectral display of amplitude, covering the frequency range 3–12 MHz. The attenuation spectra for steel samples demonstrated that this method is capable of detecting differences in average grain size. Gericke suggested that one might utilize this technique to examine specimens for proper heat treatment.

Ultrasonic spectrum analysis was used by Mercier (083, 178) to study the effect of average grain size on the attenuation spectrum. The coefficients B_1 and $(A_4 \overline{V})$ (Table 5) in the expression of attenuation (in the region $\lambda > 2\pi \overline{D}$) were tabulated for several alloys.

Canella and Monti (027), using a pulse–echo system, compared ultrasonic spectra of first and second backwall echoes. They studied stainless steel and carbon steel with the aim of devising a quality-assurance test of heat-treated specimens.

Ultrasonic Backscattering

Fay (095, 496), Koppelmann and Fay (171), Cousins (241), Sigelmann and Reid (264), Goebbels and Höller (482), and Fay, Brendel, and Ludwig (495) have outlined a method for deducing absorption and scattering losses from measurement of the ultrasonic backscattering from within a material. In contrast to the attenuation spectra from backwall echoes measured in other techniques, which describe the average interactions over a volume of material—the backscattering technique allows separation of attenuation into absorption and scattering coefficients, as a function of distance from the ultrasonic transducer. Access to opposite sides of the specimen is required for complete determination of absorption and scattering losses. However, for materials in which scattering is the dominant loss, the scattering coefficient may be measured from pulse–echo data acquired from one side of the sample.

Frequency-Dependent Attenuation

As an ultrasonic wave propagates through a material the amplitude of the wave decreases. The energy losses may be attributed to geometrical effects (beam spreading) and intrinsic effects (interactions of the wave with the material).

Intrinsic Energy Losses

Intrinsic losses due to scattering and absorption may both be incorporated in the linear attenuation coefficient. That is,

$$\alpha(f) = \alpha_s(f) + \alpha_a(f) \tag{123}$$

where α_s includes all losses from scattering and α_a includes the losses from any mechanisms producing absorption. For the case of polycrystalline materials the major contribution to the scattering coefficient is grain boundary scattering. The largest absorption loss is that due to thermal conductivity. Note that both the scattering and absorption coefficients are functions of frequency. Although some authors have presented curves of total attenuation coefficients versus frequency for some materials, few have given the relative contributions of scattering and absorption.

Scattering. Scattering occurs at boundaries of differing acoustic impedance, where the mismatch can be the result of changes in density or elastic modulus. The magnitude of the scattering loss, as well as the mechanism, is dependent on the relationship of ultrasonic wavelength to the dimensions of the impedance discontinuities. Table 6 lists the predominant scattering mechanism for a given wavelength (λ) to grain diameter ($\overline{D}$) range, as well as the functional dependence of attenuation upon frequency.

For scatterers much larger than the ultrasonic wavelength, the losses are caused by specular reflections which change the direction of the ultrasound incident on the boundary. The attenuation depends on the mean free path (distance between grain boundaries), so as grain size increases relative to wavelength the losses decrease. As the wavelength approaches the dimensions of the discontinuities, the interactions become complex—a combination of specular reflection and diffraction. For this range the attenuation coefficient is proportional to the product of grain size and the square of the frequency. Since velocity in single crystals depends on orientation, and the distribution of grains in a polycrystalline material is random, the velocity of ultrasonic waves passing through each grain is slightly different. The summation of these phase differences results in a loss of energy. In the limiting case where the boundaries are much smaller than the wavelength, energy is scattered isotropically. The amplitude of the scattered waves depends not on shape but only on the volume of the scatterer.

A large number of fairly uniformly sized particles distributed homogeneously in an ultrasonic beam will scatter the ultrasound uniformly if the particle size is less than one-tenth the wavelength of the incident ultrasound. For this Rayleigh-type scattering, the attenuation coefficient is given by

$$\alpha_s(f) = K\overline{V}f^4 \tag{124}$$

K is a constant for a particular material. It depends on the ratio of the characteristic

Table 6

Mechanisms Involved in the Scattering of Ultrasonic Waves

Wavelength to grain diameter range	Mechanism	Frequency dependence of attenuation
$\lambda \gg \overline{D}$	Rayleigh	Vf^4
$\lambda \approx \overline{D}$	Stochastic (phase)	$\overline{D}f^2$
$\lambda \ll \overline{D}$	Diffusion	$1/\overline{D}$

impedances of the suspended particles and the medium in which the ultrasound is propagating. This type of loss is predominant in emulsions and polycrystalline materials.

Absorption. While scattering accounts for the largest losses in some materials, absorption is the predominant loss in others. The major contributions to attenuation in many liquids are the classical losses due to viscosity and heat conduction. Relaxation mechanisms cause absorption in some solutions and viscoelastic materials; however, since these materials would probably not be employed in coupling the transducer and sample, they will not be discussed here.

As an ultrasound travels through a material the viscosity of the material tends to oppose the motion of the particles disturbed by the wave, and absorption of energy occurs. Energy lost in this manner is degraded directly to heat. The absorption coefficient representing losses due to viscosity is

$$\alpha_a(f) = y_2 \frac{c}{c_0^2} \frac{(2\pi f)^2 t_1^2}{1 + (2\pi f)^2 t_1^2} \tag{125}$$

where η is the coefficient of viscosity, c and c_0 the velocity at the frequency of interest and the velocity as the frequency of the ultrasound approaches zero, and $t_1 = (4/3)(\eta/\rho_0 c_0^2)$. In materials where the absorption is due to viscosity, attenuation increases as the square of the frequency. This frequency dependence can be demonstrated for water and some solutions, at frequencies in the megahertz range.

When an ultrasonic wave passes through a medium an increase in temperature will be produced at points where the material is in compression. The heat will flow from the high-temperature region into regions of lower temperature. If the flow is rapid enough, the temperature differences will tend to be smoothed out, resulting in energy losses from the ultrasonic wave. Assuming a fluid where the rigidity can be neglected, the absorption coefficient is

$$\alpha_a(f) = \frac{(2\pi f)^2 K_h(\gamma - 1)}{2\rho_0 c^3 C_p} \tag{126}$$

where K_h is the coefficient of thermal conductivity, C_p is the specific heat of the material at constant pressure, C_v is the specific heat of the material at constant volume, and $\gamma = C_v/C_p$. Again, the attenuation demonstrates a frequency-squared dependence.

In solids, the rigidity of the material cannot be neglected. Thermal conductivity leads to an absorption coefficient given by

$$\alpha_a(f) = \frac{(2\pi f)^2}{2\rho_0 c_0^2} \left(\chi + 2\eta + \frac{\lambda^a - \lambda^i}{\lambda^a + 2\mu} \frac{K_h}{C_v} \right) \tag{127}$$

where χ and η are the compressional and shear viscosities and λ^a and λ^i are the adiabatic and isothermal Lamé constants.

Polycrystalline materials can also demonstrate thermoelastic attenuation (Zener effect). The strains experienced by individual crystals differ because of their random orientation. Therefore the temperature changes in each grain will be different, leading to a random temperature distribution. Rapid heat exchange between grains causes attenuation:

$$\alpha_a(f) = \frac{\beta_h^T}{2\pi C_v} \frac{(2\pi f)^2 / \omega_h}{1 + (2\pi f / \omega_h)^2} \tag{128}$$

where β_h is the linear expansion coefficient of the specimen at temperature T, and ω_h is approximately $0.8 K_h / C_v \overline{D}^2$ ($\overline{D}$ is the grain diameter).

Loss caused by the thermoelastic effect is maximal below 100 kHz, and in many materials is small because of the wide variety of grain sizes found in most materials.

Measurement Procedures

Specimen preparation is extremely important for attenuation measurements. The sample must have two parallel and polished sides. At high frequencies the surfaces must be finished to optical tolerances. For most measurements the transverse sample dimensions should be very much larger than the width of the transducer (although a method using guided waves has been outlined by Mason).

The continuous-wave (cw) resonance technique of Bolef and DeKlerk has been summarized by Mercier (082). It is based on the frequency response of standing waves in a compound oscillator comprised of the ultrasonic transducer, coupling medium, and sample. The system is operated at or near the resonant frequency of the transducer and of the specimen. The Q of the compound oscillator is related to the linear attenuation coefficient at the resonant frequency.

Several pulse train measurement methods have been developed. A transmitting transducer produces an ultrasonic pulse with a duration which is short compared with the transit time in the sample. The amplitude of the reverberation echoes are monitored. In one technique, a pulse at the same frequency as that propagating through the specimen is transmitted through the same receiver electronics (amplifier, etc.). A calibrated attenuator is used to adjust the amplitude of the reference pulse to that of each of the echoes. Although the measurement process is slow, one is able to measure very high attenuation.

A second method utilizes an exponentially decreasing electrical signal displayed along with the returning echo signals. The curve is adjusted until matched to the echo envelope and α is read from an indicator on the exponential-signal generator. Although the method is practical and rapid, it cannot be used on materials with high attenuation.

Radiation Coupling Corrections. If the reverberation echo signals from either of the pulse train measurement systems just described is displayed, it will be apparent that the decrease in echo amplitude is not exponential. The deviation from expected behavior is caused by radiation coupling effects and energy losses at the specimen–transducer interface. Each time the ultrasonic pulse reaches the transducer a small portion is transmitted through the coupling layer into the receiver. Corrections can be made to the experimental data if the reflection and transmission coefficients at the transducer-coupling medium–specimen interfaces are known. A buffer rod technique, used by Papadakis (111), simplifies computation of the reflection coefficient correction. Lynnworth (024) developed a nomogram to speed the calculations.

Radiation coupling losses may account for a large part of the measured attenuation in samples with low attenuation and at low frequencies. This "apparent" attenuation is the result of diffraction effects (beam spreading) and phase variation over the area of the receiving transducer. Rayleigh–Sommerfield diffraction theory is used to find the radiation field produced by a piston-type circular source. Then, the receiver response is found by integrating the pressure field over the receiving transducer.

Seki, Granato, and Truell (015) have computed the radiation coupling losses for a system of two coaxial disks. As a rough approximation, they suggested that a correction of 1 db/(a^2/λ) was appropriate (a is the transducer radius). However, one should note that this approximation applies to sample thickness less than 1.8(a^2/λ). The disk-to-disk radiation coupling function derived by Rhyne (254) also may be used to calculate the "diffraction correction." Figure 97 illustrates the radiation coupling correction factor (in dB) for a range of normalized sample lengths (normalized by a^2/λ). Papadakis (025) has considered the difficult problem of diffraction losses in anisotropic materials.

Broadband spectroscopic attenuation measurements have been reported by Papadakis, Fowler, and Lynnworth (016). A buffer rod is interposed between a broadband transducer and the sample. A short ultrasonic pulse, emitted from the transducer, produces a series of echoes. Echo A arises from the buffer–specimen interface, while the two remaining echoes arise from the back reflection (b) and a reverberation (C). Each of these three signals is gated out and the spectra ($A(f), B(f), C(f)$] determined. The normalized distance (S) the ultrasonic wave has traveled from the transducer is calculated for the individual echoes:

$$S = \frac{z}{a^2/\lambda} = \frac{zc}{a^2 f} \tag{129}$$

where z is the path length in the material and c is the velocity of ultrasound propagation. The amplitude at each frequency is corrected (using Figure 97) for diffraction based on the normalized distance and using the wavelength appropriate for each frequency. If the amplitudes of the corrected spectra are represented by

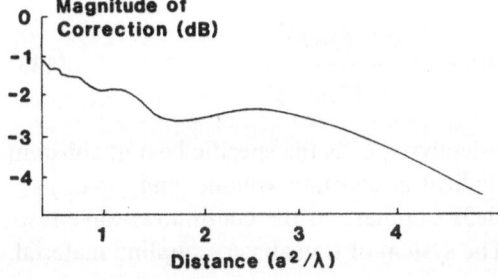

Figure 97. Radiation coupling correction factor as a function of normalized sample length (a is the transducer radius, λ the ultrasonic wave length in the sample) (after Rhyne, 254).

$A'(f)$, $B'(f)$, and $C'(f)$, then the reflection (R) and linear attenuation coefficients may be calculated from

$$R=\{A°(f)C°(f)/[A°(f)C°(f)-1]\}^{1/2} \tag{130}$$

$$\alpha(f)=\frac{\ln[-R/C°(f)]}{2l} \tag{131}$$

where $A°(f)=|A'(f)|/|B'(f)|$ and $C°(f)=|C'(f)|/|B'(f)|$. Attenuation coefficients obtained utilizing this spectroscopic method have been shown to agree well with rf-burst measurements where a quartz transducer was bonded directly to the sample.

Velocity Dispersion

Materials which demonstrate frequency-dependent velocity variations are referred to as dispersive materials. In these substances a distinction is made between the group and the phase velocities. The group velocity being the rate at which the point of maximum amplitude in an ultrasonic pulse (containing many frequency components) propagates through the material. Phase velocity is the velocity of a continuous sinusoidal wave (one frequency) in the medium. Pulse transit time measurement is adequate for calculation of group velocity; however, one must resort to "spectroscopic" techniques for finding phase velocity.

Mechanisms Causing Dispersion

The classical mechanisms giving rise to frequency-dependent absorption in a material are also accompanied by dispersion. The relative velocity dispersion occurring when viscosity is the predominant cause of loss (in liquids) is

$$\left(\frac{c_0}{c}\right)^{1/2}=\frac{1}{2}\left\{\frac{1}{[1+(2\pi f)^2 t_1^2]^{1/2}}+\frac{1}{1+(2\pi f)^2 t_1^2}\right\} \tag{132}$$

where η is the coefficient of viscosity, c_0 is the velocity as the frequency of the ultrasound approaches zero, and $t_1=(4/3)(\eta/\rho_0 c_0)$. In those fluids where the predominant absorption mechanism is thermal conductivity, the dispersion is given by

$$\left(\frac{c_0}{c}\right)^2=1+(\gamma-1)(5-\gamma)\frac{K_h^2}{C_p}\frac{(2\pi f)^2}{(2\rho_0 c^2)^2} \tag{133}$$

where K_h is the coefficient of thermal conductivity, C_p is the specific heat at constant pressure of the material, C_v is the specific heat at constant volume, and $\gamma=C_v/C_p$.

Measurement Techniques. Mercier (083) summarized the continuous-wave resonance technique of Bolef and de Klerk. The system of transducer, coupling material, and specimen is treated as a compound oscillator. As the frequency of the cw oscillator is raised, successive resonances occur. An expression involving the frequency spacing of the resonance peaks, the sample thickness, and properties of the transducer and bond may be used to find the velocity of wave propagation.

Continuous-wave phase comparison techniques (083, 253) may be utilized for finding the phase velocities at a number of frequencies. A variable-frequency cw ultrasonic signal is propagated along a sample of fixed length (L). As the frequency is slowly swept upwards, the frequencies at which input and output waves are in and out of phase are recorded. Phase velocity is

$$\text{phase velocity} = \frac{L}{N/f} \tag{134}$$

where N is the number of cycles of delay of the signal due to passage through the sample and f is the ultrasonic frequency. It should be noted that the transducers, electrical networks, and coupling to the specimen introduce frequency-dependent phase delays which must be taken into account.

The disadvantage of the cw method is its inability to discriminate between reflections, mode conversions, or other interferring signals. This limitation may be overcome by using a pulsed system where the extraneous signals are separated in time from the main signal. Martin (098) described a narrowband, sine-wave burst system for measurement of phase velocity. He used a cosine-squared modulation of the carrier frequency to minimize the amplitude of the sidebands. An expression is derived which gives the effect of dispersion on the shape of the burst. Phase comparison is performed, as in the cw method, utilizing cycles near the center of the burst since these cycles are exactly or nearly in phase with continuous waves after traveling the same path.

Pao and Sachse (221) have shown that it is possible to determine the phase and group velocities of elastic waves from the phase spectra of broadband pulses. Two wideband transducers are affixed to opposite sides of a polished sample. The received waveform is analyzed digitally to give the phase spectrum, $\angle V_2(f)$. A reference spectrum, $V_2^0(f)$, is acquired with the transducers directly in contact. The phase and group velocities may be calculated from

$$\text{phase velocity}(f) = \frac{\omega}{k} = \frac{2\pi fL}{\angle V_2(f) - \angle V_2^0(f)} \tag{135}$$

$$\text{group velocity}(f) = \frac{d\omega}{dk} = L \bigg/ \left[\frac{d\angle V_2(f)}{df} - t_0 \right] \tag{136}$$

where L is the sample length and t_0 is the time at which the input wave enters the specimen [i.e., $\angle V_2^0(f)/f$].

Questionnaire

The purpose of the questionnaire (Figure 98) was twofold: to update a listing of investigators and to ask their views on the state-of-the-art. During the early stages of the literature survey a list of authors was assembled. An attempt was made to find their current addresses. To each of these 130 individuals we sent a letter, a questionnaire, and the list of researchers. Fifty-four investigators responded, returning the questionnaires, listing pertinent publications, and providing names and addresses for persons not included on the preliminary list. The tabulated responses can be found as entries in large type on the listing of the questionnaire.

Several items which could be considered to be presently limiting the use of ultrasonic spectroscopy by industry were listed on the second page of the questionnaire. A number of respondents commented on these possible limitations. Tables 7–12 summarize their remarks.

On page 1 of the questionnaire, investigators were asked to list their present research activities in ultrasonic spectroscopy. Their replies are contained in Table 13.

When the major part of the literature survey had been completed, the list of investigators was reviewed and was amended to include only those with research interests or publications dealing directly with ultrasonic spectroscopy. The revised address list is contained in Table 14.

May 24, 1978

Gentlemen:

   Our Ultrasonics Group at The University of Tennessee is under
contract to assess the current state-of-the-art in ultrasonic
spectroscopy and to develop a comprehensive literature survey.

   We are aware of your contributions to the field of ultrasonic
spectroscopy, and would like to solicit your help in making our survey
complete. We would like to be as thorough as possible and include all
the important aspects of your work. We would appreciate it if you
could send a listing of your publications and presentations.
Additionally, we would like to request that you fill out the enclosed
questionnaire. We have enclosed a listing of persons involved with
ultrasonic spectral analysis. If you are aware of others we have not
listed, please let us know.

   We sincerely appreciate your cooperation.

                                    Sincerely,

                                    Laszlo Adler
                                    Associate Professor and
                                    Principal Investigator

Distinguished Past...   Dynamic Future...

Figure 98

ULTRASONIC SPECTROSCOPY QUESTIONNAIRE

The intent of sending this questionnaire to you is to gather information enabling an assessment of the current state-of-the-art in ultrasonic spectroscopy.

Ultrasonic spectroscopy is generally described as the field of study dealing with the extraction of information from an ultrasonic wave after interaction with a material. Ultrasonic spectroscopy has a wide variety of applications which include:

  Determination of microstructure in metals
  Assessment of surface texture of materials
  Thickness gauging
  Testing of adhesive bonds
  Characterization of defects as to size, shape, and composition
  Tissue characterization
  Study of relaxation mechanisms
  Transducer characterization
  Experimental "verification" of acoustical theories

What is your present research activity in ultrasonic spectroscopy?_____

_____

_____

_____

Number of research personnel involved with this work?___134_____

Approximate level of present funding for this research? TOTAL $1.4 M TO $2.3 M

___ < $10K, ___$10K-25K, ___$25K-50K, ___$50K-100K, $_____Other

Do you have any interest in researching ultrasonic spectroscopy if funding were made available? _38_ Yes _0_ No

What particular area of ultrasonic research would you pursue?_____

_____

_____

_____

What general level of funding would you request?

____ < $10K, ___$10K-25K, ___$25K-50K, ___$50K-100K, $_____Other

TOTAL = $1.6 M TO $2.9 M

Figure 98 (*continued*)

2

Which of the items listed below do you consider as limitations to the study and
application of ultrasonic spectroscopy?

	Development at present is satisfactory	A limitation at present	Comments
Transducers	6	38	
Signal Recording	22	17	
Analysis (Hardware and Software)	22	18	
Theory	13	32	
Acceptance by Industry	7	28	
Qualified Personnel	7	31	

Figure 98 (*continued*)

3

Would anyone from your group be interested in a symposium on ultrasonic spectral signal information and its uses?

Attendance: <u>46</u> Yes <u>2</u> No

<u>64</u> No. of Persons

Presentation of Papers: <u>32</u> Yes <u>11</u> No

Publications, Presentations, and Patents relevant to ultrasonic spectroscopy (use additional pages if necessary):

Name_____

Address (if incorrect on the list)

_____

_____

_____

_____

Please return this questionnaire to:

Professor Laszlo Adler
Department of Physics
The University of Tennessee
Knoxville, Tennessee 37916   USA

Figure 98 (*continued*)

TABLE 7

TRANSDUCERS

Number of Responses	Response
15	Limited bandwidth
3	Poor reproducibility
3	Flatter frequency spectrum
1	Better characterization of beam pattern
1	Need for phase-insensitive transducers with high overall efficiency
2	Lack of calibration procedures (frequency response and absolute sensitivity) (need for reference)
3	Lack of precise correlation of response with physical parameters - displacement, stress, particle velocity, etc.
2	Unknown effects of coupling
1	Use of arrays
1	Need for transducers with higher sensitivities
1	Theoretical knowledge not applied commercially
1	Point source of high intensity over a broad frequency range
1	Manufacture more an art than a science
1	Broad band shear-wave transducers not commercially available
1	Use laser transmitters
1	Use capacitive detectors

TABLE 8

SIGNAL RECORDING

Number of Responses	Response
6	Digitization techniques not satisfactory for frequencies above 10 MHz
1	Tradeoff between time resolution and real-time capability
1	Simpler electronics for practical applications
3	Number of samples (when digitizing) is restricted (in transient recorders)
4	Limited amplitude resolution of digitizers (transient recorders, 8-bit)
2	Expense of digitizing systems
1	NDE equipment behind the state-of-the-art of other research fields
1	Receiving amplifier and gating circuitry for frequencies greater than 100 MHz are needed
1	Better signal-to-noise ratio and dynamic range in receiver electronics is required
1	Amplifiers distort signals

TABLE 9

ANALYSIS
(Hardware and Software)

Number of Responses	Response
3	Need for more pattern recognition hardware and software
1	Expensive equipment required, in relation to conventional ultrasonic inspection systems
2	Source deconvolution software needed
1	Real time inverse processing and adaptive filtering
1	FFT on one chip would be desirable
1	Analysis techniques too complex for interpretation by field operators
1	FFT computer - frequency range 0.01 to 100 MHz
1	Data management cumbersome - multiple angles and transducer angles required
1	Data processing lags behind theory (one step in right direction is training classifiers from theory)
1	Effects of windowing on U/S data not completely understood
1	Software to ease interpretation by field operators
1	Limitations of presently available software and hardware need to be made clear

TABLE 10

THEORY

Number of Responses	Response
7	Scattering of elastic waves by an object of arbitrary shape - solution will be extremely complex (problem of application)
1	Volumetric reflectors need to be considered (present theory OK for flat reflectors)
1	Small flaw in heterogeneous material
1	More phase relations (so phase information can be used)
1	Near-field analysis
1	Finite-element approach
1	Experimental verification of simplified inversion theories
2	Theories for tissue differentiation
1	Understanding of theory amongst practitioners is poor
1	Source deconvolution
2	Attenuation in structures with preferred orientation
4	Inverse problem
1	Incorporate U/S diffraction in transducer and scattering problems and inversion
1	Correlation of spectroscopy with microstructure
1	Experimental verification of theories
1	Surface texture in high frequency range
1	Most work proceeds empirically
1	Studying caustics appears promising
1	Approaches to inverse scattering problem should be simplified

TABLE 10 (continued)

Number of Responses	Response
1	Theory is the "most important problem area . . . ," "it should be given high priority"
1	Multiple defects
1	Concentrate on more limited objectives - inverse problem may take too long to develop

TABLE 11

ACCEPTANCE BY INDUSTRY

Number of Responses	Response
2	Limited by the degree to which the interpretation can be simplified
1	Many uses are laboratory techniques
1	Not convinced ultrasonic spectroscopy is really needed
2	Expensive
2	Instrumentation too complex
1	Too time consuming
1	"Automation and demonstration that it works will be needed"
1	Complex data analysis required
1	Verification of validity of defect sizing required for code acceptance
1	Requires machined surfaces for frequencies exceeding 5 MHz
1	Not ready yet
1	No identified market (as some industrial groups accept it - ultrasonic spectroscopic techniques will become "tested" and so, proven)
1	Industry has not been exposed to its capabilities
1	Little evidence of practical problems which can be solved by ultrasonic spectroscopy
1	Very little work being directly applied to industry problems
2	Complex data presentation
1	"Empirical operation and methods abound - very little is procedure derived from understanding of the physical processes involved"

TABLE 11 (continued)

Number of Responses	Response
1	Will only be accepted when shown to be better than conventional techniques - studies of conventional versus ultrasonic spectroscopy, false positive, false negative, etc.
1	"Will come about when a few (even limited) successes are demonstrated"
1	Used by specially trained staff - for work in the nuclear power industry (in United Kingdom)

TABLE 12

QUALIFIED PERSONNEL

Number of Responses	Response
2	Insufficient technical and scientific background
2	More training necessary
1	"Level II NDE technician would probably have difficulty performing inspections with existing defect evaluation systems"
1	Only a few active researchers - in the United Kingdom
1	"Has little status in the United Kingdom"
10	Too few qualified personnel
1	Not taught to engineers
1	Qualified persons are only in the laboratories
1	"At present a playground for Ph.D's
1	Go-No Go implementation would ease need for qualified personnel
1	Personnel are few because technology is just developing
1	"There is a need for special graduate programs in the field of NDE"
1	Large gap between theory and verification and the eventual implementation of practical procedures
1	". . . damage has been done by so-called research organizations in many industries"
1	"Reduce operator skill required"
1	"Encompass variability in operator skill"
1	Used by specially trained staff - for use in the nuclear power industry (in the United Kingdom)

TABLE 13

PRESENT RESEARCH ACTIVITIES OF INVESTIGATORS
RESPONDING TO THE QUESTIONNAIRE
(Alphabetical Order)

Investigator	Present Research Activities
J. D. Achenbach	Analytical studies on the diffraction of ultrasonic waves by cracks. Computation of the scattered field as a function of frequency and scattered angle.
L. Adler	Defect characterization; wave propagation at interfaces; ultrasonic spectroscopy; elastic wave scattering.
G. Alers	Analysis of adhesive bonds; measurement of elastic constants of crystals; characterization of composites; analysis of acoustic emissions; scattering from bulk and surface defects.
C. W. Anderson	Transducer characterization, attenuation measurements versus frequency.
J. R. Birchak	Thickness gauging, defect characterization.
N. Bleistein	Characterization of defects.
J. W. Brophy	Flaw characterization (modelling of wave interactions).
A. F. Brown	Defects in composites; adhesively-bonded joints; surface cracks; acoustic emission; instrumentation design; modelling studies.
J. A. Brunk	Determination of microstructure, characterization of defects.
G. Canella	Transducer characterization.
P. L. Carson	Tissue characterization; transducer characterization and experimental verification of acoustical theories.
R. C. Chivers	Material characterization, including human tissue and tissue models; transducer characterization.
R. L. Crane	Damage accumulation in composites.

TABLE 13 (continued)

Investigator	Present Research Activities
N. E. Dixon	Transducer fabrication and characterization; pulse-shape generation techniques; process measurement and control applications (e.g., mining, pulp and paper, food processing).
P. A. Doyle	Theory of defect characterization.
D. G. Eitzen	Absolute calibration technique for transducers, also determination of defect shape from received spectra.
B. Fay	Determination of the local structure variation in metals and model substances.
D. W. Fitting	Ultrasonic spectroscopic systems design, tissue characterization, flaw characterization, pattern recognition.
J. R. Frederick	Deconvolution.
O. R. Gericke	Analysis of transmission spectra in bonded layered media.
R. S. Gilmore	Use of spectroscopy to study thermal and stress damage in composites; bond integrity in semiconductor power devices; flaw characterization in large metal forgings; transducer acceptance criteria.
P. Greguss	Tissue differentiation.
J. E. Gubernatis	Theoretical studies of defect characterization.
N. F. Haines	Study of layered media and reflections from surfaces
C. R. Hill	Differential diagnosis of malignant disease.
B. Kato	Defect characterization in welded joints.
G. Kino	Acoustic imaging.
M. Klein	Characterization of defects.
R. A. Kline	Evaluation of adhesive bond strength (particularly interfacial effects).

TABLE 13 (continued)

Investigator	Present Research Activities
M. W. Moyer	Study of material microstructure, bond testing, thickness gauging and transducer characterization.
E. Nabel	Defect sizing utilizing a deconvolution technique.
W. G. Neubauer	Determination of body or reflector geometric and elastic properties as manifested in the acoustic echo; theoretical evaluation of exact closed-form solutions as well as approximate theories and the experimental verification of both; also, body fluid constituent and contamination identification.
V. L. Newhouse	Flaw to grain signal enhancement, deconvolution.
Y. H. Pao	Calculation and interpretation of theoretical ultrasonic spectroscopy.
E. P. Papadakis	Transducer characterization, determination of microstructure in metals, testing adhesive bonds.
G. Quentin	Assessment of surface texture; characterization of surface defects, tissue characterization; experimental verification of acoustical theories.
K. J. Reimann	Scattering from rough surfaces, defect characterization, signature analysis of acoustic emission signals, experimental verification of acoustic theories.
T. L. Rhyne	Ultrasonic signature studies of lung-surface reflection.
D. E. Robinson	Tissue characterization and transducer characterization.
W. R. Scott	Study of laminated anisotropic materials with application to composite materials.
R. A. Sigelmann	Tissue characterization; transducer characterization; experimental verification of acoustical theories.
D. E. W. Stone	Analysis of adhesive bonds.
R. D. Strong	System design, including equipment, computer hardware and software.

TABLE 13 (continued)

Investigator	Present Research Activities
B. R. Tittmann	Characterization of defects; transducer characterization; experimental "verification" of acoustical theories; assessment of surface texture of materials.
A. Vary	Ultrasonic evaluation of material properties (strength).
G. Wade	Acoustic imaging for oceanic search and for medical diagnosis.

TABLE 14

INVESTIGATORS ACTIVE IN ULTRASONIC SPECTROSCOPY

J. D. Achenbach
Dept. of Civil Engineering
Northwestern University
Evanston, IL 60201

L. Adler
Dept. of Welding Engineering
Ohio State University
Columbus, OH 43210

E. E. Aldridge
Nondestructive Testing Centre
Harwell, Berks, England

G. A. Alers
Science Center
Rockwell International
Thousand Oaks, CA 91360

K. V. Ammirato
Knolls Atomic Power Lab
ERDA
Schenectady, NY 12345

C. W. Anderson
Code G-53
U.S. Naval Surface Weapons Center
Dahlgren, VA 22448

R. W. Andrews
General Electric Company
Neutron Devices Dept.
P.O. Box 11508
St. Petersburg, FL 33733

B. A. Auld
E.L. Ginzton Laboratory
Stanford University
Palo Alto, CA 94305

V. M. Baborovsky
TI Research Labs
Hinxton Hall
Hinxton, Saffron Waldon, England

J. C. Baboux
d'Ultrasons Laboratoire Bat.
502 I.W.S.A.
69621 Villeurbanne
Lyon, France

J. C. Bamber
Institute of Cancer Research
Clifton Ave.
Belmont, Sutton, Surrey, England

F. E. Barber
Harvard Medical School
Dept. of Radiology
Boston, MA 02115

F. L. Becker
Battelle Memorial Institute
Pacific Northwest Laboratories
P.O. Box 999
Richland, WA 99352

J. R. Birchak
Babcock & Wilcox
Lynchburg Research Center
P.O. Box 1260
Lynchburg, VA 24505

N. Bleistein
Denver Applied Analytics
255 S. Ivanhoe Place
Denver, CO 80222

E. Borloo
ISPRA
Varese, Italy

J. W. Brophy
Babcock & Wilcox
Lynchburg Research Center
Lynchburg, VA 24502

A. F. Brown
Department of Physics
City University
St. John St.
London EC1V4PB, England

TABLE 14 (continued)

J. A. Brunk
Bendix Corp., Kansas City Div.
P.O. Box 1159, Mail Code ME49
Kansas City, MO 64141

G. Canella
Centro Sperimentalle Metallurgico
Rome, Italy

P. Carson
University of Colorado Medical Ctr.
C278
4200 E. 9th Ave.
Denver, CO 80262

F. Chang
General Dynamics
P.O. Box 748
Fort Worth, TX 76101

R. C. Chivers
Physics Dept.
University of Surrey
Buildford, Surrey, G025XH
England

W. R. Clipson
Research & Development Labs
C. A. Parsons & Co. Ltd.
Newcastle-on-Tyne, England

F. Cohen-Tenoudji
Universite Paris VII
Groupe de Physique des Solides
De L'Ecole Normale Superieure
Tour 23-2, Place Jussieu
75221 Paris Cedex 05/ France

J. Couchman
General Dynamics
P.O. Box 748
Fort Worth, TX 76101

R. L. Crane
Materials Laboratory
AFWAL/MLLP
Wright-Patterson AFB, OH 45433

S. K. Datta
Dept. of Mechanical Engineering
University of Colorado
Boulder, CO 80302

M. DeBilly
Universite Paris VII
Groupe de Physique des Solides
De L'Ecole Normale Superieure
Tour 23-2, Place Jussieu
75221 Paris Cedex 05/ France

N. E. Dixon
ENTEC Corp.
P.O. Box 2822
Tri Cities, WA 99302

E. Domany
Weizman Institute
Rechovst, Israel

P. A. Doyle
Aeronautical Research Laboratories
Box 4331
CPO Melbourne, Victoria 3001
Australia

D. G. Eitzen
National Bureau of Standards
Sound A147
Washington, DC 20234

R. K. Elsley
Rockwell International Science Center
1049 Camino Dos Rios
Thousand Oaks, CA 91360

B. Fay
Physikalisch-Technische Bundesanstalt
2200 Braunschweig, Germany

D. W. Fitting
Department of Welding Engineering
Ohio State University
Columbus, OH 43210

TABLE 14 (continued)

J. R. Frederick
Dept. of Mechanical Engineering
University of Michigan
2046 E. Engineering Bldg
Ann Arbor, MI 48104

H. J. Fullbright, M-1, MS912
Los Alamos Scientific Lab
P.O. Box 1663
Los Alamos, NM 87545

D. J. A. Garcia-Poggio
INTA
Tottejonde Ardoz
Madrid, Spain

O. R. Gericke
NDT Advanced Research Branch
U.S. Army Materials & Mechanics
Research Center
Watertown MA 02172

S. Gilmore
Bldg. 37, Room 5032
General Electric
Corporate Research & Development
Schenectady, NY 12345

K. Goebbels
Fraunhofer-Institut fur
Zerstorungsfreie Prufahreu
D-6600 Saarbrucken
Universitat, West Germany

R. Couzze
INSA
Lyon, France

P. Greguss
Applied Biophysics Laboratory
Technical University of Budapest
Kruspe'r U. 2-4
H-1111 Budapest, Hungary

J. E. Gubernatis
Los Alamos Scientific Laboratory
T-11, MS 457
Los Alamos, NM 87545

N. F. Haines
Applied Physics Div.
Central Electric Generating Board
Berkeley Nuclear Labs.
Berkeley
Gloucester, England

S. D. Hart
Naval Research Laboratory
Code 5831
Washington, DC 20375

R. C. Heyser
Jet Propulsion Laboratory
4800 Oak Grove Drive
Mail Station 183/901
Pasadena, CA 91103

P. J. Highmore
T2 IRD3/RNPDL
UKAEA, Risley
Warrington, Cheshire
United Kingdom

C. R. Hill
Institute of Cancer Research
Royal Cancer Hospital
Clifton Ave.
Sutton, Surrey, England SM2 5PX

J. D. Hislop
Divisional Quality Engineer (NDT)
Rolls Royce, Ltd.
Aero Engine Division
Derby, England

P. Hollen
Fraunhofer-Institut fur
Zerstorungsfreie Prufahreu
D-6600 Saarbrucken
Universitat, West Germany

TABLE 14 (continued)

A. E. Holt
Babcock & Wilcox
Lynchburg Research Center
P. O. Box 1260
Lynchburg, VA 24505

P. Jenhensen
ISPRA
Varese, Italy

A. Jungman
Universite Paris VII
Groupe de Physique des Solides
De L'Ecole Normale Superieure
Tour 23-2, Place Jussieu
75221 Paris Cedex 05/ France

A. C. Kak
School of Electrical Engineering
Purdue University
West Lafayette, IN 47907

B. Kato
University of Tokey
Tokyo, Japan

R. J. Kazys
Kaunas Antanas Snieckus Polytechnic
Institute
Kaunas
Lithuania, USSR

J. C. Kennedy
Boeing Co.
Aerospace Group
Seattle, WA 98101

G. S. Kino
Ginzton Laboratories, M108
Stanford University
Stanford, CA 94305

M. Klein
Institut fur Zerstorungsfreie
Prufverfahren
Saarbrucken, Germany

R. A. Kline
General Dynamics
P.O. Box 748, MS 5984
Fort Worth, TX 76101

B. K. Lakatosh
National Science Foundation
Special Foreign Currency Science
Information Program
Washington, DC 20234

F. Lakestani
d'Ultrasons Laboratoire Bat.
502 I.W.S.A.
69621 Villeurbanne
Lyon, France

K. M. Lakin
Ames Laboratory
Iowa State University
Ames, Iowa 50011

M. Linzer
Institute of Materials Research
National Bureau of Standards
Washington, DC 20234

F. Lizzi
Riverside Research Institute
80 West End Ave.
New York, NY 10023

E. A. Lloyd
Dept. of Physics
City University
St. John Street
London, EC1V4PB, England

S. J. Mech
Hanford Engineering Development
Laboratory
P.O. Box 1970, 300 Area
Richland, WA 93352

T. Michaels
Hanford Engineering Development
Laboratory
P.O. Box 1970
Richland, WA 99352

TABLE 14 (continued)

J. B. Miller
Washington University
Dept. Of Physics
Skinker and Lindell Blvds.
St. Louis, MO 63130

R. F. Mitchell
Philips Research Laboratories
Redhill, Surrey, England

T. J. Moran
Materials Laboratory
AFWAL/MLLP
Wright-Patterson AFB, OH 45433

L. L. Morgan
British Gas Corp.
Engineering Research Station
(O.L.I.)
Killingworth, Newcastle-Upon-Tyne
England

M. W. Moyer
Union Carbide Nuclear
P.O. Box Y
Oak Ridge, TN 37830

A. N. Mucciardi
Adaptronics, Inc.
Westgate Research Park
1750 Old Meadow Road
McLean, VA 22102

E. Nabel
Bundesanstadt fürMaterialprüfung
NDT Group
Unter den Dichen 87
1000 Berlin 45
West Germany

W. Neubauer
Code 8130
Physical Acoustics Branch
Naval Research Laboratory
Washington, DC 20375

V. L. Newhouse
School of Electrical Engineering
Purdue University
West Lafayette, IN 47907

D. Nicholas
Institute of Cancer Research
Clifton Avenue
Belmont, Sutton, Surrey England

J. Obraz
National Research Institute for
Machine Design
190 00 Praha 9
Bechovive Czechoslovakia

Y. H. Pao
Dept. Of Theoretical & Applied
Mechanics, Thurston Hall
Cornell University
Ithaca, NY 14850

E. Papadakis
Ford Motor Co.
Manufacturing Processes Lab
24500 Glendale Ave.
Detroit, MI 48239

H. Z. Pawlowski
Vl, Czeska 1M3,
03-902
Warszawa, Poland

M. Perdrix
d'Ultrasons Laboratoire Bat.
502 I.W.S.A.
69621 Villeurbanne
Lyon, France

G. J. Posakony
Battelle Northwest
P.O. Box 999
Richland, WA 99352

TABLE 14 (continued)

J. Pouliquen
Laboratoire d'Ultrasons
de la Faculte Libre des Sciences
et du Polytechnicum de Lille
3 rue Francois Baes
59046 Lille Cedex, France

G. Quentin
Universite Paris VII
Groupe de Physique des Solides
De L'Ecole Normale Superieure
Tour 23-2 Place Jussieu
75221 Paris Cedex 05/ France

J. M. Reid
Institute of Applied Physiology
and Medicine
556 18th Ave.
Seattle, WA 98122

K. Reifshider
Dept. of Engineering Mechanics
Virginia Ploytechnic Institute
Blacksburg, VA 24060

K. F. Reimann, D212-E217
Argonne National Laboratory
9700 South Case Ave.
Argonne, IL 60439

P. M. Reynolcs
British Non-Ferrous Metals
Research Association
London, N.W. 1.2EU England

T. J. Rhyne
Bolt, Beranek & Newman, Inc.
50 Moulton St.
Cambridge, MA 02138

D. E. Robinson
Ultrasonics Institute
5 Hickson Rd.
Millers Point, NSW2000
Australia

J. L. Rose
Mechanical Eng. and Mechanics Dept.
Drexel University
Philadelphia, PA 19104

W. Sachse
Dept. of Theoretical and Applied
Mechanics, Thurston Hall
Cornell University
Ithaca, NY 14853

W. R. Scott
ACSTD Code 6063
Naval Air Development Center
Warminster, PA 18974

S. Serabian
Dept. of Mechanical Engineering
Lowell University
Lowell, MA 01853

J. G. Sessler
Technical Consultant
121 Jean Ave.
Syracuse, NY

J. A. Seydel
University of Missouri
Dept. of Mechanical Engineering
Roila,MO

R. A. Sigelmann
Dept. of Electrical Eng., F7-10
University of Washington
Seattle, WA 98195

M. G. Silk
Nondestructive Testing Centre
AERE Harwell
Didcot, Oxon, United Kingdom

W. A. Simpson
Oak Ridge National Laboratory
P.O. Box X
Oak Ridge, TN 37830

TABLE 14 (continued)

D. E. W. Stone
Structures Dept.
Royal Aircraft Establishment
Farnborough, Hants England

R. D. Strong
Los Alamos Scientific Laboratory
P.O. Box 1663
Los Alamos, NM 87543

J. Szilard
University of Technology
Loughborough, Liecestershire
England

B. Thompson
Ames Laboratory
Iowa State University
Ames, Iowa 50011

B. R. Tittman
Science Center
Rockwell International
1049 Camino Dos Rios
Thousand Oaks, CA 91360

A. Vary
Lewis Research Center
NASA
Cleveland, OH 44135

R. C. Waag
Dept. of Electrical Engineering
University of Rochester
River Stateion
Rochester, NY 14627

G. Wade
Dept. of Electrical Engineering
and Computer Science
University of California
Santa Barbara, CA 93106

P. N. T. Wells
Bristol General Hospital
Dept. of Medical Physics
Guinen St.
Bristol, BSI 654
United Kingdom

H. L. Whaley
Babcock & Wilcox
Research and Development Div.
Lynchburg Research Center
P.O. Box 1260
Lynchburg, VA 24305

R. M. White
University of California
Berkeley, CA 94704

J. H. Williams, Jr.
Room 30360
Massachusetts Institute of Tech.
77 Massachusetts Ave.
Cambridge, MA 02139

W. G. W. Yee
Convair Aerospace Division of
General Dynamics
P.O. Box 748
Fort Worth, TX 76101

5

Abstracted Bibliography

Work on our assessment of the state-of-the-art in ultrasonic spectroscopy was initially directed toward performing a comprehensive literature survey. We began the task by compiling references from our own files and the libraries of the University of Tennessee and Oak Ridge National Laboratory. Several reviews of ultrasonic spectroscopy were found; however, most were either limited in scope or outdated. Searches of computerized data bases were made in an effort to disclose work performed under government contracts, some of which had never been published other than in technical reports. Over 500 pertinent references have been identified, dealing with theoretical work, ultrasonic spectroscopic systems, and the application of spectroscopy to problems in nondestructive evaluation. These citations appear in this report as an abstracted bibliography.

Each entry in this bibliography has been given a serial number for ease of cross-reference with the Subject and Author Indexes. The serial number is comprised of two parts: the year of publication and an assigned number. Items cited in the text of this report are so indicated by a plus sign ($+$) following the serial number. The ordering of entries in the bibliography is completely arbitrary; no rating of importance is implied.

No claim is made for the merit of items included in this bibliography or for the lack of merit of works not included.

1973-001+

Frederick, J. R. and Seydel, J. A., "Improved Discontinuity Detection
    Using Computer-Aided Ultrasonic Pulse-Echo Technique," <u>Bulletin
    Number 185 of the Welding Research Council</u> (July, 1973).

    "The purpose of this project has been to investigate means for
obtaining improved characterization of the size, shape and location of
subsurface discontinuities in metals. This has been done by applying
computerized data processing techniques to the signal obtained in conventional ultrasonic pulse-echo systems. The principal benefits are
improved signal-to-noise ratio, and resolution.
    The received ultrasonic pulse from a discontinuity is combined
with a reference function in order to preserve both the phase and amplitude of the received pulse. Processing is performed on the power spectrum
of the resultant signal. The reference function is also used to compensate for some of the frequency-dependent scattering caused by the
material microstructure. During data acquisition, analog signal averaging
improves the signal-to-noise ratio (SNR). After data acquisition, a
digital deconvolution procedure improves the transducer longitudinal
resolution.
. . . .
    Possible applications of this data processing system to current
problems in nondestructive testing are discussed. The testing of
materials with large amounts of internal scattering and frequency
dependent attenuation appers to be particularly promising. Also, the
enhanced resolution gained by the system can be used to reveal previously
undetectable discontinuities in thin plates or those that are near a
reflecting surface such as in immersion testing." (Author)

    The author presents what we believe to be the first analysis of an
ultrasonic system (including defect) as a linear time-invariant system.
Each element of the system may be represented by its impulse response
(in the time domain) or its frequency response (in the frequency domain).
    The method of acquiring the waveform data is unique and may provide
a way in which an investigator could update an analog system. 10 refs.

1976-002 +

Simpson, W. A., "A Fourier Model for Ultrasonic Frequency Analysis,"
    Materials Evaluation 34 (12), 261-264 (Dec., 1976).

    Fourier transform techniques are applied to the interpretation of
ultrasonic frequency analysis, yielding results in basic agreement with
previous interference models, but capable of explaining anomalies
unaccounted for by such models. The modulated spectrum obtained when
two signals (separated in time or overlaping) are gated into a spectrum
analyzer is explained simply. The author notes that the frequency
spacing of maxima may vary according to the slope of the envelope.
Spectra are presented to demonstrate the principles which are discussed.
    The effects of attenuation and phase shifts are treated. The
sharpening of maxima when multiple signals are gated into the analyzer
is explained. The present approach indicates that both attenuation and
phase-shift measurements may be obtained directly from the displayed
spectrum, either separately or simultaneously. 6 refs.

1976-003+

Vary, Alex, "Correlations among Ultrasonic Propagation Factors and
    Fractures Toughness Properties of Metallic Materials," NASA TM
    X-71889, Technical paper presented at the Spring Conference of
    the American Society of Nondestructive Testing (Los Angeles,
    March 1976) (NTIS #N76-21319/8ST).

    "Empirical evidence was developed to show that a close relation
exists among fracture toughness, yield strength and ultrasonic
attenuation properties of metallic materials. The evidence was obtained
by ultrasonic probing of specimens of two maraging steels and a titanium
alloy." (Author)

    Relevant data were extracted by a frequency spectrum analysis
procedure over the frequency range between 10 MHz and 50 MHz. The
effect of microstructural factors on ultrasonic attenuation was evaluated
in terms of the slope of the attenuation coefficient versus frequency
curve characteristic of each specimen material. It was found that the
plane strain fracture toughness and also the yield strength of the
materials examined were strongly related to the slope at the attenuation
versus frequency curve.
    This paper gives an account of the methodology that was used to
extract the required ultrasonic attenuation information. Problems
associated with the methodology are discussed. 19 refs.

1977-004+

Elsley, R. K.; Richardson, J. M.; Thompson, R. B. "Determination of
    Fracture Mechanics Parameters from Elastic Wave Scattering Measure-
    ments at Low Frequencies," Project I, Unit C, Task 5, SC595.325A,
    pp. 104-141.

    Considerable information can be derived from ultrasonic scattering
measurements in which the wavelength is large with respect to the flaw
size. A certain crucial parameter in fracture mechanics (the stress
intensity factor K_1) from longitudinal-to-longitudinal scattering data
extrapolated to low frequencies. The estimate is made under the simplifying assumption that the defect is a single spheroidal void. The size,
shape, and orientation, as well as critical fracture parameters of the
defect are determined.

    The advantages and disadvantages of the low-frequency approach are
discussed. It is noted that a Ka < 0.5 is required for the analysis.
In an expansion of the scalar scattering amplitude in frequency, it is
demonstrated that the frequency-squared term dominates (for the Ka < 0.5).
The coefficient of this term, measured for several locations of the
transducer about the defect, may be used to estimate the fracture parameter
K_1. 7 refs.

1969-005+

Whaley, H. L. and Cook, K. V., "Ultrasonic Frequency Analysis," Oak Ridge
National Laboratory, Technical Report ORNL-TM-2655; also presented
at the 29th ASNT Meeting, Philadelphia (Oct., 1969).

A system was developed which can determine the frequency content of
reflected or transmitted ultrasonic pulses. The system is composed of
commercially available electronic equipment and can be used for either
contact or immersion testing. The authors utilized the system to
examine the frequency effects of a number of ultrasonic-test variables.

The effects of transducer "resonant frequency" and type of
piezoelectric material on the bandwidth of the ultrasonic pulse was
studied. It was also noted that the type of pulser (tuned or untuned)
used to excite the transducer profoundly affected the spectrum. Changes
in angulation of transducer away from normal to a flat plate caused a
"dropout" of the high frequencies. Effects of beam divergence on the
spectra were noted by changing the distance between the transducer and a
flat plate.

It was found that the inducter-resister tuning devices used on
some search tubes dramatically altered the spectrum. "Tunable peaking
coils" narrowed the spectrum. Masks (or apertures) sometimes used to
limit the ultrasonic beam diameter caused a shift of the spectra to
lower frequencies. The magnitude of the shift was greater for smaller
diameter apertures.

1963-006+

Gericke, O. R., "Determination of the Geometry of Hidden Defects by
Ultrasonic Pulse Analysis Testing," J. Acoust. Soc. Am. 35, 364-368
(March, 1963).

Difficulties with conventional ultrasonic NDE techniques are said
to be overcome by use of ultrasonic pulse analysis. In this procedure,
ultrasonic signals are utilized which contain a broad band of frequencies,
and in analogy to optics, can therefore be considered as "white" ultrasonic pulses. The form and spectral energy distribution, or "color," of
such ultrasonic pulses is influenced by the geometry of a defect from
which they are reflected. Hence, an analysis of the defect echo yields
information on the defect configuration. The successful application of
the ultrasonic pulse analysis method for differentiating between flaws
of various configurations is illustrated by several examples.

The spectra of ultrasonic pulses reflected from flat-bottomed holes,
side-drilled holes and angled notches in aluminum are presented.
Differences in the spectra were noted for various size defects, as well
as for the various shapes of defects. (It is worth noting that modulated
spectra are obtained from reflections off the angled notches.)

A brief discussion is included of electrical pulse shape on the
spectrum of ultrasound emitted by the transducer.

The frequency band under study ranged from 5 MHz to 14 MHz. 1 ref.

1966-007+

Gericke, O. R., "Defect Determination by Ultrasonic Spectroscopy,"
   J. Metals 18 (8), 932-937 (1966).

   Conventional, amplitude-based ultrasonic techniques are compared to
ultrasonic spectroscopic methods, for defect characterization. Three
sets of simulated flaws in aluminum blocks were used for the studies. One
set contains flaws whose surfaces were parallel to the test surface.
In the second set of blocks, the defect surface was oriented at an angle
of 10° with respect to the test surface. The third set was comprised of
side-drilled holes. The sizes of the defects in each group were 5/8",
5/16", 5/32", and 5/64".
   It was noted that amplitude measurements were ambiguous (e.g., equal
amplitudes for a 5/64" flat-bottomed hole at 0° and for a 5/16" side-
drilled hole). However, differences in the spectra (3.5 MHz-10 MHz) of
pulses returned from the defects gave clues to its geometry.
   The spectra for ultrasound normally incident on the defect demonstrated larger high frequency components (relative to low frequencies)
for the 5/32" and 5/64" defects. A decrease in low-frequency reflectivity
as the ultrasonic wavelength approached the defect dimension was postulated
as the cause.
   Lower frequency amplitudes were dominant in the signals from the
flaws (especially the larger sizes) oriented 10° from the test surface.
The lower frequencies are not as effectively deflected away from the
transducer by the angled defect as the better collimated medium and high
frequencies. (The spectra demonstrated no modulation.)
   Small side-drilled holes return signals weak in medium and high
frequencies.
   The author comments that defect-echo spectrum is a function of both
defect size and orientation and indicates the possibility of characteristic
spectral signatures. 3 refs.

1965-008+

Gericke, O. R., "Ultrasonic Spectroscopy of Steel," Mater. Res. Std. 5 (1),
   23-30 (1965).

   The mechanical properties of steel are, to some extent, dependent upon
the average grain diameter. The author suggests ultrasonic spectroscopy as
method for nondestructive inference of grain size. A broadband (3-16 MHz)
spectroscopic system is described which employs an analog spectrum
analyzer. A single transducer (lead zirconate niobate) operated as both
transmitter and receiver.
   Seven steel specimen, each with a different grain size (2 to 8 on
ASTM scale) were examined with the spectroscopic system. The back-wall
echo was gated into the analyzer and a spectrum was recorded for each
steel. Since the back-wall echo was always adjusted to the same level
(using amplifier gain), the low frequency region of all the spectra was
almost identical. However, the amplitude of the high frequencies varied
noticeably among the steels. Comparison of the spectra for the steels,
with a spectrum for an aluminum block, demonstrated a noticeable increase
in attenuation with decreasing grain dimensions. The spectra also indicated that attenuation increases as frequency increases. 5 refs.

1972-009+

Seydel, J. A., "Computerized Enhancement of Ultrasonic Non-Destructive
    Testing Data," Ph.D. Dissertation, Univ. of Mich., Dept. of Elec.
    Eng. (1972).

A summary of the work completed by the author may be found in ref.
(1973-001). However, a great deal more detail is included in this dis-
sertation.
    Although this manuscript does not deal with ultrasonic spectroscopy
directly, the approach to the problem and elements of the author's solution
are relevant to spectroscopy.

1970-010+

Cox, C. W., and Renken, C. J., "The Application of Signal-Processing
    Techniques to Signals from Electromagnetic Test Systems," Mater.
    Eval. 28 (8), 173 (1970).

"The demand for more accurate quantitative data directly relatable
to specimen characteristics has made necessary the development of signal-
processing methods that operate on the raw output signal, changing it to
a form that allows decisions to be made by simple computer programs.
The success of signal processing depends to a large extent on the amount
of information produced by the test system. Broadbanding transducer
filters do not increase the information present in the signal, but do
correct for the imperfect resolution of the transducer and aid in the
successful use of additional selective filters, and in pattern-recognition
techniques. . . . The practical effects of several types of processing
on the signals produced by a pulsed electromagnetic system employed in the
inspection of tubing (is) discussed." (Author)

Although this publication is primarily concerned with E-M processing
techniques, it is valuable for defining properties of ultrasonic systems.
Important signal and instrument characteristics—time-bandwidth product,
transducer bandwidth and its effects on noise and resolution—are
quantitatively defined (in the terminology of signal processing and
information theory). Techniques for broadbanding of transducers are presented and the approximations which must be made are outlined. Use of
matched filters for enhancement of signal-to-noise ratio is proposed.
11 refs.

1968-011+

Serabian, S., "Implications of the Attenuation-Produced Pulse Distortion
    upon the Ultrasonic Method of Nondestructive Testing," Mater. Eval.
    26 (9), 174 (1968).

"The use of the ultrasonic method of nondestructive testing in highly
attenuating, nondispersive materials presents difficulties uniquely non-
germane to nominal(ly) attenuating materials. The difficulties arise from
the preferential attenuation of the high frequency components of the
pulse. Under this condition, the pulse is continuously changing as it

propagates through the medium. The end result is that the pulse
experiences not only a decrease in amplitude but a gradual shift of the
frequency content to the low end of the frequency spectrum. Since there
is a greater sensitivity (to defects) at the higher frequencies, the
overall effect of preferential attenuation is a general degradation (in
accuracy) of the ultrasonic method." (Author)

The influence on pulses propagating in highly attenuating materials
is investigated. Effects on defect detectability are discussed as well
as implications involved in velocity and attenuation measurements.

A good set of figures clearly illustrates the effects discussed by
the author. Plots of attenuation as a function of frequency (for G-1
and G-2 graphite and aluminum) are included in the illustrations. 9 refs.

1971-012+

Whaley, H. L. and Adler, L., "Flaw Characterization by Ultrasonic Frequency
     Analysis," Mater. Eval. 29 (8), 182-188 (1971).

"A new method was developed for the characterization (determination
of the size and orientation) of a reflector by ultrasonic spectral analysis.
This technique can be used to determine nondestructively the nature of
flaws detected in a material by means of ultrasonic examination. It is
shown that it is feasible to characterize a flaw by this technique in spite
of its composition (i.e., crack void or inclusion) or shape and without
the need for a calibration standard. This technique is free of several
limitations inherent in techniques based on amplitude. The results presented
here are based on a series of reflection experiments in which the ends of
solid rods immersed in water and machined discontinuities in metal samples
were used as reflectors. Broadband ultrasonic pulses were analyzed after
reflection from the interface of interest. A physical and analytical model
was developed based on an interference mechanism that results from the
superposition of spherical wavelets emitted from opposing extremes of the
reflector. This model explains very well the experimentally observed
spectral variations in terms of size and three-dimensional angular orientation
of the reflector, distance between the transducer and reflector, and displacement of the reflector from the axis of the transducer for a flat
reflector of any composition. Based on this technique, a detailed procedure can be outlined for characterizing a randomly oriented natural
flaw in a plate. The feasibility and practicality of using this technique
to test various materials and in automated testing systems were considered."
(Author). 1 ref.

1974-013+

Simpson, William A., "Time-Frequency-Domain Formulation of Ultrasonic
     Frequency Analysis," J. Acoust. Soc. Am. 56 (6), 1776-1781 (1974).

The interpretation of the frequency spectra of multiple (two or
more) time-series signals is given, using Fourier analysis techniques.
The following conclusions are reached.

"1. The characteristic modulation of the observed spectrum arises
as a consequence of the time delay between input signals, whatever their

origin, and hence may be utilized to make timing measurements of any
kind." (Even if the signal overlaps.) "Thus, the separation between
maxima in the frequency domain is determined by signal delay in the time
domain."

"2. From the spectrum, one may determine the attenuation of one
signal with respect to the other; in particular, the attenuation as a
function of frequency may be obtained in a single test, using broad-band
input signals."

"3. The phase shift of one signal with respect to the other may be
obtained from the spectrum. For ultrasonic purposes, this implies the
ability to measure relative acoustic impedances, detect nonbonds, and
possibly, to discriminate between cracklike flaws and inclusions." (Author)

Whenever possible, measurement of "delta f" should be made in a
region of the spectrum where the envelope is horizontal; or, if the
minima extend to zero, then the measurements should be made at the
minima. 3 refs.

1975-014

Adler, L. and Lewis, D. K., "Scattering of Broad-Band Ultrasonic Beams
    from Circular Disks," J. Acoust. Soc. Am. 57 (Sup. 1) S59 (1975)
    (Abstract only).

"In an attempt to develop models for nondestructive testing for
flaw characterization, we have simulated "flaws" by using metal circular
reflectors in water. A broad-band ultrasonic beam was incident on the
reflectors. The scattered wave amplitudes were measured as a function of
frequency for fixed angles and as a function of angles for several frequencies. In our previous work [Laszlo Adler and H. L. Whaley, J. Acoust.
Soc. Am. 51, 881 (1972)], the frequency distribution of the scattered
sound wave was related to the size and orientation of the reflector by an
approximate theory. This theory assumed that the wavelets originating
from the extreme edges of the reflector will interfere. Now we have
adopted a theory from the diffraction of electromagnetic waves by
aperture [Keller, Appl. Phys. 28, 426 (1957)]. An expression for the
diffracted amplitude was calculated as a function of frequency and angle
for several reflector sizes. The result compares favorably with our
experiments." (Authors)

1956-015+

Seki, H., Granato, A., and Truell, R., "Diffraction Effects in the Ultra-
    sonic Field of a Piston Source and Their Effects Upon Accurate
    Ultrasonic Attenuation Measurements," J. Acoust. Soc. Am. 28, 230-
    238 (1956).

"A study is made of the ultrasonic field produced by a circular
quartz crystal transducer and the integrated response of a quartz crystal
receiver with the same dimensions as the transducer. The transducer and
receiver are taken to be coaxial, and it is assumed that the transducer
behaves as a piston source while the integrated response is proportional
to the average pressure over the receiver area. Computations are made
for the cases of interest in the megacycle frequency range (Ka = 50 to

1000; a = piston radius; λ = wavelength; k = $2\pi/\lambda$). The results contain features of use in identifying and correcting for diffraction errors. These features which apparently have been missed in previous investigations are compared with available experimental data. Finally correction formulas to account for diffraction effects in the accurate measurement of attenuation are discussed. It is shown that the order of magnitude of the diffraction attenuation is given by one decibel per a^2/λ." (Authors)

1973-016+

Papadakis, E. P., Fowler, K. A., and Lynnworth, L. C., "Ultrasonic
    Attenuation by Spectrum Analysis of Pulses in Buffer Rods: Method
    and Diffraction Corrections," J. Acoust. Soc. Am. 53, 1336-1343
    (1973).

"A method has been presented for measuring ultrasonic attenuation as a function of frequency by analyzing the spectra of broad-band echoes in a buffer/specimen system. Integral to this method is a new technique for correcting for ultrasonic diffraction. The correction technique has also been presented. It may also be used for rf bursts in buffer/specimen systems.
    The buffer rod method with diffraction corrections has been used in three experiments to demonstrate the following:
    1. That the method itself is as accurate as the regular method using rf bursts and quartz crystals bonded directly for specimens;
    2. That the use of a buffer rod on liquid buffer column is to be preferred over the use of a damped NDT transducer coupled directly to the specimen, because the latter configuration gives rise to higher erroneous attenuation readings;
    3. That damped NDT transducers may be almost critically damped, radiating a relatively flat spectrum from zero up to a frequency well below the nominal frequency to which the piezoelectric elements are cut, showing a small peak at the top of the flat band, and falling rapidly in efficiency above that peak." (Authors). 17 refs.

1975-017

Vary, Alex, "The Feasibility of Ranking Material Fracture Toughness by
    Ultrasonic Attenuation Measurements," NASA Technical Memorandum,
    NASA TM V-71769 (Sept. 1975) (NTIS #N75-29241/7ST).

"A preliminary study was conducted to assess the feasibility of ultrasonically ranking material fracture toughness. Specimens of two grades of maraging steel for which fracture toughness values were measured were subjected to ultrasonic probing. The slope of the attenuation coefficient versus frequency curve was empirically correlated with the plane strain fracture toughness value for each grade of steel." (Author) 18 refs.

1971-018

Kolsky, H. and Rader, D., "Stress Waves and Fracture," in Fracture--
   An Advanced Treatise, Chap. 9, H. Liebowitz, ed., Vol. 1, 533-567,
   Academic Press, New York (1971).

1973-019

Kolsky, H., "Recent Work on the Relations between Stress Pulses and
   Fracture," Proceedings of an International Conference on Dynamic
   Crack Propagation, George C. Sih, ed., 399-414, Noordhoff Int. Publ.,
   Leyden, The .Netherlands (1973).

1973-020+

Rose, J. L. and Meyer, P. A., "Ultrasonic Procedures for Predicting
   Adhesive Bond Strength," Mat. Eval., June 1973.

   "Ultrasonic immersion and spectroscopic procedures are presented that
are applicable for predicting and evaluating bond strength in an aluminum-
to-aluminum step lap Scotch-Weld 2216 structural adhesive joint. A
consideration of the nonlinear shear stress distribution in the bond and
the effects on evaluating a programmed flaw are included in the paper.
It is shown that it is possible to establish a reasonable correlation
between bond strength and some nondestructive testing parameter.
Limitations of the methods are reviewed." (Authors). 14 refs.

1965-021+

Papadakis, Emmanuel P., "Ultrasonic Attenuation Caused by Scattering in
   Polycrystalline Metals," J. Acoust. Soc. Am. 37, 711-717 (1965).

   "Ultrasonic attenuation measurements have been made on fine-grained
specimens of several metals. The grain-size distributions and ultrasonic
velocities in these metals were also determined. The experimental
attenuation is in good quantitative as well as qualitative agreement
with current theory. Nickel and three iron alloys, one 30% nickel
reported previously, the second 12% chromium (type 416 stainless steel)
and the third 17% chromium and 1% carbon (type 440-C stainless steel), all
gave good results. Brass also gave good results, but copper showed
much twinning, which as yet is unaccounted for." (Author). 18 refs.

   The following information is provided for each of the metal
specimens examined: (1) photomicrograph, (2) grain-image area histogram
and grain-size distribution, (3) theoretical and experimental graphs of
longitudinal and shear wave attenuation as functions of frequency.

1952-022

Roderick, R. L. and Truell, R., "The Measurement of Ultrasonic Attenuation in Solids by the Pulse Technique and Some Results in Steel," J. Appl. Phys. 23, 267-279 (1952).

"Pulse techniques for the measurement of attenuation in solids have been extended and refined sufficiently to obtain dependable measurements over a frequency range from 5 to 50 megacycles. Understanding of the relative importance of beam spreading, geometrical boundaries, and method of coupling has been improved. Coupling by means of water buffer and direct mounting is discussed in detail.
 Attenuation measurements in the frequency range from 5 to 50 megacycles have been made on chrome molybdenum steel specimens, and these measurements have shown large differences in ultrasonic attenuation for samples of the same chemical composition but different heat treatment. The resulting differences are connected with anisotropy which appears in the photomicrographs. The attenuation measurements are quite sensitive to heat treatment and other factors. The application of the methods to metallurgical problems and the physics of solids is suggested." (Author)

1973-023

Papadakis, E. P., "The Measurement of Small Changes in Ultrasonic Velocity and Attenuation," Critical Reviews in Solid State Sciences, 3 (4), 373-418 (1973).

"Several methods for measuring small changes in ultrasonic velocity and attenuation are reviewed. The discussion is restricted to traveling wave methods. In-depth coverage is given to five velocity methods: twin-specimen interferometer, sing-around, pulse super position, pulse-echo overlap, and π-point systems. Of these, the pulse-echo-overlap method provides a combination of high precision, versatility and absolute accuracy unexcelled for many applications. Phase and group velocity are both treated. Concerning attenuation, methods using rf bursts and broad-band pulses (with or without spectrum analysis) are examined. Various techniques for comparing echo amplitudes are mentioned, including AVC monitoring systems. While emphasis is on the measurement of small relative changes throughout, the techniques and corrections needed for absolute measurements are also covered." (Publisher) 95 refs.

1974-024+

Lynnworth, L. C., "Attenuation and Reflection Coefficient Nomogram," Ultrasonics 12, 72-73 (1974).

The author presents a brief (one page) summary of the buffer-rod method for determining the attenuation coefficient of a sample. A nomogram is included for rapid determination of sample attenuation and the pressure reflection coefficient at the buffer-sample interface.
 Refer to reference (1973-016) for a more detailed description of the buffered-rod technique. 4 refs.

1966-025+

Papadakis, E. P., "Ultrasonic Diffraction Loss and Phase Change in
    Anisotropic Materials," J. Acoust. Soc. Am. 40, 863-874 (1966).

   "The loss and phase change in progressive longitudinal waves from a
finite circular piston source radiating into certain anisotropic media
are calculated in this paper. Three-dimensional calculations are
presented for propagation along 3-, 4- and 6-fold symmetry axes. Relationships sufficient to permit the performance of the calculation for 2-fold
axes are derived. The anisotropy is introduced as a term in the spatial
part of the phase in the integral for the pressure in the field of the
piston. Geometry proper for pulse-echo measurements was assumed in the
calculation. The diffraction loss fluctuates with $S = z\lambda/a^2$ (s is the
distance into the crystal, λ is the wavelength, a is the transducer radius)
before becoming monotonic increasing as the logarithm of S. The positions
of the peaks in the loss are functions of the anisotropy of the medium.
The phase change, on the other hand, increases monotonically to a limit of
$\pi/2$ from S = 0 to infinity. However, the phase has plateaus 4-, and 6-fold
axes in crystals agree quantitatively with theory. Some data on 2-fold
axes are also presented. The methods for correcting attenuation and
velocity measurements for diffraction are given." (Author) 14 refs.

1974-026+

Lloyd, E. A., "Nondestructive Testing of Bonded Joints," Nondestructive
    Testing (London), 331-334 (Dec., 1974).

   "A quantitative method for the testing of adhesively bonded lap-joints
is currently being developed. (This technique should permit one to monitor
and measure) the adhesive properties of the interface between the adhesive
and adherent layer and the cohesive properties of the bonding material
itself (e.g., thickness, porosity, elastic modulus, etc.)." (Author)

   The three-layer lap-joint is modeled as a lumped parameter mass-
compliance-mass system. The compliant material (adhesive) between the
two masses (metal plates) produces a "resonance" dependent on its thickness
and mechanical properties. Results are presented for tests on several lap
joints.

1976-027+

Canella, G. and Menti, F., "Spectrum Analysis: A New Tool for Quality
    Control by Ultrasonics," NDI International [Nondestructive Testing
    (London)], 187-192 (August, 1976).

   "This paper presents the results of ultrasonic measurements performed
with spectrum analysis on forged Cr-Ni-Mo steel and on carbon steel produced by continuous casting, both subsequently subjected to various heat-
treatment cycles. The measurements have confirmed the possibility of
ascertaining structural differences in these steels, even though these may
not be immediately evident from metallography, simply by comparing two
ultrasonic pulse spectra with an appropriately selected frequency band.
These methods of inspection can be applied in industry without any great
difficulty." (Author) 2 refs.

1976-028

Gericke, O. R. and Monagle, B. L., "Detection of Delaminations by Ultrasonic Spectroscopy," IEEE Trans. Son. Ultrason., SU-23 (5), 339-345
(1976).

"A newly developed instrument for ultrasonic spectroscopy has been
evaluated on various types of laminated structures to assess its effectiveness for detecting delaminations in composites. Presented are ultrasonic
echo spectra obtained for different types of disbond conditions. In
addition, a comparison of these spectra with computer-generated
reflection data is made. Results show that in most of the cases
examined the detectability of a disbond can be predetermined by computer analysis." (Author) 11 refs.

1976-029+

Thompson, R. B. and Evans, A. G., "Goals and Objectives of Quantitative
Ultrasonics," IEEE Trans. Son. Ultrason., SU-23 (5), 292-299 (1976).

"The ultimate objective of most nondestructive evaluation (NDE)
studies is to develop a capability for predetermining the inservice
failure probabilities of a structural component, with the best possible
confidence. The role that ultrasonics can be expected to play in the
failure prediction process is reviewed. Primary emphasis is placed on
situations in which fracture is controlled by the slow growth to fracture
of a single crack. Included are discussions of the present models for
failure prediction, the manner in which flaw data can be incorporated
in them, and a summary of the ultrasonic techniques which can generate
quantitative failure prediction data. The role of ultrasonics in
situations where fracture is controlled by defect initiation is also
indicated. The paper keynotes the need for close interaction between
the ultrasonic and material scientist to ensure that the optimum use is
made of the information available in any particular situation." (Author)
39 refs.

1973-030+

Brown, A. F., "Materials Testing by Ultrasonic Spectroscopy," Ultrasonics,
202-210 (Sept., 1973).

"Ultrasonic spectroscopy is a development of the pulse-echo technique
which uses broadband (0.5-10 MHz) ultrasound and analyses the spectra of
the echo pulses. It has already proved its value in cases where the
geometry is essentially two-dimensional. Examples include the determination of grain size in metals and quality control in carbon-fibre composites and glued joints, while a new development has opened up the
possibility of measuring the depth and width of surface cracks down to
sub-millimetre size. When the problem is essentially three-dimensional,
as in characterization of discrete defects in metals, full interpretation
of the echo signal cannot as yet be achieved except in a few special
cases.

This paper is the first of two reviews on ultrasonic spectroscopy. The second will cover the generation and reception of wideband ultrasound." (Author) 18 refs.

1974-031+

Papadakis, E. P., "Ultrasonic Spectroscopy Applied to Double Refraction in Worked Metals," J. Acoust. Soc. Am. 55 (4), 783-784 (1974).

"An experiment has been performed showing that ultrasonic spectroscopy can be used to study texture in worked metals and other textured poly-crystals. Echoes from a broadband shear transducer polarized at 45° between two characteristic directions in a rolled aluminum specimen showed nulls in their spectrum corresponding to differences of $(2 n-1)\pi$ in phase between the components of the wave amplitude resolved along the characteristic directions. From the frequencies of the nulls, the fractional velocity difference $\Delta v/v$ was computed to be 0.021, a reasonable value for aluminum with moderate texture. Ultrasonic spectroscopy is particularly useful to study texture when the first echo contains several nulls, i.e., $n > 1$ in the phase difference $(2n-1)\pi$." (Author) 6 refs.

1975-032+

Greguss, P. and Waidelich, W., "Ultrasonic Holographic Fourier Spectros-copy via Optical Fourier Transforms," IEEE Trans. Comp., C-24 (4) (April, 1975).

"Holographic Fourier spectroscopy has been extended to ultrasonic radiation. A recently developed acoustical-to-optical conversion cell (AOCC) was used to display the spectroscopic halogram. Real-time optical Fourier transformation of the ultrasonic interferogram can be obtained." (Author) 13 refs.

1975-033+

Seydel, James A., "Methods Development for Non-Destructive Measurement of Bond Strength in Adhesively Bonded Structures," AFML-TR-75-151 Technical Report for the Air Force Materials Laboratory (1975).

A system for ultrasonic spectroscopy of bonded structures is described in great detail. Samples of adhesively bonded aluminum plates were investigated. The aim of the experiments reported in this reference was to measure reflectivity as a function of frequency and to identify, for various samples, differences in the magnitude and phase spectra. Later work was to include investigation of spectral differences in relation to bond strength.
The acquisition system digitizes the ultrasonic waveform in "equiva-lent time." The digitized data is stored on magnetic disk for later processing by a minicomputer. The processing algorithms utilized in this project are briefly discussed. Waveforms, as well as magnitude and phase spectra, are presented for the bond samples tested. 8 refs.

1975-034+

Rose, Joseph L. and Meyer, Paul A., "Ultrasonic Procedures for the
    Determination of Bond Strength," (NTIS AD-A015 291) AFOSR-73-2480A
    Technical Report for the Air Force Office of Scientific Research
    (1975).

Adhesive bondline defects such as chemical segregation, variation
in cure, gas entrapment or inadequate surface preparation are responsible
for many bond failures. Analytical models have been developed that can
be used to relate these defects to the manner in which they affect the
reflection of an ultrasonic pulse from such a bondline. Computer analysis
(listings are included in the appendix) of these models has shown that
these variations occur in the form of amplitude changes in the time
domain and spectral shifts in the frequency domain. The experimental
phase of the program indicated that a variation in substrate surface
preparation does give reflected amplitude changes predicted by the
models.
    The authors provide an excellent overview of adhesive bonding
procedures. They also note the steps in bonding where "flaws" are
generated. An extensive bibliography is included in the report. 63 refs.

1970-035+

Whaley, H. L. and Cook, K. V., "Ultrasonic Frequency Analysis," Mat. Eval.
    28 (3), 61-66 (1970).

This publication is essentially the same as reference (1969-005).
An ultrasonic frequency analysis system is described (in great detail),
and its potential uses in NDT are investigated. 3 refs.

1970-036+

Whaley, H. L. and Adler, L., "Model for the Determination of the Size
    and Orientation of Reflectors from Ultrasonic Frequency Analysis,"
    (Summary), J. Acoust. Soc. Am. 48 (1), 102 (1970).

Refer to reference (1971-012) for a complete description of this
interference model.

1970-037+

Whaley, H. L. and Adler, L., "A New Technique for Ultrasonic Flaw
    Determination by Spectral Analysis," Tech. Memo., Oak Ridge
    National Laboratory, ORNL-TM-3056 (1970).

A method is introduced for characterization (determination of the
size and orientation) of a reflector by ultrasonic spectral analysis.
Flaws (cracks, voids or inclusion) may be characterized, without the
need for a calibration standard.
    "The results presented in this report are based on a series of
reflection experiments in which the ends of solid rods immersed in water

and machined discontinuities in metal samples were used as reflectors.
Broadbanded ultrasonic pulses were analyzed after reflection from the
interface of interest." (Authors)

A physical and analytical model is developed based on the
assumption of interference of spherical wavelets diffracted from opposing
extremes of the reflector.

"This model explains the experimentally observed spectral variations
in terms of size and three-dimensional angular orientation of the
reflector, distance between the transducer and reflector and displacement of the reflector from the axis of the transducer from a flat
reflector of any composition. Based on this technique, a detailed
procedure can be outlined for characterizing a randomly oriented natural
flaw in a plate. The feasibility and practicality of using this
technique to test various materials and in automated testing systems
were considered." (Author) 2 refs.

1971-038+

Whaley, H. L. and Adler, L., "Determining the Size of Flaws in Materials,"
    Instrum. Contr. Sys. 44 (10), 98-102 (1971).

"If an ultrasonic pulse containing a broad range of frequencies is
used to detect a discontinuity, spectral analysis of the reflected waveform can provide size information. The finite size of a discontinuity
causes interference patterns in the reflected signal. The frequency
difference between peaks in the observed spectrum can be geometrically
related to the discontinuity dimensions." (Author)

The ends of circular and noncircular brass reflectors were examined
with the technique. The authors found good agreement between the physical
dimensions of the reflectors and the dimensions measured by their
technique. 3 refs.

1972-039+

Whaley, H. L. and Adler, L., "A New Approach to the Old Problem of
    Determining Flaw Size," Int. J. Fracture Mech., 112-113 (1972).

Conventional ultrasonic tests estimate the size of a detected
defect from the amplitude of an ultrasonic pulse reflected from it. The
inaccuracy of these techniques is reviewed and an alternate method using
a broadband ultrasonic pulse is proposed. The results are said to be
independent of the amplitude of the reflected signal.
The accuracy of the spectral analysis technique was investigated
by studies on a steel test block. A variety of flat-bottomed holes of
differing sizes was machined into the plate at various angles. The
amplitude-sizing method failed to give accurate size information; however, the spectral analysis technique accurately (within 10%) sized the
defects as well as indicating the orientation of each flaw. 3 refs.

1972-040+

Adler, L. and Whaley, H. L., "Interference Effect in a Multifrequency
    Ultrasonic Pulse Echo and Its Application to Flaw Characterization,"
    J. Acoust. Soc. Am. 51 (3), 881-887 (1972).

    "The dependence of spectral variations with a reflected broadband
ultrasonic pulse on the size and orientation of the reflector was
determined experimentally. An analytical model is developed on the
assumption that interference of the waves received from the edges of the
reflecting surface is responsible for the variations of the received
frequency spectra. It explains the experimental results very well. The
feasibility of determining the size and orientation of hidden flaws in
metals by this method is demonstrated." (Author) 3 refs.

1972-041+

Adler, L. and Whaley, H. L., "Ultrasonic Flaw Determination by Spectral
    Analysis," U.S. Patent 3,662,589 (1972).

    "A method of quantitatively determining the size and orientation
of flaws within a material. A broadband pulse having a frequency width
of at least several MHz is used for an input pulse from a transducer
which is directed toward the test material which may contain a flaw,
and the average frequency interval between interference maxima in the
reflected spectrum is relatable to both the size and orientation of
the flaw. An experimental method using two known angle changes allows
determination of the size and angle of randomly oriented discontinuities
within metal specimens." (Authors)

1973-042+

Whaley, H. L. and Adler, L., "Ultrasonic Flaw Determination by Spectral
    Analysis" (to U.S.A.E.C.) U.S. Patent 3,776,026 (1973).

    "A method of quantitatively determining the size and orientation of
flaws within a material. A broadband pulse having a frequency width of
at least several MHz is used for an input pulse from an ultrasonic
transducer which is directed toward the immersed test material which may
contain a flaw, and the frequency interval between two consecutive frequency maxima of the reflected spectrum due to the difference in distance
between each end of the flaw and a receiving transducer is relatable to
both the size and orientation of the flaw." (Author)

1973-043

Whaley, H. L., Adler, L., Cook, K. V. and McClung, R. W., "Some Appli-
    cations of Spectral Analysis in Ultrasonic Testing," p. 45 in
    Fracture and Flaws, Proc. 13th Ann. Symp., Mar. 1-2, 1973, pt. 2,
    Am. Soc. of Mech. Engr. and Am. Soc. for Metals.

1975-044+

Whaley, H. L., Cook, K. V., Adler, L. and McClung, R. W., "Applications
of Frequency Analysis in Ultrasonic Testing," Mat. Eval. 33 (1),
19-24 (1975).

Applications of spectral analysis in several areas of ultrasonic
nondestructive testing are discussed. These applications involve control of ultrasonic test variables (transducers and alignment of an
ultrasonic beam normal to a surface), identifying and sizing flaws and
improving resolution for thickness measurements and detection of near-
surface flaws. It is demonstrated that frequency analysis may be used
to differentiate between spherical and crack-like flaws, as well as
between pores and bursts in electron beam welds.
    A formula is developed to relate the thickness of a thin sheet (or
tube) to changes in frequency "interference" minimum. Measurements of
thickness are claimed possible with material of 0.05 mm (in aluminum,
with a 25 MHz transducer). Resolution of near-surface flaws (0.025 mm)
is also postulated.
    The authors assert that advantages of the spectral analysis
technique are reproducibility, precision and simplicity. 8 refs.

1975-045+

Whaley, H. L., Adler, L., Cook, K. V. and McClung, R. W., "Measurement
of Flaw Size in a Weld Sample by Ultrasonic Frequency Analysis,"
Oak Ridge National Laboratory Technical Memo, ORNL-TM-4800 (1975).

The ultrasonic frequency analysis technique was applied to the
characterization of flaws in an 8-inch-thick heavy-section steel
specimen. Quantitative determination of flaw dimensions and orientation
was possible.
    A multitransducer system was developed for frequency analysis. A
study of the system geometry is presented. The transmitting transducer
is located directly over the flaw, normal to the specimen surface. Two
receiving transducers are placed in a plane with the transmitter and
angled toward the defect. The spacing of frequency maxima in the two
received signals, as well as a knowledge of material properties, enter
into the calculation of defect dimension and orientation. A computer
program (listed in the appendix) was written to systematize the required
calculations. 7 refs.

1977-046+

Adler, L. and Lewis, D. K., "Diffraction Model for Ultrasonic Frequency
Analysis and Flaw Characterization," Mat. Eval. 35 (1), 51-56
(1977)

"Ultrasonic frequency analysis is developing into a powerful tool
in nondestructive evaluation. In this technique an incident broadband
pulse hits the target (flaw) and the scattered pulse is spectrum
analyzed. The interpretation of the spectrum from various targets and
the identification of targets from their spectra presents a problem
amenable to solution. In our earlier work the frequency distribution

of the scattered wave was related to the size and orientation of the
target by an approximate theory. This theory assumed that the wavelets
originating from the edge of the target will interfere. By realizing
that any flaw will diffract sound waves like an aperture diffracts
light, we have adopted a geometrical theory of diffraction of flaw
characterization problems. An exact expression was derived which
relates the amplitude and the frequency of the scattered wave to the
size and orientation of a flaw. Computer calculations of this diffraction
theory agree very well with the experimentally obtained spectra from
frequency analysis measurements. This approach is applied to the
characterization of flaws in metals." (Author) 9 refs.

1977-047

Adler, L., Cook, K. V., Whaley, H. L. and McClung, R. W., "Flaw-Size
    Measurement in a Weld Sample by Ultrasonic Frequency Analysis,"
    Mat. Eval. 35 (3), 44-50 (1977).

"An ultrasonic frequency-analysis technique has been developed and
applied to characterize flaws in an 8 in (203 mm) thick heavy-section
steel weld specimen. The technique applies a multitransducer system.
The spectrum of the received broad-band signal is frequency analyzed
at two different receivers for each of the flaws. From the two spectra,
the size and orientation of the flaw are determined by the use of an
analytic model proposed earlier." (Author) 4 refs.

Also refer to the abstract for reference 1975-045.

1975-048+

Scott, W. R., "Ultrasonic Spectrum Analysis for NDT of Layered Composite
    Materials," NADC-75324-30, Report for the Naval Air Dev. Ctr.
    (NTIS #AD-A028 856/3ST).

A theoretical model has been devised to describe the interactions
of ultrasonic plane waves with laminated structures. Using this model,
measurable parameters such as the frequency dependent transmission and
reflection coefficients can be calculated for laminates consisting of
layers of elastic monolithic materials. A computer program (LAYER) was
written, incorporating the relations of the model. However, the program
was not listed.
    Experiments and calculations for some simple finite periodic
laminates have demonstrated:

"1. In general, the spectra are not simply periodic.
  2. If a finite periodic laminate is modeled assuming an infinite
      laminate—the frequency regions of high attenuation are
      accurately predicted. Detailed transmission and reflection
      spectra cannot be predicted.
  3. For wavelengths longer than the period of the laminate (ℓ),
      behavior is similar to a dispersive monolithic material.
      Anomalous dispersion and strong attenuation occur for wave-
      lengths given by integer multiples of (ℓ).

    4. Graphite/epoxy, fiberglass and B/Al composites have properties
       which are similar to the simple laminates modeled." (Author)

Acoustic amplitudes of waves propagating near the resonant (thickness)
frequencies can be used to modulate c-scan intensities, thereby generating
maps of a sample which reveal small changes in either modulus or thickness. 7 refs.

1968-049+

Hartman, G. S., Lewis, D. H., Cressman, R. N. and Bautz, W. J.,
    "Evaluation of Ultrasonic Equipment for Rotor Forging Inspection,"
    A report prepared for ASTM Committee A-1 on Steel, Subcommittee V1
    on Forgings, Special Task Group on Large Turbine and Generator Rotor
    Forgings, Task Group on ASTM A-418-64 (Oct. 15, 1968).

Pulsers, receivers and transducers from a number of manufacturers
were analyzed for applicability to rotor inspection. The purpose of the
examinations was to obtain information to explain the differences in
ultrasonic responses of various instruments to natural defects. A
large part of the work involved spectrum analysis.
    The committee noted a number of factors which affect the amplitude
of ultrasonic indications obtained from discontinuities in rotor
forgings. Among these factors which affect amplitude are the frequency
responses of the components of test systems. A large number of spectra
are presented to document their work. The committee also recognized that
the type, shape, size and orientation of the defect have profound effects
on amplitude (as well as frequency response) of indications. Recorded
spectra illustrate their contention.

1966-050+

Gericke, Otto R., "Experimental Determination of Ultrasonic Transducer
    Frequency Response," Mat. Eval., 409-411 (1966).

A swept-frequency (0.5 MHz to 12 MHz) system is described for
determining frequency response characteristics of ultrasonic transducers.
The method involves an actual transmission (through a 1" thick aluminum
cylinder) and reception (backwall echo) of ultrasonic signals and yields
the loop response curve of a transducer. In examining the frequency
response of a number of commercially made ceramic transducers, large
variations were found. Deviations of the maximum response from the
frequency indicated as well as wideband response were noted. The author
suggests that a transducer response curve be included in the manufacturer's
specifications.
    Examples are given of how differences in frequency response can
affect amplitude and travel-time measurements. The effects of frequency
response on beam collimation are also briefly discussed. 5 refs.

1966-051+

Lopilate, S. A. and Carter, S. W., "Unbond Detection Using Ultrasonic
    Phase Analysis," Mat. Eval. 24, 690 (1966).

"The development of an ultrasonic phase sensitive test procedure is
described for the inspection of adhesive-bonded structures. The application of this technique depends upon the principle that the polarity of
an ultrasonic wave is influenced by the relative acoustic impedance involved
when it intercepts and reflects from the boundary of two materials. For
the investigations conducted, unbonds as small as 1/4" in diam were
resolved, independent of acoustic variables, i.e., velocity attenuation,
frequency, and coupling, which one normally encounters for those
materials involved in aerospace design. The development effort leading
to the practical application of the physical principle and instrumentation
is described." (Author)

1978-052+

Tittmann, B. R., "Ultrasonic Measurements for the Prediction of Mechanical
    Strength," NDT International 17-22 (February, 1978).

"Results are presented on the scattering of ultrasonic waves from
defects embedded in metals and on the influence of these defects on the
mechanical properties. A simple correlation has been discovered between
an empirical expression for the ductility of a sample containing a single
void and the back-scattered ultrasonic power derived from scattering
theory. An expression is derived which relates the ductility (reduction
in cross-sectional area at tensile fracture) with the high-frequency
(15 MHz) back-scattered ultrasonic signal in an approximately linear
fashion. The data obtained on diffusion bonded Ti-alloy samples are in
good agreement with the predictions. These results open the door to
the prediction of ductility from ultrasonic measurements." (Author)
17 refs.

1973-053

Tattersall, H. G., "The Ultrasonic Pulse-Echo Technique as Applied to
    Adhesion Testing," J. Phys. D; App. Phys. 6, 819-832 (1973).

"Flaws are commonly detected by their ability to reflect a burst
of ultrasound back to a piezoelectric transducer, and a flaw is taken
to be found if an ultrasonic reflection is detected. In this form,
relying simply upon the existence or otherwise of a signal, the pulse-
echo technique is not suited to the detection of weak bonding between
an adherend and an adhesive, because both perfect and imperfect interfaces between dissimilar materials always reflect ultrasound. In 1971
Rolls-Royce sponsored an investigation into a highly developed form of
pulse-echo technique, in which all the information from the reflected
pulse is used, expressed in terms of accurate amplitude and phase
measurements. It has been found possible to detect "slackness" at an
interface, in situations where this slackness is not severe enough to
amount to an actual unbond." (Author)

1974-054+

Lloyd, E. A., "Non-destructive Testing of Bonded Joints—A Case for
    Testing Laminated Structures by Wide-Band Ultrasound," Non-Destr.
    Test. (London), 7 (6), 331-334 (1974).

"The author describes a technique of testing adhesively bonded lap
joints of a wide range of materials. It is designed to help setting up
quality control for manufacture. The author believes it can be developed
into a cheap, light instrument." (Author)

1947-055

Mason, W. P. and McSkimin, H. J., "Attenuation and Scattering of High
    Frequency Sound Waves in Metals and Glasses," J. Acoust. Soc. Am.
    19 (464-473 (1947).

"By using a pulse method, attenuation and velocity measurements
have been made for aluminum and glass rods in the frequency range from
2 to 15 megacycles. The sound pulses are generated by crystals waxed to
the surface of the rod. This wax joint limits the band width of the
transmitted pulse and measurements are made using long pulses which
approach steady state conditions. The reflected pulses show evidence of
several normal modes which can be minimized by using specially shaped
electrodes. Longitudinal waves show delayed pulses of smaller magnitude
that are caused by the longitudinal wave breaking up into reflected
longitudinal and shear waves at the boundary. This effect is small if
the diameter of the rod is 20 wave-lengths or more.
    The measured losses for aluminum rods show a component proportional
to the frequency and another component proportional to the fourth power
of the frequency. The first component is the hysteresis loss found for
most solid materials. The component proportional to the fourth power
of the frequency is caused by Rayleigh scattering losses which are the
result of differences in the elastic constants between adjacent grains
caused by changes in orientation. Calculated scattering losses agree
quite well with the measured values. The fourth-power scattering law holds
quite well until the grain size is equal to one-third of a wave-length.
For higher frequencies the scattering loss increases more nearly with
the square of the frequency. Glasses and fused quartz have a loss
directly proportional to the frequency, showing that any irregularities
must be of very small size." (Author)

1948-056

    Mason, W. P. and McSkimin, H. J., "Energy Losses of Sound Waves in
Metals Due to Scattering and Diffusion," J. Appl. Phys. 19, 940-946 (1948).

"When high frequency longitudinal and transverse sound waves are
sent through a multicrystalline rod of metal, attenuation losses result
because of scattering and diffusion of sound waves by the grains. When
the grain size is less than one-third of the wave-length, these losses
are due to Rayleigh fourth power law scattering and are proportional to
the grain volume. The scattering factor depends on the anisotropy of the
elastic constants. Two different factors are obtained, one for shear

waves and one for longitudinal waves. These factors have been evaluated
for cubic and hexagonal metals. From the measured Elastic constants the
only metals with a low loss are aluminum, magnesium, and tungsten. The
calculations indicate that the losses for aluminum and magnesium are
about equal for longitudinal waves, but for shear waves magnesium has a
very low shear loss. It has been found experimentally that magnesium
has nearly as low a loss as fused quartz.

Experiments with higher frequencies show that when the wave-length
is one-third of the grain size or less, the transmission process
becomes a diffusion process similar to the propagation of a heat wave.
The grain sizes determine the mean free path, and the loss becomes
inversely proportional to the grain diameter. An approximate formula
for diffusion losses has been obtained which agrees closely with the
experimental values." (Author)

1964-057

Gericke, O. R., U.S. Army Materials Research Agency, Technical Report
    TR 64-44 (1964).

1962-058

Gericke, Otto R. and Maguire, John L., "Determination of Defect Geometry
    by Ultrasonic Pulse Analysis Testing," Watertown Arsenal
    Laboratories, Technical Report WAL TR 830.5/3.

Test specimens containing artificial defects (side-drilled holes
and angled notches) were consecutively subjected to a conventional ultrasonic pulse echo test and a pulse reflection method employing "polychromatic"
ultrasonic energy. Results are compared from the viewpoint of the
effectiveness of these two methods to differentiate among the various
flaw geometries. It was found that the polychromatic pulse method
permits a more conclusive geometry inspection for defects which are
smaller than the ultrasonic search beam diameter.

Change in the shape of the pulse envelope was used as the criteria
for identification of geometry and size of the reflector. In the case of
the angled notches, it was found (for these specimens) that the envelope
possessed two maxima. The distance between the two maxima can be
related quantitatively to the angle of the notch (at least between 20°
and 30°). This appears to be the first attempt made by anyone at
measuring defect orientation. Unfortunately a thorough analysis was not
presented. 2 refs.

1960-059

Gericke, O. R., "Spectrum and Contour Analysis of Ultrasonic Pulses for
    Improved Nondestructive Testing," Watertown Arsenal Laboratories,
    Technical Report WAL TR 830.5/1 (1960).

1960-060

Maquire, J. J., "Determination of Flaw Geometry by Ultrasonic Pulse
    Contour and Spectrum Analysis," Watertown Arnsenal Laboratories,
    Technical Report WAL TR 830.5/2.(1961).

1964-061+

Gericke, O. R., "Ultrasonic Spectroscopy of Steel," U.S. Army Materials
    Research Agency, Technical Report, AMRA TR 64-44 (1964)
    (NTIS #AD-614 113)

    This report contains the same information as reference (1965-008);
however, it is more detailed in the description of equipment and experimental technique.

    "It is possible to discriminate between grain sizes varying from
2 to 8 (ASTM scale) on the basis of ultrasonic attenuation spectra
observed. Compared with conventional multiple-back-echo attenuation
testing, the spectroscopic method offers the advantage that measurements can be carried out much faster. A further feature of the technique
is that the back-echo amplitude is not involved in the measurement and
thus can be used as a control indication to continuously monitor
overall equipment sensitivity and transducer coupling conditions."
(Author). 5 refs.

1968-062

Gericke, O. R., "Ultrasonic Pulse-Echo Spectroscopy," Presentation at the

    The experimental apparatus used by the author is reviewed and the
importance of transducer selection is stressed. Examples of the spectra
of narrowband and wideband pulses which had propagated through samples
of aluminum and copper demonstrated the additional information available
with spectroscopic methods. The spectrum of narrowband pulses which
passed through the aluminum was essentially the same as that of the pulse
through the copper. The high frequency components (above 8 MHz) of
the wideband pulse through the copper were strongly attenuated while the
ultrasound through the aluminum was transmitted with little attenuation.
    The use of spectroscopy for determining grain size in steel is
mentioned. Differential spectroscopy (as reported in reference 1967-063)
was used for comparison of two spectra." (Author). 3 refs.

1967-063

Gericke, O. R., "Differential Ultrasonic Spectroscopy for Defect and
    Microstructure Identification," Army Materials Research Agency,
    Technical Report AMRA TR 67-07 (1967).

1970-064+

Gericke, O. R., "Ultrasonic Spectroscopy," Chapter 2 in Research
    Techniques in Nondestructive Testing, 31-62 (1970).

The author discusses theory, applications and future of ultrasonic
spectral analysis in NDT. A brief discussion of the techniques used,
prior to the advent of spectroscopy, for obtaining frequency information
from ultrasonic pulses is given.
    The author notes the importance of pulser and transducer in the
spectroscopic system. The spectra of various pulses is presented (calculated from the Fourier series). The effects of modulation on sine-waves
are also explored. The transducer is modeled as a forced damped
oscillator. It was noted that a large bandwidth implies a low Q. The
possibility of matching the pulse length and shape to the transducer,
for maximum bandwidth, is mentioned. Proper coupling of the transducer
to the samples, to prevent resonance in the coupling layer, is stressed.
    The electronic systems required for ultrasonic spectroscopy are
given in block diagrams. The advantages and disadvantages of frequency
modulation (swept frequency), pulse (with spectrum analyzer) and pulsed
frequency modulation (pulsed version of swept frequency) systems are
presented.
    Applications of ultrasonic spectroscopy are examined. Frequency
analysis may be used for the determination of transducer frequency
response, measurement of specimen thickness, the examination of microstructure in polycrystalline metals and amorphous materials and also
for the inference of the defect geometry of single or multiple defects.
    Future improvements in spectroscopic equipment are postulated.
18 refs.

1964-065

Gericke, O. R., "Dual Frequency Ultrasonic Pulse-Echo Testing," J. Acoust.
    Soc. Am. 36, 313-322 (1964).

"The problem of defect-size determination by means of ultrasonic pulse-
pulse-echo testing and the benefits accrued from the utilization of a
system that permits simultaneous testing at two different frequencies
are discussed. The design of a dual-frequency pulse-echo test system
employing only a single ultrasonic transducer is described, and its
usefulness is examined for discriminating between dislike discontinuities
(flat-bottom holes) of various diameters. In addition, experiments are
described that indicate that the frequency dependence of ultrasonic-energy
losses and the longitudinal-wave velocities in materials can be
determined faster and with a higher degree of reliability by dual-frequency
testing than by conventional methods. Also reported is an experimental
investigation of the frequency dependence of transducer coupling in
contact testing involving interposed layers of liquid couplant." (Author)

1970-066+

Legge, R. D., "Frequench Characteristics of Ultrasonic Pulses Used for
    Flaw Detection," Ultrasonics for Industry Conference Papers 1970,
    31-36 (1970).

"In order to obtain good range resolution in flaw detection, short
pulses are generally employed, which do not have the same frequency
characteristics as longer pulses. The connection between pulse length
and frequency content is discussed. The effects of coupling to, and
attenuation in, the test sample, are considered. The advantages and
disadvantages of short and long pulses are described; these include
discussion of resolution (in range and in angle), problems associated
with size-estimation of flaws, sensitivity, and frequency analysis of
flaw echoes." (Author). 38 refs.

1967-067

Gericke, O. R., Physics and Nondestructive Testing (W. D. McGonnagle,
     ed.), pp. 257-282, Gordon and Breach (1967).

1972-068+

Papadakis, E. P., Fowler, K. A. and Lynnworth, L. C., "New Uses of
     Ultrasonic Spectrum Analysis," Proceedings of the 1972 IEEE
     Ultrasonics Symposium, IEEE Press, 81 (1972).

     Ultrasonic spectrum analysis has been extended and improved to
yield quantitative data on ultrasonic attenuation and on broadband
transducer loop response as functions of frequency. Both new results
depend upon diffraction (beam spreading) corrections which are applied
to the received echo amplitudes in the frequency domain. In both cases,
the transducer is coupled to one end of a buffer rod with plane parallel
ends. The amplitude versus frequency for the first echo after correction
for diffraction and input pulse spectrum is the transducer loop response
(for two transductions). For attenuation measurements, a specimen with
plane parallel faces is bonded to the opposite end of the buffer rod,
and three echoes are analyzed: the interface echo and the first two
reverberations in the specimen. After correction, the amplitudes at
various frequencies are used to compute the reflection coefficient and
the attenuation. Measurements agree with earlier data taken with quartz
plates bonded directly to the specimens. Responses of some broadband
transducers are essentially flat down to zero frequency, indicating
critical damping." (Author). 20 refs.

1977-069+

Adler, L., Cook, K. V. and Simpson, W. A., "Ultrasonic Frequency Analysis,"
     Chap. 1 in Research Techniques in Nondestructive Testing, Vol. 3,
     1-50 (1977).

     This chapter is an organized summary of the authors' work in ultra-
sonic frequency analysis through 1975.

     "A new ultrasonic technique to characterize flaws has been discussed.
The technique is based on the use of a broadband ultrasonic pulse and the
analysis of the frequency content of the scattered pulse from a defect.
Hence, the technique is called frequency analysis. The amplitude of the

scattered energy from known defects is related to the frequency of the
scattered pulse and the size and orientation of the defects is derived
from diffraction theory. Two alternatives to the exact treatment,
(a) the Interference Model and (b) the Fourier Transform Model, have been
given and each of the models experimentally verified. Further work is
required to solve the diffraction problem for other than circular-shaped
flaws as well as for three-dimensional flaws. An inclusion of mode
conversion at the defects will also make the analysis more realistic.

   The frequency analysis technique has been successfully applied to
determine the size of planar circular flaws with arbitrary orientation
in aluminum and carbon steel; to identify flaw shape in titanium; to
measure thickness in tube walls; and to characterize flaws in a heavy
section steel weldment.

   With further experimental and theoretical development, the ultrasonic
frequency analysis method will take its place amongst the other comple-
mentary nondestructive testing methods applied in industry." (Author)
11 refs.

1973-070

Neumann, E., Würstenburg, H., Nabel, E. and Mundry, E., "Analyse von
    Ultraschalleches durch Impulsspektometrié; Vortragstagung 1973;
    Zerstörungsfreie Materialprüfung" der DGZFP und der Eisenhütte
    Österreich in Salzburg am 18. und 19.10.1973.

    "Analysis of Ultrasonic Echoes by Impulse (Wide-Band) Spectroscpy"

1974-071

Nabel, E., Neumann, E., Mundry, E. and Wustenburg, H., "Analysis of Ultra-
    sonic Echoes by Means of Pulse Spectroscopy," V. Konferenz fur
    Zeitgemässe Benessung, VI. Kongress für Materialprufung, Budapest,
    28.10-1.11.1974, Erschienen im Tagungsband, Verlag: Adamediai Kiado,
    Budapest.

1976-072

Nabel, E. and Neumann, E., "Evaluation of Flaw Indications by Ultrasonic
    Pulse Amplitude and Phase Spectroscopy," 8. WCNDT, Cannes,
    6.-11.9.76, Preprint 3 H 6.

1976-073

Nabel, E., "The Importance of Phase for Ultrasonic Spectroscopy and
    Deconvolution," Seminarvortrag, City University, London, 27.10.1976.

1977-074

Nabel, E., "Analyse von Ultraschallanzeigen durch Entfaltung," Vortrags
    veranstaltung "Neuere Verfahren zur Analyse von Ultraschall-
    Befunden" der DGZFP-Fachausschüsse Sonderprufverfahren und
    Ultraschall-Prüfverfahren, BAM Berlin, 24.2.1977.

    "Analysis of Ultrasonic Indications by Unfolding (Deconvolution)"

1977-075

Nabel, E., Mundry, E., "Die Entfaltung von Ultraschallechos,"
    DGZFP-Jahrestagung, Bremen, 16.-18.5.1977.

    "Unfolding of Ultrasonic Echoes"

1977-076

Mundry, E., Kutzner, J., Nabel, E., Wustenburg, H., Neumann, E. and
    Schmitz, V., "Flaw Size Determination by Ultrasonic Pulse
    Spectroscopy," Acoustical Holography and Focussing Probes—A
    Critical Comparison, IIW/IIS Colloquium Commission V, Kopenhagen,
    3.-8.7.1977.

1977-077

Wustenburg, H., Mundry, E., Kutzner, J., Nabel, E., Neumann, E., and
    Schmitz, V., "Vergleich der Fehl ergrössenbestimmung mit Ultra-
    schallimpulsspektroskopie, akustischen Holographie und
    fokussierenden Prüfköpfen"; VDEh-Gemeinschaftssitzung, Düsseldorf,
    29.9.1977.

    "Comparison of the Determination of the Enlargement of Defects
with Ultrasonic Pulse Spectroscopy, Acoustical Holography and Focused
Test Probes."

1978-078

Nabel, E. and Mundry, E., "Versuche zur Deutung von Ultraschallechos durch
    Entfaltung," DGZfP-Jahrestagung, Mainz, 24.4-26.4.1978.

    "Experiment on the Interpretation of Ultrasonic Echoes by Unfolding
(Deconvolution)."

1977-079

Nabel, E., "Bewertung von Echoanzeigen durch Ultraschall-Impulsspektrometric
    nach Amplitude und Phase," Material prüfung 19 (2), 58-64 (1977).

"Evaluation of Flaw Indications by Ultrasonic Pulse Amplitude and Phase Spectroscopy."

Until now in nondestructive testing with ultrasound there are no criteria which allow to answer on the questions "What? Which size? and Which orientation?" in a sufficient unambiguous manner. The echo amplitude which is used in many cases as the sole measure for evaluation depends on orientation and size of the reflector, as well as on the frequency and the bandwidth of the transducer. In practice these dependencies cannot be taken into account. The described results were found during researches on methods for unambiguous evaluation of echoes by spectrum analysis according to amplitude and phase and by deconvolution of the signals." (Author)

1977-080

Nabel, E., "Analyse von Ultraschallanzeigen durch Entfatung,"
    Berichtband der DGZFP "Neuere Verfahren zur Analyse von Ultraschall-
    Befunden," enschienen November 1977.

"Analysis of Ultrasonic Indications by Unfolding (Deconvolution)."

1977-081

Nabel, E., "Die Impulsantwort als Ergelonis der Ultraschallspektrometrie -
    Versuche zur exakten Bewertung von Ultraschallechos," Materialprüfung
    19 (12), 496-499 (1977).

"The Impulse Response as a Result of Ultrasonic Spectroscopy—
Attempts of Exact Evaluation of Ultrasonic Echoes."

"This paper will report some effects, which are of fundamental importance for ultrasonic spectroscopy and deconvolution. In many cases the importance of these effects is not recognized, although they are responsible for the most aberations between experimental results and theory. The paper deals with the improvements during analysis of echo indications by calculating the impulse response of a reflector (deconvolution of ultrasonic echoes). The importance of phase information is emphasized as well as the influence of the low frequencies, or low diffraction orders, on the experimental results is shown. As an example the results of a simulated convolution will be demonstrated with respect to the theory of replica pulses, published by Freedman." (Author) 9 refs.

1976-082+

Mercier, Noëlle, "Nondestructive Control of Materials by Ultrasonic Tests,"
    A technical translation (from French) of a report from the Office
    National d'Etudes et de Recherches Aérospatiales (NTIS #N7626555).

This report is a translation of a thesis and published as a technical note by ONERA.

A bibliographic study of nondestructive test methods for solids and of the radiation field produced by a transducer of finite dimensions is presented. Much of the work describes techniques for velocity and attenuation measurement. The problems encountered, and corrections for their effects is discussed. The application of these techniques for nondestructive detection of local inhomogeneities is addressed.

Experimental verification of two generalized testing methods was carried out. The first method employs analysis of the ultrasonic signal (amplitude and phase) on passage through a metal specimen of constant thickness. Analysis of the amplitude and phase of the signal gives specimen attenuation and ultrasonic propagation velocity. A good spatial resolution is obtained with the use of 1 mm diameter probes.

The second method, utilizing a test rig equipped with broad-frequency band electrostatic transducers, is used to determine the attenuation characteristics of the specimens as a function of the frequency (5-15 MHz) and permit classification of these specimens as to their grain dimensions. The capacitative transducers (described in moderate detail) demonstrated a wide frequency-band loop response.

This report gives a good overview of ultrasonic techniques and provides some experimental work relating sample attenuation to grain size in metals. 50 refs.

1974-083+

Hudgell, R. J., Morgan, L. L. and Lamb, R. F., "Nondestructive Measurement
    of the Depth of Surface-Breaking Cracks Using Ultrasonic Rayleigh
    Aves," Br. J. NDT 16, 144-149 (September, 1974).

"In ultrasonic testing, defect size has traditionally been estimated from the measurement of echo amplitude and/or shadowing such as the 20 dB drop method. Certain techniques utilizing ultrasonic Rayleigh waves give the possibility of estimation of crack depth by observing change in the time domain. Three Rayleigh wave techniques are considered—transmission around the defect, mode conversion at the defect and spectroscopy. This paper describes the techniques, outlines some of the experimental work which has been done, and discusses the present state of the art and possible routes for future development." (Author) 10 refs.

1975-084+

Goebbels, K., "Gefuegebeunteilung Mittels Ultraschall-Streuung,"
    Materialpruefung 17 (7), 231-233 (1975).

"Assessment of the Structure of Steel, Using Scattered Ultrasonic Pulses."

"The conditions for a quantitative assessment of the structure of steel by scattered ultrasonic pulses, particularly grain-size determinations, are defined. Measurements, using 6 and 10 MHz (transverse waves), on 29 ferrite and austenitic steels of different chemical compositions and structures are discussed. Satisfactory agreement between ultrasonically and metallographically determined grain sizes was obtained for the grain diameter range of 10 to 100 micrometers. Different ultrasonic frequencies must be used for grain sizes outside this range." (NTIS) (In German).

1974-085+

Morgan, L. L., "The Spectroscopic Determination of Surface Topography Using Acoustic Surface Waves," Acustica 30, 222-228 (1974).

"Rayleigh waves reflected from features in a surface may be used to characterize the feature geometry and, by manipulation of the spectrum of the reflected signal, the information of interest (the impulse response of the features) can be separated from other spurious information. Two methods for effecting this are described. The first, using manipulation in the complex frequency domain, has been verified using a slot as a target and two-digit interdigital transducers as generator and detector, having bandwidths in excess of 1 to 10 MHz. The second method uses the cepstrum of the signal and gives similar results but has wider practical application." (Author) 6 refs.

1975-086+

Kennedy, J. C. and Woodmansee, W. E., "Signal Processing in Nondestructive Testing," J. Test. Eval. 3 (1), 26-45 (1975).

"In testing electron beam welds the delay of an electronic gate has been synchronized to the transducer motion to improve the detectability of a tight interface crack in a tensile specimen. Cross-correlated techniques and multiple transducer arrays have been used to improve the signal-to-noise ratio of artificial flaws in welded panels. Signal averaging has been used to reduce random noise in the through-transmission ultrasonic inspection of a honeycomb composite. Megacycle range ultrasonic flaw information has been recorded on a low frequency FM tape recorder by rapidly sweeping an electronic gate through the time interval of interest." (NTIS) 6 refs.

1975-087+

Sachse, Wolfgang and Chian, Chian-Thang, "Determination of the Size and Mechanical Properties of a Cylindrical Fluid Inclusion in an Elastic Solid," Mat. Eval. 33 (4), 81-86 (1975).

"The size and mechanical properties of a fluid inclusion in an elastic solid are determined from an analysis of the pulse time-record and corresponding Fourier power spectra of broadband ultrasonic pulses scattered by the inclusion. The density ratio of the fluid to the solid investigated ranged from 10^{-4} to 1.4, and the wave speed ratio ranged from 0.05 to 0.92. The size of the cylinder and the wave speed of the fluid contained in it are determined from measurement of the pulse arrival times, which are interpreted in terms of the theory of geometric acoustics. The usefulness and limitations of ultrasonic pulse spectroscopy measurements in determining these characteristics of the

inclusion are also described. The density of the included fluid can be
determined from an amplitude analysis of the scattered pulse." (NTIS)
11 refs.

1975-088

Silk, M. G., "Better Ultrasonic Resolution by Data Unfolding," Non-Destr.
     Test. (London) 8 (2), 94-99 (1975).

     "Conventional ultrasonics cannot give good resolution of defects
in thick steel because of the effect of the spreading of the beam. Ultrasonic holography overcomes this difficulty at the expense of time and
complex computer processing. The authors propose an improvement of
conventional ultrasonics by using data unfolding on the results, using a
small on-line computer. Their preliminary investigation shows that
improved resolution is possible by this means under certain conditions.
There are limitations but the authors recommend further work to establish
the method with certainty." (NTIS) 4 refs.

1975-089+

Fleischmann, P., Rouby, D., Lakestani, F. and Baboux, J. C., "Spectrum
     Analysis of Acoustic Emission," Non-Destr. Test. (London), 8 (5),
     241-244 (1975).

     "The authors describe a data acquisition system to determine the
frequency spectrum between 500 KHz and 3 MHz of the initial acoustic
emission. They measured the transfer function of the piezoelectric
transducer and of the specimen for their results using a computer. They
present an application of this device to the study of the emission of a
low-carbon steel." (NTIS) 7 refs.

1974-090+

Graham, L. J. and Alers, G. A., "Broadband Ultrasonic Attenuation
     Measurements in Unusual Materials," IEEE Sonics and Ultron. Symp.,
     Milwaukee, Nov., 1974, pp. 703-706, IEEE Press (1974).

     "In applying ultrasonic nondestructive testing methods to cases
where the ultrasonic path length is very long (acoustic emission) or
where the material under test is highly attenuating (composites, honeycomb, insulating foam), the attenuation characteristics of the structure
determines the ultrasonic frequency range to be used. In order to
establish attenuation characteristics quickly over a broad frequency range,
an acoustic instrument consisting of a white-noise generator and a damped
piezoelectric receiver has been assembled for the frequency range of
0-3 MHz. By varying the path length, the frequency dependence of the
attenuation can be obtained. The system used to make measurements are
described and results for a variety of structures and materials are
presented." (NTIS) 7 refs.

1974-091+

Sachse, W., "Density Determination of a Fluid Inclusion in an Elastic
    Solid from Ultrasonic Spectroscopy Measurements," 1974 IEEE Sonics
    and Ultrason. Symp., Milwaukee, Nov., 1974, pp. 716-719, IEEE Press.

"An ultrasonic method is presented by which the density of the
fluid contained in a cylindrical cavity in an elastic solid can be
determined. The method depends on the measurement of the reflection
coefficient of a broadband ultrasonic pulse incident on the inclusion;
the reflection coefficient is determined from the spectral analysis of
each of the scattered pulses. Also discussed are the results of
measurements made on three sizes of cavities, 1/32 in., 1/16 in., and
1/8 in. in diameter and containing various density fluids." (NTIS)
8 refs.

1974-092+

Mast, P. W. and Rose, J. L., "Signature Techniques for Defect
    Characterization," 1974 IEEE Son. and Ultrason. Symp., Milwaukee,
    Nov., 1974, pp. 720-725, IEEE Press.

"Several integral transformation techniques are presented that
produce feature extraction magnification for various ultrasonic defect
signals. A class of fast tensor transformations is studied. Such
integral transform techniques as Fourier, Mellin, Hilbert, and Hadamard are
also used to examine selected ultrasonic response functions from model
defects. Ultrasonic response transforms are examined along with such
modeled flaw problems as wall thinning and delamination detection,
determining dispersive effects due to density variations or geometries,
and attenuation resulting from absorption of grain boundary scattering.
Several ultrasonic response functions are presented along with the
various transform functions. The structural aspects of various transforms
are also studied and approaches presented for examining larger classes of
transforms than those that are normally considered in signal processing
work today." (NTIS) 9 refs.

1975-093

Rose, J. L., Mast, P. W. and Niklas, L., "Potential of 'Simulearning'
    in Flaw Characterization," B. J. NDT 17 (6), 176-181 (1975).

"The concept of 'simulearning' is introduced as a technique used
to integrate various aspects of analytical wave mechanics, flaw
characterization analysis, learning machine philosophies, and various
mathematical pattern recognition techniques. A 'simulearner' can be
described as an automated logic system that examines the response from
simulated training cases and ultimately remembers their characteristics."
(NTIS) 42 refs.

1975-094

Mucciardi, A. N., Shankar, R. and Buckley, M. J., "Applications of
    Adaptive Learning Networks to Nondestructive Evaluation Technology,"
    IEEE Proc. Nat'l Aerosp. Electron. Conf., Dayton, June 1975,
    pp. 430-469, IEEE Press.

    "The overall objective of this work was to demonstrate feasibility
of adaptive nonlinear signal processing techniques, from cybernetics and
bionics, applied to characterization of ultrasonic nondestructive testing
waveforms for accurate inferences of flat-bottom hole sizes. The
classified waveforms were ultrasonic pulse echoes obtained from two
different sets of 7075-T6 aluminum area-amplitude test blocks and three
different transducers. The eight flat-bottom hole defect sizes ranged
from 1/64- to 8/64-inch in steps of 1/64-inch." (NTIS) 20 refs.

1976-095

Fay, B., "Ermittlung der Korngroesse von Stahl mit dem Verfahren der
    Ultraschallrueckstreuung." Arch. Eisenhuettenwesen 47 (2), 119-126
    (1976).

    "Determination of the Grain Size of Steel Using the Ultrasonic Back-
Scatter Method."

    "This paper deals with the theories and mathematics of the application
of the ultrasonic back-scattering method in nondestructive determination
of the grain size of steel specimens. It reports plotting of back-scatter
curves at 5 to 19 MHz on specimens with different structures of different
steels produced by differing methods, and determination of the grain size
values from the amplitudes of scattered waves and the scatter coefficients.
The effect of multiple scatter of the acoustic pulses on the shape of the
back-scatter curves is discussed. Good agreement was obtained between
the results of this and conventional methods." (NTIS) 10 refs. In
German.

1976-096

Kraut, E. A., "Review of Theories of Scattering of Elastic Waves by
    Cracks," IEEE Trans. Son. Ultrason. SU-23 (3), 162-167 (1976).

    "The detection of cracks with the aid of ultrasonics is an important
nondestructive evaluation (NDE) technique. The corresponding theoretical
problem of the scattering of elastic waves by cracks has also been studied
by scientists working in many different fields. Contributions have come
from such diverse areas as geophysics, applied mathematics, electrical
engineering, and continuum mechanics. Many of the results obtained by
workers in those fields are also of interest to the NDE community and for
that reason a review is presented here of current results and profitable
directions for future research." (NTIS) 44 refs.

1976-097

Haines, N. F., "Ultrasonic Spectroscopy," Physics in Technology 7 (3),
     108-115 (1976).

     "Development of techniques in ultrasonic spectroscopy for NDT should
provide the inspection engineer of the 1980s with precise measurement on
structural integrity. A review of progress over the past decade is
presented." (Author)

     The author provides a good overview of several uses for ultrasonic
spectroscopy. Experimental and theoretical works are summarized for
the following uses:

     Flaw characterization
     Inspection of layered media
     Measurement of the depth of surface-breaking cracks
     Grain-size determination
The author speculates about the future of ultrasonic spectroscopy.
12 refs.

1976-098+

Martin, A. G., "Phase Velocity Measurements in Dispersive Materials by
     Narrow-Band Burst Phase Comparison," Presented at the Denfense Conf.
     on NDT (23rd), Sept., 1974, San Francisco (NTIS #AD-A033903/6ST).

     "A new method for the measurement of phase velocity of ultrasonic
waves by phase comparison of narrow-band bursts is described. The method
was developed for the determination of properties such as elastic moduli
of composites or other materials in which phase velocity varies with frequency. In such dispersive materials, individual cycles within a burst
travel at a velocity different than for the burst as a whole. The new
method is based on the phase comparison of entering and emerging
sinusoidally-shaped bursts. Two wideband transducers and a buffer are
used for reference. The specimen is interposed between a set of transducers and buffer identical with the reference set. Bursts are fed to
the two assemblages and the phases of the signals after transmission are
compared. The signal is varied from low to high ultrasonic frequencies
while in-phase and out-of-phase conditions and frequencies are recorded.
From the data, phase velocity is calculated at each recorded frequency.
The method and some results are discussed." (Author)

1976-099

White, R. M., "Some Device Technologies Applicable to Nondestructive
     Evaluation," IEEE Trans. Son. Ultrason. SU-23 (5), 306-312 (1976).

     "A brief review is made of some technologies which might usefully
be applied in nondestructive evaluation (NDE). The technologies (or
devices) are adaptive networks, analog/ditigal converters, charge-transfer
devices, commercial signal-processing systems, lasers, medical imaging
systems, memories, microprocessors and large-scale integration, microwave
sources and system, picture processing, piezoelectric thin films, scanners

(acoustical and optical), surface acoustic wave devices, time-shared
computing, and tomography." (NTIS) 35 refs.

1976-100

Buck, O., "Harmonic Generation for Measurement of Internal Stresses as
     Produced by Dislocations," IEEE Trans. Son. Ultrason. SU-23 (5),
     346-350 (1976).

     "The present knowledge on the state of microscopic internal stresses,
as produced by dislocations, and the relation of these internal stresses
to materials properties such as flow stress and the state of fatigue are
briefly reviewed. It is shown that ultrasonic harmonic generation is
sensitive to the dislocation density and loop lengths and thus to microscopic internal stresses. Some experimental data on harmonic generation
in a deformed and fatigued aluminum specimen, which are in agreement with
the above theory, are presented. Thus the technique of ultrasonic
harmonic generation is shown to be a new and potentially powerful
technique for the nondestructive evaluation of technologically important
materials." (NTIS) 35 refs.

1976-101

Yee, B. G. W. and Couchman, J. C., "Application of Ultrasound to NDE of
     Materials," IEEE Trans. Son. Ultrason. SU-23 (5), 299-305 (1976).

     "A review of state-of-the-art applications of ultrasonics to the
nondestructive inspection of materials for defects is given along with
some of the limitations associated with the present state-of-the-art.
These limitations include (1) the inability to determine defect dimensions
quantitatively, (2) the lack of suitable reference standards, (3) the
general lack of equipment for the inspection of parts with complex
geometries, (4) the nonreproducibility of transducer characteristics, and
(5) the lack of a sound theoretical foundation for ultrasound-discontinuity
interaction. Some of the current research and development work in the use
of digital computers for scanning parts with complex geometries and in
signal processing to increase inspection reliability and defect detection
is also described." (NTIS) 33 refs.

1976-102

Lakin, K. M. and Fedotowsky, A., "Characterization of NDE Transducers and
     Scattering Surfaces Using Phase and Amplitude Measurements of Ultra-
     sonic Field Patterns," IEEE Trans. Son. Ultrason. SU-23 (5), 317-322
     (1976).

     "The characterization of transducers for quantitative NDE applications
requires that the radiation pattern, conversion efficiency, and bandwidth
be accurately determined. These quantities may, in principle, be determined if the transducer's construction and constituent parts are
independently known. However, most often the internal details of the

transducer are unknown and subject to statistical variations and aging. A measurement technique and system for characterizing transducers based upon external measurements is described, which does not rely upon knowledge of the transducer's construction." (NTIS) 11 refs.

1975-103

Pao, Y. H. and Mow, C. C., "Theory of Normal Modes and Ultrasonic Spectral Analysis of the Scattering of Waves in Solids," Rand Corp. Report R-1891-RC, Dec. 1975. (NTIS #AD-A026674/2ST)

"Spectral analysis of ultrasonic pulses in elastic solids has attracted wide attention in recent years as a tool of the quantitative nondestructive test methods. So far, a complete analysis of the power spectra of the scattered pulses of any geometry is still lacking. This has hindered the general understanding and the application of ultrasonic spectroscopy to the detection of inclusions of flaws in solids. This report presents a theory of the spectral analysis of the scattering of elastic waves and illustrates it with numerical results for the scattering by a circular cylindrical fluid inclusion in a solid. When the spectral frequencies are nearly equal to the real parts of the principal frequencies of the fluid inclusion in free vibration, the power spectrum of the scattered pulses undergoes a rapid rise and fall in magnitude because of the selective transmission of an incident wave. The conspicuous peaks and valleys of the backward and forward scattering spectra can be identified with the overtone frequencies of the two lowest normal modes of the cylinder, from which the characteristics of the fluid inclusion, the ratio of the wave speed to radius, can be determined. An application of spectral analysis to quantitative nondestructive testing of materials is discussed." (Author) 12 refs.

1975-104

Shcherbinskii, V. G. and Belyi, V. E., "New Informative Index for the Nature of Flaws in Ultrasonic Inspection," Sov. J. NDT 11 (3), 279-288 (1975).

"A new informative index—the shape factor—is proposed for the nature of a flaw. Formulas are presented for computing it and also the results of an experimental check." (NTIS)

1976-105

Yee, B. G., Couchman, J. C. and Chang, F. H., "Resonance Effects on Energy Flux Coefficients for Ultrasonic Interactions with Thin Gap Discontinuities in Titanium," Mat. Eval. 34 (11), 245-250 (1976).

"This paper presents calculations of the reflected, mode-converted, and transmitted energy coefficients from a parallel interface separation when a plane shear or longitudinal wave is incident on the separation. The calculations treat the idealized case of monochromatic, unbounded waves incident on discontinuities with infinitely large parallel surfaces. The analyses of results of these idealized calculations clarify several points

in the interpretation of signal response from shear wave and Delta-Scan
inspection of cracks on crack-like flaws in solid materials. The problems
related in this paper considered longitudinal and shear waves obliquely
incident on (1) air- and (2) water-filled gaps in titanium." (NTIS)
11 refs.

1976-106

Meyer, P. A. and Sutton, S., "Aspects of Model Analysis and Pattern
    Recognition in Adhesive Bond Inspection," Nucl. Metall. 20 (p + 1-2),
    1976, Proc. of the Int. Conf. on Comput. Simul. for Mater. Appl.,
    Gaithersburg, Apr. 1976, p + 2, pp. 1097-1106.

"This paper discusses the application of a technique called
simulearning to the ultrasonic inspection of adhesive bonds. A simulearner can be described as an automated logic system that searches for
data classification or performance potential correlation possibilities
making use of theoretical techniques to generate large numbers of data
sets required in learning machine analysis. A sample problem of classify-
ing the attenuation characteristics of adhesive bondlines is presented.
The analytical procedures for generating the ultrasonic response
functions is discussed." (NTIS) 8 refs.

1976-107

Meyer, P. A. and Rose, J. L., "Modeling Concepts for Studying Ultrasonic
    Wave Interaction with Adhesive Bonds," J. Adhes. 8 (2), 107-120
    (1976).

"Recent research has shown that such adhesive bondline defects as
chemical segregation, variation in cure, gas entrapment or inadequate
surface preparation are responsible for many adhesively bonded structural
failures. Analytical models have been developed in this work that can be
used to relate these "flaws" to the manner in which they affect the
reflection of an ultrasonic pulse from such a bondline. The results of
this study provide a substantial resource base for extended research
through which ultrasonic inspection can become a reliable NDT technique
for bond strength determination." (NTIS) 12 refs.

1976-108

Meyer, P. A. and Rose, J. L., "Ultrasonic Determination of Bond Strength
    Due to Surface Preparation Variations in an Aluminum-to-Aluminum
    Adhesive Bond System," J. Adhes. 8 (2), 145-153 (1976).

"This paper reports on a study which was undertaken to demonstrate
the feasibility of the use of analytical models in the experimental
ultrasonic evaluation of interface conditions in an aluminum-aluminum
adhesive bond system. The results of the study show that a variation in
bond strength due to surface preparation can be detected ultrasonically
through careful inspection and signal processing analysis." (NTIS) 7 refs.

1976-109

Birchak, J. R. and Gardner, C. G., "Comparative Ultrasonic Response of
    Machined Slots, and Fating Cracks in 7075 Aluminum," Mat. Eval. 34
    (12), 275-280 (1976).

"To assess the suitability of machined flaws as standards for
ultrasonic inspection for fatigue cracks, reference blocks were developed
for quantitative comparison of pulse echo amplitudes produced by machined
flaws and true fatigue cracks of corresponding size and configuration.
Flaw shapes investigated were rectangular, triangular, and half penny,
representing the most commonly encountered fatigue crack configurations.
For each shape, electric discharge machined (EDM) and saw slits were
produced; actual fatigue cracks were grown in low cycle and in high cycle
conditions. Ultrasonic parameters investigated were frequency, bandwidth
and wave mode. The ultrasonic signals produced by low cycle fatigue
cracks averaged 5 dB stronger than those from high cycle cracks of corresponding size and configuration." (NTIS) 7 refs.

1975-110

Gray, D. H. and Darby, D. M., "Automated System for Testing," Met. Eng.
    Quant. 15 (1), 1-8 (1975).

"A commercial ultrasonic test facility was interfaced to a medium-
sized multiuser computer, and programs were written to accomplish data
taking and analysis for all presently used test methods. The system can
supply a digitized ultrasonic RF waveform to the computer for Fourier
analysis. All test results can be plotted on a local X-Y plotter or
strip-chart recorder. Surfaces of revolution and flat plates have been
inspected with good results. The transducer position may be changed
between data scans via the computer. The capability exists for tracking
a complex geometric shape with the transducer." (NTIS) 1 ref.

1968-111+

Papadakis, E. P., "Buffer-Rod System for Ultrasonic Attenuation Measure-
    ments," J. Acoust. Soc. Am. 44, 1437-1441 (1968).

"A method has been developed for making ultrasonic attenuation
measurements in a large sample by means of a solid buffer and temporarily
bonded to the sample. A piezoelectric transducer is permanently bonded
to the other end of the buffer rod. The ratios of the amplitudes of
certain echoes are used to find the ultrasonic attenuation in the sample
and the reflection coefficient at the buffer-sample interface at various
frequencies across the band of the transducer. The echo-amplitude ratios
are not affected by operating the transducer away from its resonant
frequency, because none of the echoes considered arise from reflections
at the buffer-transducer interface. A numerical analysis of the echo
amplitudes is presented in which the reflection coefficient is assumed
constant and the ultrasonic attenuation is taken as a power of the frequency. Experimental data are given on shear waves in a glass buffer rod
on a fused-silica sample. The reflection coefficient varies little while

the ultrasonic attenuation increases with frequency, fluctuating near
30 MHz." (Author) 8 refs.

1965-112+

Papadakis, E. P., "Revised Grain Scattering Formulas and Tables," J.
    Acoust. Soc. Am. 37, 703-710 (1965).

"The current theory of Rayleigh and stochastic scattering in polycrystalline materials is reviewed and compared with (1) former theory
and (2) experiment. Rayleigh scattering giving ultrasonic attenuation
equal to STF^4 (S is the Rayleigh scattering factor, T the average scattering volume, f the frequency) occurs when $\lambda > 2\pi\bar{D}$ (λ is the wavelength,
$\bar{D}$ the average grain diameter); stochastic scattering yielding ultrasonic
attenuation equal to $\Sigma\bar{D}f^2$ (Σ is the stochastic scattering factor) occurs
when $\lambda < 2\pi\bar{D}$. The average scattering volume and average grain diameter
must be evaluated by taking their averages over the grain-size distribution in the metal. When this is done, the current theory accounts rather
well for the scattering component of the ultrasonic attenuation in polycrystalline metals. Former theory underestimated the scattering. A
tabulation is made of the scattering factor S and Σ is various materials.
The computed scattering factors show that polycrystalline samples of the
following materials should have low attenuation: aluminum, chromite,
chromium, magnesium, magnetite, silicon, strontium nitrate, tungsten
vanadium and YIG." (Author) 47 refs.

1967-113+

Serabian, S., "Influence of Attenuation Upon the Frequency Content of a
    Stress Wave Packet," J. Acoust. Soc. Am. 42, 1052-1059 (1967).

"This paper presents experimental data concerning the distortion
of the frequency spectrum of a stress wave packet as produced by attenuation
effects in graphite." (Author) 15 refs.

The results presented in this paper are summarized in the abstract
for reference 1968-011.

1968-114+

Gericke, O. R., "Ultrasonic Pulse-Echo Spectroscopy," Report AMMRC-TR-
    68-15 (NTIS #AD-675 465).

"Basic requirements for the use of spectroscopic techniques in the
field of ultrasonic pulse-echo inspection are discussed. Specialized
electronic instrumentation developed for ultrasonic spectroscopy and
examples illustrating the practical usefulness of the equipment are
described. It is shown that the microstructure of metals, for instance,
can be determined on the basis of the spectral signature of the ultrasonic back-echo and that the interpretation of defect indications in terms
of actual defect geometry can be improved by spectrum analysis."
(Author) 3 refs.

1972-115+

Fang, T. C., "Scattering of Acoustic Waves by a Spherical Inclusion in
    a Solid," Grumman Report RE-440 (NTIS AD-749 411).

    "An analysis is presented of surface deformation and scattered
energy due to scattering of a high frequency plane longitudinal wave by a
spherical inclusion in an isotrophically elastic solid medium. The
inclusion is taken to be an isotropically elastic sphere of different
material or a spherical cavity. The average rate of total scattered
energy has been derived in terms of the expansion coefficients of the
scattering waves. Exact and approximate, for both Rayleigh scattering
and finite small inclusion, expressions for the expansion coefficients
have been obtained." (Author) 20 refs.

1966-116+

Gericke, O. R., "Ultrasonic Spectroscopy," Report AMRA-TR-66-38 (NTIS
    #AD-647918).

    "Various experimental difficulties discussed in the report had to
be overcome to be able to adopt spectroscopic procedures for ultrasonic
testing and to construct an ultrasonic spectroscope. The ultrasonic
pulse-echo spectroscope recently developed and its practical use for
nondestructive inspection purposes are described. Test results obtained
with the instrument indicate its usefulness for determining defect
characteristics which cannot be revealed by conventional ultrasonic
test methods. In addition, it is shown how the ultrasonic spectroscope
can be used to distinguish microstructures of steel specimens."
(Author) 3 refs.

1976-117+

Cohen, E. R., "Analysis of Ultrasonic Wave Scattering for Characterization
    of Defects in Solids," Interim Rept. #2, 16 March 1975-15 March 1976,
    Air Force Office of Scientific Research Report SC579.3IR
    (NTIS #AD-A030930/2ST) (1976).

    "Measurements have been carried out on scattering of ultrasonic
waves by a solid spherical inclusion (tungsten carbide) in titanium
alloy. Both direct scattering and mode-converted scattering angular
distributions were measured for both shear and compressional incident
waves. The scattering from an arbitrary shape was expressed in terms of
an integral equation from which an improved Born approximation was
developed. In this formulation, the Born approximation reduces to the
Rayleigh limit at low frequencies." (Author) 6 refs.

1976-118+

Rose, J. L., "Ultrasonic Procedures for the Determination of Bond
    Strength," Air Force Office of Scientific Research Interim Report
    (NTIS #AD-A027 704/6ST) (1976).

"Substantial progress was made in developing a sound resource base
in model analysis, data acquisition, and signal processing for advancing
the state of the art in the ultrasonic inspection of adhesive bonds. A
computer program was developed with the capability of calculating the
theoretical longitudinal and shear reflections from a homogeneous isotropic adhesion bond due to an obliquely incident longitudinal wave.
A fast data acquisition system consisting of a PDP 11/05 minicomputer,
Biomation 8100 A/D converter and a Tektronix 4014 graphics display was
also developed. Attention is now being placed on examining the ultrasonic inspection potential of additional metal to metal adhesive bonds,
and separately, the examination of selected composite materials, the
procedures of which use basic concepts presented in this report." (NTIS)
2 refs.

1976-119+

Shankar, R., Mucciardi, A. N., Cleveland, D., Lawrie, W. E. and Reeves,
    H. L., "Adaptive Nonlinear Signal Processing for Characterization
    of Ultrasonic NDE Waveforms. Task 2: Measurement of Subsurface
    Fatigue Crack Size," Air Force Mat. Lab. Rept. 687 (final)
    (NTIS #AD-A031 464/1ST).

"A new NDE nonlinear signal processing system has been developed to
detect and measure small, subsurface fastener hole fatigue cracks. The
system synthesized from nondestructive evaluation (NDE) waveform parameter
inputs is capable of detecting and measuring quantitatively subsurface
fatigue cracks in the size range of 0 to 279 mils to within 70 percent
of their nominally characterized lengths. Previous investigations had
achieved a 50 percent detection rate for cracks larger than 30 mils, and
no detection capability for cracks smaller than 30 mils. However, the
fatigue crack measurement system reported herein is the first known fatigue
crack NDE system capable of detection and measurement for this wide range.
The NDE waveforms were recorded from sixteen aluminum sample specimen cracks
under two different experimental conditions. Series 1 was recorded with the
ultrasonic transducer wedge at 20 degrees to the plane of the material
surface, and Series 2 with the wedge at 30 degrees. A 10 MHz transducer
was used for both series." (NTIS) 10 refs.

1975-120+

Curtis, G. J., Joinson, A. B. and Lloyd, P. A., "The Role of Transient
    Spectrum and Damping Analysis in Assessing the Strength of Polymetric
    Adhesive Metal Bonding," Air Force Mat. Lab. (monitor)
    (NTIS #AD-A009172/8ST).

"This report describes the derivation of an acoustic impact testing
facility and some results that have been obtained using it. The primary
task has been to examine what acoustic phenomena associated with the
response to impact of an adhesively bonded joint relate to the strength
of the joint. A 'first look' system employing a 40 microsec. contact
time impactor, a broadband polymeric foil transducer and a PDP8/I on-line
computer has been used to identify the parameters of interest and the
most efficient analysis scheme. This system accesses data at rates up to

4 MHz with a data block size of 256, 10 bit words thus specifying the frequency interval between points in the computed frequency spectrum as (sampling frequency/256). The analysis has centered around the computation of the fast Fourier transform of the impact response to obtain frequency response spectra and damping information. Initial results using the 'first look' system have shown that the presence of artificial defects in rectangular bars can be detected. For adhesively bonded joints, results suggest that damping analysis may be useful in assessing glue line porosity, whereas modal frequency shift is sensitive to cohesive strength." (NTIS) 49 refs.

1973-121+

Brockelman, R. H., "Evaluation of Advanced Ultrasonic Testing Techniques for Diffusion-Bonded Titanium Alloy Aircraft Structures," Report AMMRC-TR-73-16 (NTIS #AD-760 673).

"The report describes the initial effort to develop effective nondestructive test methods for diffusion-bonded titanium aircraft components. It was demonstrated that several ultrasonic techniques have the potential for overcoming the extraneous background scattering noise normally encountered in titanium structures thereby improving the sensitivity to defect detection at the bond joint. The ultrasonic techniques examined were high resolution flaw detection, spectroscopy and compound scan." (Author) 8 refs.

1971-122+

Gericke, O. R., "Improved Ultrasonic Spectroscope for Nondestructive Inspection," Air Force Mat. Lab. Report AMMRC-TR-71-36 (NTIS #AD-734 840).

"Improved instrumentation for ultrasonic spectroscopy was developed for the purpose of advanced nondestructive inspection of artillery projectiles. The equipment employs high-gain amplifiers to enable the reception and spectral processing of ultrasonic echo signals whose amplitudes are just above the electronic noise level. By incorporating special circuitry, high sensitivity is achieved without incurring a significant reduction in near-surface resolution which normally is caused by the amplifier paralysis attributable to the transducer excitation pulse. With the newly developed equipment, it is therefore possible to overcome the detrimental effect of losses in ultrasonic signal amplitude arising from the unfavorable geometry and microstructure of projectiles." (Author) 4 refs.

1973-123+

Gray, D. H. and Darby, D. M., "Ultrasonic Nondestructive Testing System Automation," Oak Ridge Y-12 (NTIS #Y-1883).

"A commercial ultrasonic test facility has been interfaced to a medium-sized multiuser computer. The large volumes of data, formerly correlated to flaw size and flaw location manually, are now automatically accumulated and analyzed, and the results are plotted within a matter of minutes." (NTIS) 1 ref.

Also refer to the abstract for reference 1975-110.

1974-124+

Van Doren, S. L., Pond, R. B., Sr. and Green, R. E., Jr., "Acoustic Characteristics of Twinning in Indium," Air Force Office of Sci. Res. (Monitor) (NTIS #AD-A009 163/7ST).

"Since mechanical twinning is one of the major causes of fatigue failure in metals, acoustic emission measurements were made during mechanical deformation of indium single crystals. Indium was selected since upon stress application well delineated twins develop which can be made to disappear upon stress reversal. Associated with this twinning and detwinning are high intensity acoustic emissions and therefore indium affords an excellent material with which to develop techniques. By recording the acoustic emission signals on a video tape recorder it was possible to play them back as often as desired, to observe them in slow and stop motion, and to play any desired portion of the signal through a spectrum analyzer for frequency analysis. The experimental results show that a correspondence exists between the duration of the acoustic signal and the volume of twinned material as determined by cinematographic observation of the generation and growth of twin bands." (NTIS) 6 refs.

1971-125+

Mann, L., Jr. and Young, M. H., "Data Analysis and Correlation with Digital Computers-Nondestructive Testing," Air Force Prod. Agen. (Monitor) Report Bull-107 (NTIS #AD-734321).

"The purpose of this research was to ascertain feasibility of using digital computers to facilitate nondestructive testing techniques. Energy envelopes were made and were analyzed on the LSU hybrid computer which consists of the EAI 680 for the analog component and a Sigma 5 for the digital component. The sound envelopes were subjected to a filtering program making use of the Fourier analysis in order to arrive at a response frequency. The response frequency for 'good' specimens were compared against the response frequency for faulty specimens. During the course of the investigations nondestructive tests were investigated which included stress analysis, dimensional properties, faulty specimens and faults in long, thin wall tubing." (Author) 11 refs.

1975-126+

Mucciardi, A. N., Shankar, R., Cleveland, J., Lawrie, W. E. and Reeves, H. L., "Adaptive Nonlinear Signal Processing for Characterization of Ultrasonic NDE Waveforms. Task 1: Inference of Flat Size," Air Force Mat. Lab. (Monitor) Interim Rept. #687 (NTIS #AD-A013 881/8ST).

"The overall objective of this first work task was to demonstrate feasibility of adaptive nonlinear signal processing techniques, from cybernetics and bionics, applied to characterization of ultrasonic non-destructive testing (UNDT) waveforms for accurate interferences of flat-bottom hole sizes. The classified waveforms were ultrasonic pulse echoes obtained from 2 different sets of 7075-T6 aluminum test blocks and 3 different transducers. The eight flat-bottom hole defect sizes ranged from 1/64 to 8/64 inch in steps of 1/64 inch." (NTIS) 20 refs.

1973-127+

Papadakis, E. P., Fowler, K. A., Lynworth, L. C., Carnevale, E. H. and Chen, J., "Measurement of Elastic Moduli of Materials at Elevated Temperature," Panametrics, Quarterly Report #2 (NTIS #AD-759 060) (also refer to NTIS #AD-753 844—Quart. Rept. 1).

"The report contains new data on modulus and attenuation at elevated temperatures. It also contains theoretical calculations pertaining thereto. Specific topics discussed include the following: Measurements on rememdur to 1000 C; Corrections for spurious echoes in wire experiments; Forming graphite specimens for thin-wire ultrasonic measurements; Fourier spectrum analysis and synthesis of pulses." (NTIS) 5 refs.

1967-128+

Frank, L. M. and Kubiak, E. J., "Nondestructive Methods Development for the Evaluation of Thin and Ultrathin Sheet Materials," Air Force Mat. Lab. (Monitor) (NTIS #AD-832 035/0ST).

"The FM Lamb wave system was modified to permit two modes of operation, a swept frequency mode and a single frequency mode. This modification proved to be extremely useful for laboratory testing because in the single frequency mode, amplitude and frequency measurements can be made digitally with a resultant accuracy of better than 1%, while in the swept frequency mode several modes can be observed simultaneously. All the electronics of the FM Lamb wave system were packaged into a mobile console to facilitate field testing. This included rebuilding of the receiver section with a resulting increase in sensitivity and adding an additional mode of data display. In addition, a precision miniwheel scanner was designed and fabricated for inspecting large samples and for on-line field testing. The results of laboratory testing showed the system to be very sensitive to laminar defects, thickness, and anisotropic properties. An IBM 1130 computer was programmed to solve the equations describing Lamb wave propagation and with the use of an on-line plotter, the curves of incident transducer angle versus frequency-thickness product were generated for many materials. Many points on these curves were then verified experimentally." (Author) 2 refs.

1975-129+

Thompson, D. O., "Proceedings of the ARPA/AFML Review of Quantitative NDE," Air Force Material Laboratory Rept. SC595-10AR (NTIS #AD-A023 622/4ST) (1975).

"The edited transcripts of the ARPA/AFML Reivew of Quantitative NDE held on July 15-17, 1975, at the Science Center, Rockwell International, are presented in this document. Several key topics form the core of these presentations and discussions. They include quantitative ultrasonics, adhesives and composites, emissions related to the prediction of failure, and residual stress. In addition a panel discussion is presented related to the role of ultrasonic standards in the emerging quantitative ultrasonics area. It is believed that this document provides a reasonable summary of NDE research and development currently underway in the areas selected for presentation." (Author).

1974-130+

Carey, C. A., Carnevale, E. H., Chen, J., Fowler, K. A. and Larson, G. S., "Measurement of Elastic Moduli of Materials at Elevated Temperature," Panametrics (NTIS #AD-780 231/7); also see AD-763 763).

"Ultrasonic measurement of elastic properties of solids at elevated temperature comprises the central theme in the report of some twelve years of R and D under the support of the ONR Acoustics Branch. The principal methods used were momentary contact with pressure coupling, for bulk specimens of cross-sectional dimensions large compared to wavelength, and continuous contact, with welded joints, for wires and rods of cross-sectional dimensions small compared to wavelength. The bulk specimens were tested with broadband pulses of rf bursts having frequencies typically above 1 MHz. The wires were tested with pulses having 0.1 MHz as a typical center frequency. Other topics include: transducers, velocimetry, spectrum analysis, infrasonic measurements of moduli in plastic tensile test specimens, measurement of polymerization, and applications of ultrasonics such as nondestructive testing, acoustic emission, thermometry, flowmetery and process control," (Modified author abstract by NTIS). Numerous refs.

1967-131+

Gericke, O. R., "Differential Ultrasonic Spectroscopy for Defect and Microstructure Identification," Report AMRA-TR-67-07 (NTIS #AD-650 787).

"The report presents improved experimental procedures for ultrasonic spectroscopy which make it possible either to compare the spectra of two ultrasonic echo signals or to obtain a curve representing differences in the spectral amplitude distribution of these signals. It is shown that the differential method can be used to distinguish two defects from each other which, although differing in size, produce indications of identical height in an ordinary pulse-echo test. It is further shown that the new spectroscopic technique can be used to determine differences in the microstructure of materials if these differences affect the attenuation of ultrasound." (Author) 4 refs.

1974-132+

Zuckerwar, A. J., "Application of Ultrasonic Signature Analysis for
    Fatigue Detection in Complex Structures," Report NASA-CR-138113
    (NTIS #N74-22531/9).

    "Ultrasonic signature analysis shows promise of being a singularly
well-suited method for detecting fatigue in structures as complex as
aircraft. The method employs instrumentation centered about a Fourier
analyzer system, which features analog-to-digital conversion, digital
data processing, and digital display of cross-correlation functions and
cross-spectra. These features are essential to the analysis of ultrasonic signatures according to the procedure described here. In order
to establish the feasibility of the method, the initial experiments were
confined to simple plates with simulated and fatigue-induced defects
respectively. In the first test the signature proved sensitive to the
size of a small hole drilled into the plate. In the second test, performed
on a series of fatigue-loaded plates, the signature proved capable of
indicating both the initial appearance and subsequent growth of a fatigue
crack. In view of these encouraging results it is concluded that the
method has reached a sufficiently advanced stage of development to
warrant application to small-scale structures of even actual aircraft."
(Author) 15 refs.

1975-133+

Cohen, E. R. and Tittmann, B. R., "Analysis of Ultrasonic Wave Scattering
    for Characterization of Defects in Solids," Air Force Office of Sci.
    Res. Rept. SCTR-75-12 (NTIS #AD-A023 136/5ST).

    "The report summarizes the results of a program to characterize the
scattering of ultrasonic acoustic waves by defects in solids. The program
is intended to analyze and verify by direct measurement scattering by
spheroidal cavities and inclusions in a metallic matrix. The work to
date has been confined to analysis of spherical scatterers and the
measurement of scattering from spherical voids. The measured scattering
patterns are in agreement with the predictions of the theory. The
observed deviations may be understood as due to the effects of the pulse
shape and finite pulse length of the transmitting transducer compared to
the monochromatic wave form of the theoretical analysis. It is concluded
even from these limited measurements that a spherical cavity has the
characteristics necessary for use as standard defect for NDT." (Author)
19 refs.

1970-134+

Kennedy, J. C. and Woodmansee, W. E., "Electronic Signal Processing
    Techniques. Phase II. Nondestructive Testing," Report D180-10589-1,
    Boeing (NTIS #AD-716803).

    "Signal averaging was used to enhance flaw indications in the
ultrasonic inspection of electron beams welds. An electronic gate,
synchronized to the transducer motion through the use of an electrically

controllable delay was also used to enhance flaw indications. To aid in
electrical signal processing, a technique for recording ultrasonic video
information on a low-frequency tape recorder was developed. In
preparation for optical matched filtering, ultrasonic information was
recorded on a photographic transparency in a C-scan format. An XY
scanning densitometer was constructed to remove the intensity modulated
information from the film and to aid in the production of hard copy
recordings. Phaselock detection was used to perform through transmission
eddy current testing. Methods were developed for obtaining quantitative
through transmission data. Through transmission thickness measurements
were performed and procedures were developed for obtaining a linear
relationship between part thickness and eddy current signal. Applications
to chemical milling and in-motion thickness measurements were demonstrated. Phaselock detection and signal averaging were used to measure
resistance and inductance changes in conventional eddy current coils
during a scanning operation. Quantitative data was produced from bridge
unbalance signals." (Author) 10 refs.

1968-135+

Martin, G. and Moore, J. P., "Exploratory Development of Nondestructive
    Testing Techniques for Diffusion Bonded Interfaces," Air Force Mat.
    Lab. Report N.A.-68-565 (NTIS #AD843914/5ST) (1968).

"The report covers exploratory development of nondestructive testing
techniques for solid-state diffusion-bonded parts. A literature survey
describes the current inspection practices employed on diffusion-bonded
laminates and the nature of the interface and bond properties. Test
specimens of titanium, columbium, and TD-nickel were prepared for
evaluating various test methods. The specimens provided a range of
thicknesses and varying degree of bond quality as determined by mechanical
property tests. In addition, a number of diffusion-bonded specimens were
supplied by the Air Force. Conventional ultrasonic techniques and recently
developed selective interface inspection techniques clearly indicate bond
conditions where these vary from a no-bond condition to substandard bonds.
A number of methods show potential correlation between a quantitative
expression of bond quality such as shear or tensile strength. These
methods include logitudinal and shear wave velocity, ultrasonic attenuation,
and vibration analysis." (Author) 8 refs.

1975-136+

Szabo, T. L., "Obtaining Subsurface Profiles from Subsurface-Acoustic-
    Wave Velocity Dispersion," Air Force Cambridge Res. Lab. Report
    AFCRL-TR-75-0273 (NTIS #AD-A010 081/8ST) and J. Appl. Phys. 46 (4),
    1448-1454.

"Surface acoustic waves can be used to probe nondestructively subsurface gradients (caused by physical processes) by changing their
penetration depth with frequency. A perturbation-theory integral equation
that describes the influence of a known gradient on the velocity dispersion
is reviewed. The experimental situation requires solution of the inverse
problem: obtaining the profile from measured velocity data. The inverse
problem is solved exactly by the use of Laplace transforms for gradients

of the general form F(z). The nature of the solution allows observations
about the gradient-dispersion relationship, the physical interpretation
of dispersion curves, and the representation and measurement of gradients
to be made. Six commonly occurring gradient functions and their dispersion curves are compared by use of an equivalent-area parameter formulation. The solution agrees well with published experimental results for
residual stress in an aluminum block and for damaged layers on YZ LiNbO3
substrates. For cases in which the analytic form of neither the dispersion
nor the gradient is known, a method is presented for obtaining the
gradient from raw data. This technique is applied to published results
for quench-hardened-steel cylinders, and the experimental implications
of gradient determination are discussed." (Author) 21 refs.

1975-137+

deBilly, M., Doucet, J. and Quentin, G., "Angular Dependence of the Back-
    scattered Intensity of Acoustic Waves from Rough Surfaces," Ultra-
    sonics International 1975, Conf. Prof; IPC Science & Technology
    Press, London (1975).

    "The authors have studied how the intensity of ultrasonic waves back
scattered by different rough surfaces varies with the angle between the
normal to the surface and the incident beam ($\theta = 0, 20°$). The ultrasonic
frequency was varied from 5 to 25 MHz. The roughness properties of the
surfaces were known; the r.m.s. roughness (h) of the samples ranged from
6 to 92 μm. The results obtained are compared in order to examine the
influences of the mean roughness of the scatterers and of their correlation length (L). The possibility is then discussed of assessing, as in
radar experiments, a 'signature' for the target." (Author) 20 refs.

1975-138+

Quentin, G., deBilly, M., Cohen-Ténoudji, F., Doucet, J. and Jungman, A.,
    "Experimental Results on the Scattering of Ultrasound by Randomly
    and Periodically Rough Surfaces in the Frequency Range 2 to 25 MHZ,"
    1975 IEEE Ultrasonics Symp., Proc., 102-106 (1975).

    "Preliminary results of backscattering experiments at three discrete
frequencies (5, 15 and 25 MHz) have been described . . . It was suggested
that the diagrams of the angular dependence of the ultrasonic intensity
backscattered by randomly rough surfaces could be used to obtain precise
information about the rms roughness $\underline{h}$ and its correlation distance $\underline{L}$.
A new set of experiments have been performed and the 'ultrasonically'
estimated value of $\underline{h}$ is obtained with a precision of the order of one
micron.
    We have also begun a study of the scattering from rough surfaces,
using ultrasonic spectroscopy. In these experiments the insonifying
ultrasonic pulse has a quite large bandwidth (2 to 9 MHz). With these
two methods we can study the dependence of the backscattered ultrasonic
intensity versus the frequency and the angle of incidence. We shall
discuss . . . the results obtained with three kinds of rough surfaces
used as scatterers: randomly rough surfaces; one-dimensionally periodic

surfaces (with a spatial period in the range of some hundreds of microns); and two-dimensionally periodic grid.

A special attention will be given to the information about surface profiles which can be deduced from these experiments." (Author) 10 refs.

1976-139+

deBilly, M., Cohen-Ténoudji, F., Jungman, A. and Quentin, G., "The Possibility of Assigning a Signature to Rough Surfaces Using Ultrasonic Backscattering Diagrams," IEEE Trans. Son. Ultrason., SU-23 (5), 356-363 (1976).

The authors "had previously reported experimental results on the backscattering of ultrasound waves by surfaces either randomly, or periodically rough. Two different series of experiments were performed to study the surfaces. In the first one, we used narrow-band ultrasonic pulses, and our experimental device automatically records the variations of the backscattered intensity versus the grazing angle. The operating frequencies are 2, 5, 15 or 25 MHz. In the second (series of experiments), the insonifying pulse is broadband (2-9 MHz within 3 dB), and the frequency spectrum of the echoes is plotted for discrete values of the grazing angle. For randomly rough surfaces, the ultrasonic value of the rms roughness (h) is now automatically deduced from the backscattering measurements, and the range in which good precision is obtained has been extended towards smaller and higher values of the roughness (3 μm < h < 100 μm). The 'secondary' influence of the autocorrelation distance of the profile will also be discussed. In a second part we present experimental results showing the ultrasonic spectroscopy is able to give useful information about spatial periodicities in one or two dimensions, and examine the correlation of defects in the gratings with 'ghosts' appearing in the spectrum as in optical diffraction experiments. We finally present an original procedure used to evaluate the periodicity of more typical surfaces where it is hidden in a random background." (Author) 10 refs.

1977-140+

Jungman, A., Cohen-Ténoudji, F. and Quentin, G., "Diffraction Experiments in Ultrasonic Spectroscopy; Preliminary Results on the Characterization of Periodic or Quasi-Periodic Surfaces," Ultrasonic International 1977 Conf. Proc.; IPC Science and Technology Press (1977).

"In order to obtain quantitative information about the profile of periodically or quasi-periodically rough surfaces, a series of backscattering experiments have been carried out, using a broad-band insonifying pulse. Ultrasonic spectroscopy is shown to be a very powerful tool for the measurement of spatial periodicities even when the main component of the surface profile is random. For diffraction gratings exhibiting the same periodicity, but different profiles, we studied the influence of the exact profile on frequency spectra. A comparison between the ultrasonic spectral data and the theoretical values computed by using the Fourier transform of the surface profile is performed." (Author) 6 refs.

1974-141+

Thompson, D. O., "Proceedings of the Interdisciplinary Workshop for
    Quantitative Flaw Definition," Air Force Mat. Lab. (Monitor),
    Report SC595.21TR (NTIS #AD/A-003 672/3ST).

    "Contents: Economic motivation and potential impact of NDE:
Fracture control for high performance ships and related NDE requirements;
State-of-the-art NDE in quantitative inspection; Factors affecting ultrasonic waves interacting with fatigue crack fasteners; Interface waves on
interference-fit fasteners; Acoustic wave scattering from a sphere; Impulse
analysis of ultrasonic indicia; Adaptive nonlinear modeling for ultrasonic
signal; Studies of electromagnetic sound generation for NDE; Laser
generated ultrasonic beams; Advanced magnetic methods of flaw detection;
Advanced composites status review; Microscopic description of bond
strength, mechanics and processes; Ultrasonic procedures for predicting
adhesive bond strength; Applicability of ultrasonic resonance spectroscopy
to measuring the strength of bonded materials; NDE prediction of adhesive
bond failure areas; Kinetics of moisture degradation in advanced composites;
Residual stress effects on the properties and characteristics of engineering; Physical origins of residual stress and present physical techniques
for measurement; Acoustic emission—a summary of current understandings;
Acoustic emission study of twinning in indium crystals and Pb-Sn alloys;
Exo-electron emission from metals; Energy distribution of photo-stimulated
electron emission and fatigue specimens; The effect of surface roughness
on the frequency of surface plasmons." (NTIS). Numerous refs.

1975-142+

Johnson, D. M., "Model for Predicting the Reflection of Ultrasonic Pulses
    from a Body of Known Shape," Con. Elec. Gen. Board (UK)
    (NTIS #RD/B/N-3273).

    "The reflection of broad-band ultrasonic pulses from an object can
result in modifications to both the frequency spectrum and phase information in the pulse. The changes are largely controlled by the shape and
nature of the surface of the object. This has been the subject of many
recent experimental investigations from which it is apparent that there
is a need for a simple, unifying theoretical model to account for the
observations.
    Three approximate methods for determining the scattered field have
been reviewed, one of which has been developed in a general form and
should have considerable applications. From the known shape of the object
surface, the response function can be evaluated numerically and the effect
on the incident pulse predicted." (Author) 11 refs.

1976-143+

Meyer, M. W. and Oakes, R. E., Jr., "Ultrasonic and Impact Techniques
    Used to Characterize Liquid-Phase-Sintered Tungsten Alloys,"
    Conf. on High Density Alloy Penetrator Materials, Charlottesville
    (24 May 1976) (NTIS #Y/DA-6661).

"The Oak Ridge Y-12 Plant has been involved in producing liquid-phase-sintered tungsten alloys for penetrator applications. The majority of the effort has been centered on two materials—a 90 W-7 Ni-3 Fe alloy with 25 percent cold work and a 95 W-3.5 Ni-1.5 Fe alloy with 18 percent cold work. Fracture toughness concepts, along with instrumented Charpy impact tests, were used to measure toughness differences and temperature effects on toughness for several alloys. The instrumented Charpy was useful in screening alloys having different processing parameters. Ultrasonic techniques can be used to detect conditions of matrix depletion and incomplete sintering. Spectral analysis shows high attenuation of all frequencies in areas of matrix depletion and very low attenuation of energy between 10 and 15 MHz for conditions of incomplete sintering when compared to normal material. Efforts are underway to develop a higher density alloy. Of major concern in the higher tungsten alloys is cracking which may occur during swaging. Preliminary ultrasonic techniques to detect cracks are discussed." (Author)

1975-144+

Wallace, S. A., "Acoustic Emission: An Overview," Oak Ridge
     (NTIS #Y-DA-6376).

"The mechanisms and applications of acoustic emission testing are reviewed. A bibliography of selected literature covering the period 1967 to 75 is presented." (NTIS) Numerous refs.

1976-145+

Adler, L. and Lewis, D. K., "Scattering of a Broadband Ultrasonic Pulse by Discontinuities," IEEE Trans. Son. Ultrason. SU-23 (5), 351-356 (1976).

"In order to obtain quantitative information from ultrasonic waves scattered from hidden defects, a series of scattering experiments were carried out using broadband ultrasonics. The scattering amplitudes were measured as a function of frequency and scattering angle from various reflectors in water and from flat-bottomed holes in metals. A geometrical theory of diffraction was applied to obtain analytical expressions for scattering of both longitudinal and shear waves in crack-like flaws. The agreement between theoretical predictions and experimental results is good. Scattering experiments were also carried out from various shaped cavities embedded in diffusion-bonded titanium samples. The wave is mode converted and scattered as both shear and longitudinal waves. These two waves are time separated, gated out electronically, and the signal spectrum analyzed separately. The resultant spectrum is a characteristic of the cavity shape." (Author) 7 refs.

1974-146+

Rogers, W. S., Jr., "Data Acquisition and Processing System for Ultra-sonic Research," Air Force Inst. of Tech., Wright-Patt. AFB, Ohio, School of Engr., Thesis Report GE-EE-74-33 (NTIS-785 127/2).

The Air Force Materials Laboratory, Ultrasonic Research Facility
at Wright-Patterson Air Force Base, Ohio, is investigating the use of
ultrasonics as a tool for the nondestructive evaluation of aerospace
materials. The development of a computerized data acquisition and processing system for the facility's ultrasonic research is described in
this paper. The system provides the capability to sample and record
electrical representations of ultrasonic wave trains reflected from a
material sample under test." (Modified author abstract, by NTIS.)

Because this report was also the author's Master's thesis, the system
(especially computer programs) is described in great detail. 19 refs.

1964-147+

Redwood, M., "Experiments with an Electrical Analog of a Piezoelectric
    Transducer," J. Acoust. Soc. Am. 36 (10), 1872-1880 (196).

"An electrical analog of a piezoelectric transducer has been built
and used to demonstrate the generation and detection of acoustic waves and
the electrical characteristics of a piezoelectric resonator. The circuit
uses an artificial transmission line to represent the distributed-constant
mechanical properties of the transducer, and is therefore capable of
reproducing the behavior of a transducer under both transient and continuous wave conditions. The theory of the equivalent circuit of a
transducer is first extended to facilitate interpretation of the
physical processes of generation and detection. This is done by
developing an 'impulse sequence' that takes into account waves generated
at both faces of a transducer, and the delay in time and reflections
that they undergo in passing through the transducer. This analysis is
used to discuss examples of waveforms obtained with the analog in the
following situations (with various simulated combinations of backing
and load impedance): (1) as a generator of acoustic waves when excited
by an electrical signal in the form of (a) a short impulse, (b) a step,
(c) short trains of sinusoidal oscillations of various lengths; (2) as
a detector of acoustic waves when excited by an acoustic signal in the
form of (a) a step, (b) a train of sinusoidal oscillations. In
detection, the effect of terminating the transducer with high and low
resistances is also demonstrated. Experiments concerning the continuous-
wave response of the analog are also described." (Author) 8 refs.

1961-148+

Redwood, M., "Transient Performance of a Piezoelectric Transducer,"
    J. Acoust. Soc. Am. 33 (4), 527-536 (1961).

"The methods of transform calculus are used to solve several
problems in which the transient response of a piezoelectric transducer
is of interest. The electrical signal produced by a step function of
force is derived, both for open-circuit and resistive loading at the
electrical terminals, and the mechanical signal produced by a voltage
step is also discussed. The analyses are performed for both the plate
transducer in compressional thickness vibration and the bar in compressional length vibration, and the important differences between the

two transducers are discussed. The analyses commence with the fundamental
piezoelectric equations, and solutions are found which represent successive
time-delayed reflections of the mechanical transient between the end
faces of the transducer. The results are also discussed with reference
to the exact transmission-line electrical equivalent circuits of the
transducers, whose development is outlined briefly. Simple equivalent
circuits which do not involve lines are described; these make it possible
to determine many features of the transient without recourse to the full
theoretical analysis." (Author) 11 refs.

1974-149+

Sachse, W., "Ultrasonic Spectroscopy of a Fluid-Filled Cavity in an
    Elastic Solid," J. Acoust. Soc. Am. 54, 891-896 (1974).

    "The scattering of wide-band ultrasonic pulses from a circular,
cylindrical, fluid-filled cavity in an elastic solid is investigated with
a digital ultrasonic-spectroscopy technique. Based on the data from
three sizes of cavities (1/32-in., 1/16-in., and 1/8-in. diameter), each
being filled with five different fluids, we find that either the diameter
of the cavity or the wave speed of the fluid contained in it can be
determined from the time history of the scattered pulses. From the
spectral analysis, an empirical formula is established relating the incre-
mental spectral frequencies (the intervals between successive frequency
maxima or minima of the spectra) to the diameter of the cavity and the
wave speed of the fluid contained in it." (Author) 13 refs.

1974-150+

Chang, F. H., Yee, G. W. and Couchman, J. C., "Spectral Analysis Technique
    of Ultrasonic NDT of Advanced Composite Materials," Non-Destr. Test.,
    194-198 (1974).

    "The authors describe an ultrasonic-frequency-spectral analysis
method of detecting flaws in advanced composite materials. This nondestructive technique depends on the phenomenon of resonance interference
of acoustical waves in materials. When the material thickness is an
integral multiple of the half wavelength of the sound waves, destructive
interference of a return echo by multiple reflections in the material
produces anti-resonant dips in the frequency spectrum for the reflected
signal. The period of these dips is related to the material thickness
normal to the beam path. Delaminations or voids in a plane perpendicular
to the direction of propagation of the sound waves may be observed through
their characteristic anti-resonant frequencies. Graphite-epoxy composite
specimens containing flat-bottom holes and small planar voids were used as
examples of the application of this technique. The authors present
analytical development, experimental procedures, and spectral analysis
results in this paper." (Author) 8 refs.

1975-151+

Bifulco, F. and Sachse, W., "Ultrasonic Spectroscopy of a Solid Inclusion
    in an Elastic Solid," Ultrasonics, 113-116 (May, 1975).

"Experiments were conducted to measure the arrival time and the power density spectra of wide-band ultrasonic pulses as scattered by a circular, cylindrical, solid inclusion in a matrix of aluminum. Cavities of 3.18 and 6.35 mm diameter were alternately filled with four different solids possessing a wide range of acoustic properties. Results show that the time history and the spectral analysis of the scattered pulses can be used to determine the size of the cavity or the wave velocity in the solid inclusion." (Author) 4 refs.

1976-152+

Blinka, J. and Sachse, W., "Application of Ultrasonic-Pulse-Spectroscopy Measurement to Experimental Stress Analysis," Exp. Mech., 448-453 (Dec., 1976).

"We apply the technique of ultrasonic pulse spectroscopy to measure the interference effects between two shear waves propagating in specimens loaded in uniaxial compression. We show that the power spectrum of an echo containing both fast and slow components of a shear wave will exhibit periodic minima. The periodicity exhibited in the spectrum is $1/\Delta\tau$, where $\Delta\tau$ is the difference in arrival time between the fast and slow waves. A change in the state of stress which produces a change in the two shear velocities results in a stress-dependent change in wave-arrival times. Because of this velocity change, the frequency at which a particular minimum occurs in the spectrum changes, and this can be used to indicate the state of stress in the material.

Our results indicate that, if the spectrum minima frequencies could be resolved to within 10 kHz, the principal-stress differences within 36 psi (0.251 MPa) could be measured in specimens of aluminum 1 in. (2.54 cm) thick. Inherent in analyzing and measuring echo-interference effects is a single-echo requirement. Thus, transducer coupling effects are minimized and measurements in highly attenuating materials or at high frequencies in normal attenuating materials are possible. This technique shows considerable promise as a means of measuring and monitoring the applied stresses in materials." (Author) 11 refs.

1977-153+

Scott, W. R. and Gordon, P. F., "Ultrasonic Spectrum Analysis for Nondestructive Testing of Layered Composite Materials," J. Acoust. Soc. Am. 62 (1), 108-116 (1977).

"In this paper, a simple model is presented which predicts the ultrasonic frequency spectra for a broad class of layered composite materials having a finite number of laminas. This model predicts spectra for arrays of glass plates in water and these spectra are experimentally verified. Precisely regular spectra are predicted for single plates, while irregular spectra are predicted for all of the arrays studied. Results relating to nondestructive testing which have emerged from this investigation include methods for predicting spectra for layered composite materials and techniques for mapping small changes in the modulus and thickness of composite materials. Also discussed is the existence of forbidden frequency bands for which ultrasound transmission is strongly attenuated in thick layered composites." (Author) 7 refs.

1971-154+

Engelson, M., <u>Spectrum Analyzer Measurements: Theory and Practice</u>,
    Tektronix, Inc., Beaverton, Oregon 97005

    "This book is primarily concerned with the problem of measurements
in the frequency domain by means of spectrum analyzers. Thus circuit
design or construction details are not considered. Basic system parameters
are, however, discussed in some detail since these have a direct bearing
on the interpretation of measurement data. Two types of signals are
treated in detail: those composed of discrete or line spectra and those
composed of continuous or dense spectra. Continuous wave (CW) or sinusoidal amplitude modulation (AM) is an example of the former, while
pulsed-RF is treated as the latter. The third class of signals comprising
random variables and requiring statistical methods are excluded from the
detailed discussion, though some applications are included.
    The discussion follows a dual approach: Part I is a mathematical-
process-oriented approach while Part II applies the theory of Part I to
specific measurement problems." (Author) 53 refs.   292 pages

1973-155+

Cheney, S. P., Lees, S., Gerhard, F. B., Jr., and Kranz, P. R., "Step
    Excitation Source for Ultrasonic Pulse Transducers," <u>Ultrasonics,</u>
    111-113 (May, 1973).

    "A circuit is described for exciting ultrasonic transducers by step
discharge through avalanching transistors. While a free-running mode is
possible, the circuit is designed for synchronous operation at 1 and 10
KHz repetition rates. During the receive mode the transducer is decoupled
from the power source. The output is obtained from a buffer amplifier to
match a 50 ohm coaxial cable. The circuit described produces a 250 V
step discharge into a 24 picofarad load in 10 nanoseconds." (Author)
2 refs.

1978-156+

Adler, L., Cook, K. V., Simpson, W. A., and Lewis, D. K., "Ultrasonic
    Flaw Detection by Spectral Analysis in Structured Material,"
    U.S. Patent Pending.

    "A method of quantitatively determining the size and location of
flaws within anisotropic materials is described. A broadband ultrasonic
pulse having a frequency width of at least several MHz is used as an
input pulse from a transducer which is directed at a specific angle
toward the potentially flaw-containing material. The average frequency
interval between interference maxima in the reflected spectrum is
relatable to the size and location of the flaw when the velocity of
sound for the specific angle of orientation is utilized in computations."
(Author)

1972-157

Dixon, N. E., "Method of Generating Unipolar and Bipolar Pulses,"
    Patent 3,656,012.

    "The method for generating bipolar and unipolar mechanical pulses
is described. The unipolar pulses can be in a form of a single unipolar
pulse or pairs of unipolar pulses of opposite polarity." (Author)

1972-158

Wang, H. S., "Multiple-Use Subsonic Waves," Edited trans. of K'o Hsueh Ta
    Chung (Mainland China) N9, 348-349 (1964) by G. Hwang.

    "The report reviews the applications of ultrasonic radiation detection
to fields of seismic detection, geophysics, underwater communications,
storm forecasting, nondestructive testing, agriculture and medicine."
(NTIS)

1967-159

Tang, M. C., "A Resonance-Type Ultrasonic Thickness Gage for Measuring
    the Wall Thickness and Corrosion of Chemical Plant Equipment,"
    Edited trans. of Hua Hsueh Shih Chieh (Chinese People's Republic) 18
    (8), 377-379 (1964).

    "A discussion is given of the principles of thickness measurement
by the ultrasonic resonance method and equipment designed for applying
this technique to the measurement of the depth of corrosion in chemical
equipment is described. Emphasis is placed on the proper use of transducers and coupling media for optimum results. In actual practice,
certain key points should be selected for testing signal identification is
best achieved with an oscilloscope, and if the resonant signal appears very
often during measurements of thickness, the accuracy can be improved by
taking the mean of several measurements. Some of the parameters of the
Chinese model HS-1 meter, which has a maximum error of 4-4%, and an
average error of 1-3% when measuring steel having a thickness of 3-30 mm,
are given. At the present time, this technique cannot be used for
measuring the wall thickness of small-diameter pipes." (Author)

1971-160

Steffens, R. W. and Zeutschel, M. F., "Electrostatic Ultrasonic Non-
    destructive Testing Device," Patent 3,577,774 (4 May 1971).

    "An ultrasonic device for nondestructively testing an electrically
conductive sample comprises an electrode mounted to said sample and
electrically insulated therefrom. Means are provided for generating a
pulsed potential difference between the sample and the electrode to
generate an elastic wave in the sample. A bias voltage is applied to the
electrode and an oscilloscope is used to detect, relative in time to the
applied pulsed potential difference of the electrode, changes in the
bias potential of the electrode response to the elastic wave." (Author)

1975-161+

Stoker, G. C., "Third Generation Computer Interface for Ultrasonic Pulse
    Echo Nondestructive Testing," Sandia Labs. (NTIS #SAND-75-0056).

"A third generation computer interface is described in detail.
Echo Trapp III as it is called interfaces standard ultrasonic pulse echo/
through transmission units with minicomputer systems. This approach
allows computerized data acquisition and analysis for improved NDT
results. Brief results are included." (NTIS)

1975-162

Stone, D. E. W., Clarke, B., "Ultrasonic Attenuation as a Measure of
    Void Content in Carbon-Fibre Reinforced Plastics," Non-Destr. Test.
    (London), 8 (3), 137-145 (1975).

"Describes a technique for determination of the void content of
carbon-fiber reinforced plastics (cfrp) from its effect on the attenuation
of ultrasound. The authors describe briefly the fabrication of cfrp
laminates and the defects which cause an increase in ultrasonic attenuation.
The article concentrates on the effect of voids and shows that void content
and ultrasonic attenuation are directly related. This attenuation
increases with frequency. The authors establish a close correlation
between ultrasonic attenuation and interlaminar shear strength." (NTIS)
20 refs.

1970-163

Sessler, J. G. and Weiss, V., "Improvement in Crack Detection by Ultra-
    sonic Pulse-Echo with Low Frequency Excitation," Syracuse Univ.;
    (NTIS #AD-716 642).

"The results of studies to determine the feasibility of inducing
stress in metal samples with low frequency excitation have indicated
that the effective use of this technique is limited to excitation in the
cyclic frequency range of 1000 Hz and below. Best results were obtained
with solid coupling of low frequency exciters to test material. Attempts
to induce stress with excitation at 20 KHz were not very successful with
the techniques and equipment used in these studies." (Author)

1975-164

Sessler, J. G. and Weiss, V., "Crack Detection Apparatus and Method,"
    Patent 3,867,836 (NTIS #AD-D000730/2ST).

"The patent describes an apparatus and method for nondestructively
detecting the presence and the location of crack in materials, using low
frequency mechanical vibrations. The material specimen or structure having
a crack to be detected is subjected to tensile or compressive forces due
to excitation caused by low frequency sound waves or mechanical vibrations

from a generator, thus changing the opening and thereby changing the effective size of the crack in the specimen. An ultrasonic search unit is used to follow modulations of reflected energy at the crack interface due to variations of the effective size of the crack. The search unit is controlled by an ultrasonic pulser-receiver which displays the amplitude of echo from the crack on an oscilloscope." (NTIS)

1971-165

Sessler, J. G., Weiss, V., Sengupta, M. and Barrus, L., "The Effect of Stress Fields on the Ultrasonic Energy Reflected from Discontinuities in Solids," Syracuse Univ. Res. Corp. for ARPA (NTIS #AD-719 434 Quart. Rept.).

"The challenge to the field of nondestructive testing (NDT) is to develop procedures capable of resolving defects of critical size and to accurately define defect geometry. The NDT research program now in progress under the contract is a modification of the ultrasonic pulse-echo method of flaw detection. The approach is based on the principal that when a stress field is applied to a discontinuity (defect or flaw) in a solid, a change in the geometry of the discontinuity is effected. The change in geometry can result in corresponding changes in amplitude and/or pattern of ultrasonic energy reflected from the discontinuity. It is reasonable to expect that the reflected energy response should be relatable to specific distinguishing features of a given type of discontinuity." (Author)

1970-166

Reifsnider, K. and Sawyer, S., "A Correlation between Pulse-Echo Ultrasonic Attenuation and Hardness," Virginia Polytechnic Inst., Report VPI-E-70-23 (NTIS #AD-717 099).

"Nondestructive evaluation and characterization of materials, especially in terms of parameters critical to modern design and fabrication, is a problem of growing magnitude. One example of this problem is the inability of any present nondestructive test technique to definitely establish the yield strength of a fabricated section of steel, especially if spacial gradients of material properties occur. The present report describes some results of one nondestructive method which does offer some improvement in such capabilities and indicates a direction for continued development. In particular, a theoretical correlation is established between pulse-echo ultrasonic attenuation values and hardness for alloys with variable impurity content. Experimental tests which qualitatively support the correlation are also described." (Author)

1966-167

Martin, G. and Moore, J. F., "Development of Nondestructive Testing Techniques for Honeycomb, Heat Shields, Vol. I, Final Report," Report NASA-CR-81145, NA-66-912, Vol. I (NTIS #N67-15279).

"The development of a portable nondestructive testing and scanning/ recording system, using ultrasonic techniques, for the detection of disbonds in composite honeycomb structures by inspection methods operating from one side only of the honeycomb is described. Breadboard systems of five ultrasonic techniques were developed to inspect Saturn honeycomb composites: pulse echo interference, impedance, decrement, spectrum analysis, and intermodulation." (NTIS)

1972-168

Lynnworth, L. C., DuBois, J. L., and Kranz, P. R., "Prototype Ultrasonic
        Instrument for Quantitative Testing," Report NAS8-26931, Prepared
        for Marshall Space Flight Center by Panametrics, Inc.
        (NTIS #N73-17562).

"A prototype ultrasonic instrument has been designed and developed for quantitative testing. The complete delivered instrument consists of a special pulser/receiver which plugs into a standard oscilloscope, a special rf power amplifier, a standard decade oscillator and a set of broadband transducers for typical use at 1, 2, 5 and 10 MHz. The system provides for its own calibration, and on the oscilloscope, presents a quantitative (digital) indication of time base and sensitivity scale factors and some measurement data. Performance includes a velocimetry capability of better than 0.1%." (Author)

Consideration is given to the effects of transducer capacitance, cable length and damping resistance on the wideband performance of the instrument (Panametrics 5051 P/R).

1978-169+

Lin, J. C., Yu, F. T. S. and Tai, A. M., "Ultrasonic Blood Flow Spectral
        Analysis Using Coherent Optics," IEEE Trans. Biomed. Engr. BME-25(3)
        (1978).

"The paper presents a coherent optical technique for displaying and analyzing a blood-flow-generated ultrasonic Doppler spectrum. The system is highly cost-effective and produces spectrograms on-line. Other advantages include a large, continuously variable bandwidth, an instantaneous display of velocity profile, and simultaneous display of temporal spectra. The system makes use of the Fourier transformation property of converging lens and its use of processing time signals. The spectrum obtained from the brachial artery of a normal subject compares favorably with the spectrograms obtained using electronic spectrum analyzers." (Author)

Such a system could be used to monitor flows in pipes at industrial sites.

1974-170

Lautzenheisen, C. E., Greer, A. S., Jr., Jolly, W. D. and Ying, S. P.,
        "Evaluation of Heavy Section Vessels Using Acoustic Techniques,"

Period. Insp. of Pressurized Components, Conf., Proc., London, June 1974, Publ. by Inst. of Mech. Eng., London (1974).

"During the past five years, Southwest Research Institute has maintained a very active research and development program designed to meet the continuing need to improve inspection techniques. The Institute has been especially active in the areas of acoustic emission, acoustic holography, acoustic spectrometry and advanced ultrasonics. Practical results of this research are presented along with some discussion on application of NDT technology to nuclear power systems." (NTIS) 22 refs.

1973-171

Koppelmann, J. and Fay, B., "Quantitative Evaluation of Ultrasonic Backscattering Measurements in Metals," Translation from Acustica, 29, 297-302. (NTIS $UCRL-Trans-10758) (1973).

"Measurements of ultrasonic backscatter have the advantage that sound attenuation measurements can be carried out on a side of a poly-crystalline test specimen which has no plane-parallel surfaces. A formula is obtained for quantitative evaluation of the backscatter measurements and the accuracy of the formula is illustrated by means of some sample measurements. The accuracy obtained in the measurement shows that this method of measurement is comparable with the classical multiple echo method." (NTIS)

1974-172

Jacobs, J. E., "Scanning Ultrasonic Spectrograph for Fluid Analysis," Patent APPL-512622, DHEW-G-7-73 (NTIS #PB-238 130/9ST).

"The patent application describes a method and apparatus for time delay acoustic spectrographic analysis of fluid composition, using frequency modulated ultrasonic vibrations which are transmitted through the fluid from a sending transducer to a receiving transducer. By scanning the frequency of the modulated vibrations at a sufficiently high rate standing waves which give erroneous indications of fluid component concentrations are eliminated. The frequency of the signal at the sending transducer is compared with that at the receiving transducer and a comparing means whose output is related to the frequency difference of the signal at the receiving transducer from the sending transducer measures the number of times this signal shifts 360 degrees during a scanning period. This measurement is directly related to a property of the fluid being measured." (NTIS)

1976-173

Green, R. E., Jr. and Pond, R. B., Sr., "An Ultrasonic Technique for Detection of the Onset of Fatigue Damage," Air Force Off. Sci. Res. (Monitor) (NTIS #AD-A035 578/4ST).

"The primary purpose of this research was to develop an optimum ultrasonic technique for the early detection of fatigue damage. It is concluded that the optimum ultrasonic technique for early detection of fatigue damage is a technique which permits simultaneous measurements of acoustic emission activity and ultrasonic attenuation changes." (NTIS)

1970-174

Green, D. R. and Dixon, N. E., "Thermal and Ultrasonic Nondestructive
    Testing Methods for Carbon-Carbon Composites," Report CONF-700112-1
    Am. Soc. Mech. Eng. Symp., Albuquerque (NTIS #BNWL-SA-3054),
    Battelle.

"Thermal and mechanical properties are important factors in the performance of carbon-carbon composites. Flaws and property differences in composite parts can often be detected using ultrasonic and high-speed thermal methods recently developed at Battelle-Northwest (BNW). A feasibility study demonstrating the application of these methods to carbon-carbon composites was conducted. This study made use of a new thermal image transducer method, an infrared scanning method and a high-power broad-band ultrasonic method using special transducers." (NTIS)

1976-175

Evans, A. G., Tittmann, B. R. and Ahlberg, L. A., "Failure Prediction
    in Ceramics Using Ultrasonics," Report SC5064.1TR (NTIS #AD-A033 004/
    3ST) (1976).

"The requirements for failure prediction in ceramics using ultra-sonics have been examined. These show that the absolute prediction of failure at acceptable stress levels and lifetimes requires high frequencies, in the 50-400 MHz range. The ability to utilize such high frequencies for flaw detection studies is primarily dependent on the attenuation of the material in this frequency range. Attenuation studies performed on a variety of ceramic materials have shown that the coarse grained or porous ceramics are excessively attenuating, whereas the fine grained fully dense ceramics are acceptable. An analysis of attenuation using numerical scattering cross sections and microstructural parameters has demonstrated that the attenuation is entirely predictable from the large extreme of the microstructure. The analysis has also suggested that attenuation measurements may permit the implementation of statistical failure prediction at high levels of confidence, in materials that are not amenable to absolute failure prediction using high frequency ultra-sonics." (Author)

1976-176+

Sancan, S. and Sachse, W., "Determination of the Geometry and Mechanical
    Properties of a Fluid-Filled Cylindrical Bi-Inclusion in an Elastic
    Solid," Proc. 1976 IEEE Ultrason. Symp., 54-57 (1976).

"The geometry and mechanical properties of a fluid bi-inclusion in an elastic solid are determined from analyses of the time records of broad-band ultrasonic pulses scattered by two side-drilled cylindrical cavities in blocks of aluminum. Various cavity sizes, spacings, and inclusion fluids were investigated to determine the geometric and material influences on the sequence of scattered pulses. The density ratio of the inclusion fluid to matrix solid ranged from 4.3×10^{-4} to 0.59 while the wave speed ratio ranged from 0.05 to 0.31. The wave number-inclusion radius product, αa, ranged from 0.7 to 15. It is shown how the sizes, spacing, and orientation of the inclusion and the contained fluids' wave speed can be determined from measurements of the pulse arrival-times, interpreted according to the theory of geometric acoustics. Inclusion-fluid density determinations are made from the Fourier amplitudes of the scattered pulses." (Author) 11 refs.

1976-177+

Chang, F. H., Flynn, P. L., Gordon, D. E. and Bell, J. R., "Principles and Application of Ultrasonic Spectroscopy in NDE of Adhesive Bonds," IEEE Trans. Son. Ultrason. SU-23 (5), 334-338 (1976).

"Ultrasonic spectroscopy utilizes the principle of constructive and destructive wave interferences to obtain information about ultrasound in the frequency domain. For applications in the nondestructive evaluation (NDE) of adhesive-bonded structures, destructive interference of the pulsed sound waves at the boundaries of the adhesive bond layer produces spectroscopic signals characteristic of the bond. The bondline thickness can be determined accurately from the frequency minima in the spectrum. The width of the antiresonance dips at half-power points and the amplitude of the dips are related to the acoustical properties at the interfaces of the adhesive layer. The amplitude ratio of the RF signals reflected from the boundary surfaces of the adhesive layer in the time domain is also affected by the acoustic impedance mismatch at the boundaries. A combination of these measurements in the frequency and time domains characterize the adhesive properties between the adherend and the adhesive. The experimental set-up consists of a wide-band pulser/receiver as an energy source, a highly damped piezoelectric sensor serving as the transmitting and receiving transducer, and a digital computer for waveform digitization and Fourier transform. Single lap-shear adhesive-bond specimens using three types of adhesive systems were used in a correlation study between bond strength and NDE parameters from the ultrasonic spectroscopy method. An empirical bond strength/NDE parameters relationship established from the correlation study has been justified by analytical calculation. The experimental procedures, analytical derivations of the sound wave interactions in the layered media, and the statistically valid correlation of the bond strength with NDE parameters is presented." (Author) 7 refs.

1975-178+

Mercier, N., "Ultrasonic Classification of Metals by Grain Size," Proc. 1975 Ultrasonics International, 64-67 (1975).

"This paper describes work carried out to verify the principle of a new method of ultrasonic nondestructive control of materials. Making use of a set-up equipped with broad-frequency-band electrostatic transducers, this method permits the investigation of the behavior of metallic materials as a function of frequency. For the alloys studied it has been possible to formulate a law of the form $Af + Bf^4$ for the attenuation coefficient in the range 5-15 MHz. In this range, the coefficient B is proportional to the mean grain volume, which thus allows a classification of the specimens according to their grain size." (Author) 6 refs.

1975-179+

Pouliquen, J. and Defebvre, A., "Use of Surface Acoustic Waves for the Detection of Modifications Produced by Deformation Fatigue or Corrosion on Metallic Surfaces," Proc. 1975 Ultrasonics International, 102-106 (1975).

"This paper describes a new method that uses Rayleigh waves in a feedback arrangement to study metallurgical phenomena such as micro-deformation, fatigue or corrosion. The acoustical method is two hundred times more sensitive than resistive strain gauges for measuring deformation by static flexure. It is also possible to detect the initiation and growth of fatigue failure by means of Rayleigh-wave attenuation variation. For corrosion, Rayleigh-wave frequency variations and attenuations variations seem appropriate, but this latter parameter might be characteristic of cold working." (Author) 6 refs.

1975-180+

Lloyd, E. A., "Developments in Ultrasonic Spectroscopy," Proc. 1975 Ultrasonics International, 54-57 (1975).

"Following a brief introduction to ultrasonic spectroscopy and a review of current instrumentation, the paper continues by examining problems in which ultrasonic spectroscopy has an obvious application: gauging, testing of bimetallic strip, adhesive bond testing, and honeycomb structures. The paper concludes with the modelling of pipe section and the means being developed to define the size and shape of discrete targets. Links with fracture mechanics are proposed." (Author) 2 refs.

1977-181+

Tittmann, B. R., Thompson, D. O. and Thompson, R. B., "Standards for Quantitative Nondestructive Examination," Nondestructive Testing Standards - A Review (H. Bergen, editor), ASTM Special Tech. Pub. #624, 295-311 (1977).

"In this report, the subject of ultrasonic standards is reappraised in terms of history, philosophy of calibration, and future needs. In answer to the critical need for a procedure to calibrate ultrasonic systems for quantitative nondestructive examination (NDE), a new calibration standard and procedure is proposed.
    The calibration standard proposed is the far-field sphere (cavity or inclusion embedded in a solid or a ball suspended in a liquid). As a

result of recent work, the sphere now is understood well theoretically and experimentally and can be reproduced and fabricated in a solid, for example, by diffusion bonding techniques. A major advantage of the sphere is that it has no preferential orientation, the transducer alignment is not critical, and it allows multipoint checks on a single standard block.

The backbone of the calibration procedure is an equation which relates the transmitter signal to the received signal in a quantitative way. With the help of this equation, the scattering parameters, namely, the angular dependence, frequency dependence, and amplitude of the differential scattering cross section may be determined from data obtained on a calibration standard. The results may be compared then with the invariant theoretical solution to verify the proper operation of the ultrasonic system. A key feature of the development is the G-factor which is a proposed figure of merit for a transducer. The discussion includes a technique for its simple determination and its use in the calibration procedure." (Author) 26 refs.

1967-182+

Merkulova, V. M., "Accuracy of the Pulse Method for Measuring the Attenuation and Velocity of Ultrasound," Sov. Phys.-Acoustics 12 (4), 411-414 (1967).

"The transmission of a high-frequency ultrasonic pulse through a medium with different frequency dependencies of the attenuation factors is investigated. Equations are given for estimating the errors of the pulse method for measuring the attenuation and velocity of sound." (Author) 4 refs.

1973-183+

Highmore, P. J., "Impedance Matching at Ultrasonic Frequencies Using Thin Transition Layers," Proc. 1973 Ultrasonics International, IPC Science and Technology Press, 112-118 (1973).

"This paper investigates the use of double transition layers for impedance matching at ultrasonic frequencies, based on a theoretical approach analogous to the input impedance analysis of transmission lines. It is shown how a resonator can be made using two thin layers to match semi-infinite media and, more importantly, to match a transducer into a semi-infinite medium. The practical choice of layer materials is considered in some detail together with a discussion of the layer thickness tolerances. Finally, the influence of the matching layers is demonstrated experimentally by observing the transient characteristics of a short pulse." (Author) 14 refs.

1975-184+

Kay, M., Shimmins, J., Manson, G. and England, M. E., "A Computer Interface for Digitizing Ultrasonic Information," Ultrasonics, 18-20 (January 1975).

"An interface is described which links an ultrasonic B-scanner to a
small digital computer. The interface dititizes the image signals from
the scanner to 10 levels. The 10 levels are then converted to a four-bit
binary word, four of which are packed into a 16-bit word. The 16-bit words
are then transferred to the digital computer." (Author) 3 refs.

1977-185+

Heyman, J. S. and Cantrell, J. H., Jr., "Application of An Ultrasonic
    Phase Insensitive Receiver to Material Measurements," Proc. 1977
    IEEE Ultrason. Symp. (1977).

"The theory of a phase insensitive receiver based on acousto-electric
effect is presented along with experimental characteristics of a CdS acoustoelectric converter (AEC). Since the AEC is nearly phase insensitive, it is
ideal for measurements in inhomogeneous materials and/or materials with
irregular flatness and parallelism. Through transmission ultrasonic C-scan
data of phantom flaws demonstrates a significant improvement in flaw
characterization with an AEC over that of a conventional transducer. In
addition, measurements with conventional transducers in anisotropically
stressed metal samples are shown to lead to grossly inaccurate results due
to phase sensitivity. Various other measurements are presented with data
contrasting conventional transducers with an AEC in specific NDE applications."
(Author) 13 refs.

1961-186+

Papadakis, E. P., "Grain-Size Distribution in Metals and Its Influence on
    Ultrasonic Attenuation Measurements," J. Acoust. Soc. Am. 33 (11), 1616-
    1621 (1961).

"A transformation has been derived relating the number of spheres of a
certain radius R per unit volume (the 'volume distribution of spheres') to the
number of circles smaller than a certain radius r per unit area (the 'area
distribution of circles') appearing on a plane cutting through the volume.
The transformation was applied to several hypothetical grain-size distributions for polycrystalline metals to find the resulting hypothetical area
distribution of grain images on photomicrographs. Comparison of the
hypothetical area distributions to experimentally found area distribution
gave the following conditions that the true volume distribution of grains
must meet: (1) It must be finite at R=0, and (2) it must have a nonzero
decreasing tail for large values of R. The common assumption of a single
grain diameter is insufficient to explain the experimental area distribution
of grain images. The functions $N_v(R)=R^n \exp(-kR)$ and $N_v(R)-\exp[-(\ln R/R_e)/2\sigma v]$
were judged plausible for the volume distribution function of grains, and a
correction was computed for the attenuation formulas for Rayleigh scattering
of ultrasonic waves in polycrystalline metals by taking averages of R^6 and R^3
over these functions." (Author) 7 refs.

1977-187+

Rose, J. L., "A 23-Flaw Sorting Study in Ultrasonics and Pattern
    Recognition," Mat. Eval., 87-92, 96 (July 1977).

"An establishment of a two-mode classification scheme as either a
sharp or smooth surface defect, for the pattern recognition algorithms
employed in this study, produced a 92 percent reliability value for predicting the sharp or smooth classification value. Reliability values of
separating the flaws into other useful engineering classes are also outlined in the paper. Results of this study show promise in developing
reliable ultrasonic data acquisition procedures and suitable pattern
recognition algorithms for solving the many flaw classification problems
that exist today." (Modified author abstract) 9 refs.

1974-188+

Mills, G. S., "Time-Domain Analysis of Ultrasonic Pulse Diffraction for
    Defect Characterization," Mat. Eval., 256-258 (December 1974).

"Procedures for preprocessing ultrasonic test data are introduced that
could be applied effectively to many ultrasonic test systems. The
particular problem of treating thickness measurement of thin films is
treated in detail. When pulse superposition occurs in the time domain,
making it impossible to measure layer thickness by measuring echo distance
on a time scale, it is possible to resort to a curve of peak-to-peak amplitude of the reflected pulse from the thin film as a function of the ratio
of layer thickness to input pulse duration.
    Frequency analysis will work effectively as a film-thickness measuring
device provided the characteristics of the ultrasonic transducer being
used are broadband and cover the required frequencies for proper analysis."
(Author) 3 refs.

1977-189+

Loew, M. H., Shankar, R. and Mucciardi, A. N., "Experiments with Echo
    Detection in the Presence of Noise Using the Power Cepstrum and a
    Modification," Proc. 1977 IEEE Int. Conf. of Acous., Speech and Sig.
    Process. (May 1977).

"The power cepstrum of a function is found by computing the power
spectrum of the logarithm of the power spectrum of that function. In
addition, some applications of the cepstrum employ a smoothing (windowing)
function immediately before or after the logarithm operation.
    This paper describes some experiments in echo-time detection and fre-
quency-band power measurements performed on ultrasound waveforms, both by
the standard cepstral method and by the deletion of either or both of the
window and the logarithm from that method. It is shown that echo-time
detection in the presence of noise is achieved more reliably without the
logarithm step for the waveforms of interest. Our experimental findings
compare favorably with theoretical results reported recently by Hassab and
Boucher." (Author) 7 refs.

1975-190

Winter, T. G., Pereira, J. and Bednar, J. B., "On Driving a Transducer to
    Produce Pulses Shorter Than the Natural Period of the Transducer,"
    Ultrasonics, 110-112 (May 1975).

"When an acoustical transducer is given a short electrical impulse,
the pressure wave produced is usually some sort of damped sine wave whose
exact shape is determined by the physical properties of the transducer
and the medium to which it is coupled. In order to produce pressure pulses
shorter than this, it is necessary to drive the transducer with a more
complex waveform. This paper describes how to calculate and produce the
voltage waveform for any desired pressure waveform and shows some
examples of the voltage waveforms necessary to drive a loudspeaker to
produce pulses shorter than the natural resonant period of the loudspeaker." (Author) 1 ref.

1974-191+

Simanski, J. P., Pouliquen, J. and Defebvre, A., "Loading Transducers
    for Nondestructive Testing and Signal Processing by Acoustic Bulk
    Waves," Ultrasonics, 100-105 (May 1974).

"Nondestructive testing and signal processing with acoustic bulk
waves require the use of loaded transducers having a large bandwidth.
The technique of loading quartz or ceramic transducers by backing with a
material having a relatively large acoustic impedance is well known. A
brief review of the methods and results of this classical type of loading
is given. Another method of broadening the bandwidth by backing the
transducer with one or several metallic layers of variable thickness
followed by a semi-infinite medium is then proposed. A brief mention is
also given to the technique of matching the backed transducer to the
propagating medium. Several practical examples illustrating the use of
the methods are given." (Author) 10 refs.

1974-192+

Brown, A. F. and Weight, J. P., "Generation and Reception of Wideband
    Ultrasound," Ultrasonics, 161-167 (July 1974).

"A previous article on ultrasonic spectroscopy described the applications of wideband ultrasonic techniques to materials testing. Here it is
shown how wideband (0.5-20 MHz) transducers for pulse-echo systems can be
produced. Excitation is achieved by specially-shaped electrical pulses
while, for the reception mode, high-gain wideband amplifiers and gates are
available. Frequency analysis of the ultrasonic pulses is carried out by
a novel, inexpensive spectrum analyzer. New Ideas for data processing,
both analogue and digital, are reviewed." (Author) 7 refs.

1974-193+

Curtis, G., "A Broadband Polymeric Foil Transducer," Ultrasonics, 148-154
(July 1974).

"If a 0.001 in (0.025 mm) thick foil of polyethylene teraphthalate is
metallized upon one side and a dc voltage is applied to the metallizing,
then the foil will adhere firmly by electrostatic attraction to any conducting surface. In acoustic emission experiments with such a foil as a
capacitance detector it was observed that by cycling the polarizing
voltage it was possible to increase the sensitivity of the device until it
was only 30 dB below that of a damped pzt transducer. Such an enhanced
response probe has been found to have a frequency response which is flat
from 10 kHz to 5 MHz. It also has the advantages of being lightweight;
needs no coupling agent when applied to conducting substrates; is stable
over long periods; can be made in any shape or array of shapes and will
conform to curved surfaces." (Author) 2 refs.

1973-194+

Canella, G., "The Ultrasonic Field in Water and Steel," Proc. 7th Int.
Conf. on NDT, Session H-07 (Warszawa, 1973).

"The investigations on the trend of the field of ultrasonic pressure
transmitted by a transducer normally used for nondestructive testing, both
in water and in steel, had two objectives: the first was to obtain a
direct comparison of propagation in the two media, and the second was to
study the contribution to attenuation of the reflected ultrasonic signal,
due to beam spread, as a function of the area of the reflector and the
distance of the latter from the transducer." (Author) 9 refs.

1977-195+

Shibayama, K., Matsunaka, T. and Sato, H., "Exact Design of Acoustic
Matching Networks for Ultrasonic Transducers," Proc. 7th Int. Cong.
Acoust., Session K-65 (Madrid, 1977).

"The piezoelectric transducer having a frequency response suitable
for an ultrasonic device has been strongly required. This paper presents
a synthesis of the acoustic matching networks for the transducer."
(Author) 3 refs.

1976-196+

Silk, M. G., "The Determination of Crack Penetration Using Ultrasonic
Surface Waves," Ultrasonics, 290-297 (December 1976).

"A theoretical discussion of the use of ultrasonic surface waves
for detecting and evaluating surface opening defects in materials is
presented. It is suggested that such techniques should not be confined
to laboratory use, being well-suited to testing structures with awkward
geometries. Experimental data have been obtained to determine the
accuracy of the technique." (Author)

A frequency modulation technique for varying the penetration of the surface is examined. 10 refs.

1975-197+

Rassing, J., "Ultrasonic Relaxation Spectrometry," Chemical and Biological Applications of Relaxation Spectrometry, Wyn-Jones (ed.), 1-16, D. Reidel Pub. Co. (Holland).

   "The application of elastic waves of different frequencies (20 KHz-1 GHz) to the investigation of molecules and their interactions is often referred to as ultrasonic spectrometry. The method has been employed mainly as a tool for studying fast physical and chemical reactions and offers access to the determination of time constants ranging from 10^{-5}-10^{-10} sec. for the approach to equilibrium by the reaction in question.
   For a given compound several physical and chemical reactions with different time constants may exist. These reactions may increase the absorption of the ultrasonic energy. Since the reactions have different time constants, relaxation times, one type of a reaction may predominate over a certain frequency range whereas another type may predominate in a different portion of the acoustic spectrum, thus giving rise to the concept of 'ultrasonic relaxation spectrometry.'
   The aim of the experimental technique is to provide measurements of the absorption of ultrasonic energy at different frequencies for a sound wave which travels through the system in which the reaction in question takes place." (Author) 11 refs.

1977-198+

Behravesh, M., "Frequency Dependent Ultrasonic Reflectivity from Plates in a Liquid," Proc. 9th Int. Cong. Acoust., Session K-4 (Madrid, 1977).

   "Recent theoretical investigations of the reflectivity of a bounded ultrasonic beam from a solid plate in a liquid indicate that the non-specular reflected beam effects are a function of the frequency, f, of the incident ultrasound. This suggests that for a given product fd (frequency x plate thickness) one would expect to see different reflected beam profiles depending on the particular values of f and d." (Author) 3 refs.

1974-199+

Rose, J. L. and Meyer, P. A., "Ultrasonic Signal-Processing Concepts for Measuring the Thickness of Thin Layers," Mat. Eval., 249-258 (December 1974).

   "A systematic approach is presented for developing ultrasonic signal-processing procedures that can be used to measure the thickness of thin layers. Such preprocessing procedures as time between echoes, peak-to-peak amplitude analysis, and Fourier Transform techniques are considered in the paper. The work can be extended to such areas of study as the property evaluation of thin planar defects in metals, flaw characterization analysis in an adhesively bonded system, and also applied to the fatigue crack analysis problem, including an evaluation of possible corrosion and sealer properties in the thin layer." (Author) 3 refs.

1977-200+

Alers, G. A., Flynn, P. L. and Buckley, M. J., "Ultrasonic Techniques for Measuring the Strength of Adhesive Bonds," Mat. Eval., 77-84 (April 1977).

"Ultrasonic methods for predicting both the cohesive and the adhesive strength of a metal-to-adhesive bond are investigated by theoretical and experimental techniques. The basic premise on which such a prediction capability might be expected is the argument that measurable changes in the elastic properties of the adhesive or the interface should be associated with changes in the cohesive or adhesive strength. With the aid of a digital computer to make the ultrasonic measurements rapidly and accurately, it is demonstrated that the cohesive strength of a bond can be predicted from quantitative measurements of the velocity of sound and the attenuation in the adhesive layer. It is predicted and experimentally indicated that the lowest frequency at which a standing wave resonance of the entire sandwich structure occurs can be used to predict the quality of adhesion at the adhesive-to-metal interface." (Author) 12 refs.

Ultrasonic spectroscopy is used to measure the resonance frequencies.

1972-201+

Ermolov, I. N., "The Reflection of Ultrasonic Waves from Targets of Simple Geometry," Non-Des. Test., 87-91 (April 1972).

"This paper discusses the problem of estimating the size of flaws using ultrasound. A solution to the problem, for various targets, is suggested that uses Krautkramer's method of similarity for a disc reflector. Examples are worked out for different conditions." (Author) 9 refs.

1977-202+

Sachse, W., "The Scattering of Elastic Pulses and the Nondestructive Evaluation of Materials," Mat. Eval., 83-89, 106 (October 1977).

"This paper is a review of the results of experiments and their interpretation of the scattering of broadband ultrasonic pulses from various smooth obstacles imbedded in an elastic solid. The scattering by circular and elliptical cylinders and multiple inclusions possessing a wide range of acoustic properties is described. It is shown that both arrival time and spectral analysis of the scattered signals contain information regarding the characteristics of the scatterer. The use of either to characterize an obstacle is discussed." (Author) 21 refs.

1977-203+

Gilmore, R. S. and Czerw, G. J., "The Use of Radiation Field Theory to Determine the Size and Shape of Unknown Reflectors by Ultrasonic Spectroscopy," Mat. Eval., 37-45, 56 (January 1977).

"This paper deals with the application of field theory to explain ultrasonic spectroscopy measurements. Calculations similar to those use to describe radar antenna or ultrasonic transducer radiation patterns can be applied to the evaluation of reflectors in metal. Seven reflector shapes are treated specifically—a dipole, an n tupple pole, a continuous line, a rectangle, a square, a ring source, and a circular disk, and it is shown how the calculations can be applied to evaluate reflectors whose geometry may not approximate any of these specific shapes. The results include the effects of the uniformity of both the illuminating wave and the reflector target strength." (Author) 19 refs.

1977-204+

Evans, A. G., Kino, G. S., Khuri-Yakub, P. T. and Tittmann, B. R., "Failure Prediction in Structural Ceramics," Mat. Eval., 85-96 (April 1977).

"The failure prediction requirements and the pertinent accept/ reject criteria for structural ceramics are derived, and the available failure prediction techniques are examined vis-à-vis the failure prediction relations in order to highlight the capabilities and limitations of each technique. The need for additional techniques is thereby demonstrated. The capabilities of the ultrasonic technique are extensively evaluated in order to determine its ability to satisfy the deficiencies in the existing failure prediction repertoire. The prospects are shown to be very encouraging, but the results of several key studies must be awaited before defining the ultimate role of ultrasonic failure prediction techniques." (Author) 35 refs.

1978-205+

Doyle, P. A. and Scala, C. M., "Crack Depth Measurement by Ultrasonics: A Review," Ultrasonics, 164-170 (July 1978).

"A review is given of both bulk and surface wave ultrasonic methods for the measurement of the depth of surface-breaking cracks. Research is considered which relates to techniques for measuring crack depth by studying the scattered pulse amplitude, by using time-of-flight methods, or by carrying out ultrasonic spectroscopic analysis. Measurement of the transit time of bulk waves appears most likely to provide simple and reliable depth measurement in the near future. Further work in the other two areas should also prove valuable. Some suggestions are made of promising directions for future research." (Author) 50 refs.

1977-206+

Childers, D. G., Skinner, D. P. and Kemerait, R. C., "The Cepstrum:
A Guide to Processing," Proc. IEEE 65 (10), 1428-1443 (1977).

"This paper is a pragmatic tutorial review of the cepstrum literature
focusing on data processing. The power, complex, and phase cepstra are
shown to be easily related to one another. Problems associated with
phase unwrapping, linear phase components, spectrum notching, aliasing,
oversampling, and extending the data sequence with zeros are discussed.
The advantages and disadvantages of windowing the sampled data sequence,
the log spectrum, and the complex cepstrum are presented. The influence
of noise upon the data processing procedures is discussed throughout the
paper, but is not thoroughly analyzed. The effects of various forms of
filtering the cepstrum are described. The results obtained by applying
whitening and trend removal techniques to the spectrum prior to the
calculation of the cepstrum are discussed.
We have attempted to synthesize the results, procedures, and
information peculiar to the many fields that are finding cepstrum analysis
useful. In particular we discuss the interpretation and processing of
data in such areas as speech, seismology, and hydroacoustics. But we
must caution the reader that the paper is heavily influenced by our own
experiences; specific procedures that have been found useful in one
field should not be considered as totally general to other fields.
It is hoped that this review will be of value to those familiar
with the field and reduce the time required for those wishing to become
so." (Author) 86 refs.

1974-207+

Pao, Y. H. and Sachse, W., "Interpretation of Time Records and Power
Spectra of Scattered Ultrasonic Pulses in Solids," J. Acoust. Soc.
Am. 54, 1478-1486 (1974).

"Experiments were conducted to investigate the scattering of a
wide bandwidth pulse by a fluid-filled cylindrical cavity in an elastic
solid. The pulse amplitude-time record and the corresponding Fourier
power spectra of the scattered pulses are analyzed. The sequence of
pulses received by a transducer is identified by applying the theory of
geometric acoustics and the pulse arrival times are related to the cavity
diameter and the wave speed in the fluid. The power spectrum is interpreted in terms of the resonance of the fluid inclusion, the natural
frequencies of which also depend on the size and property of the
inclusion. Investigations show that both pulse-time record and power
spectrum can be utilized to detect the size and the wave speed of a
fluid inclusion in a solid." (Author) 14 refs.

1959-208

Krautkrämer, J., "Fehlergrössenermittlung mit Utraschall," Arch.
Eisenhuettenwessen 30, 693-703 (1959).

"Defect Sizing Investigation with Ultrasound."

The author presents an amplitude-based sizing method for flaws and proposes interrogation of the defect with wideband ultrasound as a possible alternative.

1973-209+

Dory, J., "Les Possibilités d'Application de L'Analyse Fréquentielle au Contrôle Non-Destructif par Ultrasons," Proc. 7th Int. Conf. NDT, Session C-24 (Warszawa, 1973).

"The Possible Application of Frequency Analysis to Non-destructive Testing by Ultrasonics."

1974-210+

Thompson, D. O., "Requirements for and Advances in Nondestructive Evaluation," Proc. 1974 IEEE Ultrason. Symp., 642-652 (1974).

"In this paper a limited review is given of current problems in nondestructive evaluation and approaches that are being taken in some of these problem areas. It is pointed out that a serious need exists to 'bridge the gap' between NDE and the materials and engineering research communities. Because of various changes in structural design philosophy that have been introduced in recent years, new NDE techniques must be much more quantitative in nature than previously required. Several new approaches which exemplify these requirements in ultra-sonics, acoustic emission, and adhesive bond strength measurements are discussed." (Author) 25 refs.

1974-211+

Szabo, T. L., "Nondestructive Subsurface Gradient Determination," Proc. 1974 IEEE Ultrason. Symp., 565-567 (1974).

"Surface acoustic waves can be used to nondestructively probe subsurface gradients (caused by physical processes) by changing their penetration depth with frequency. A perturbation theory has been developed that describes th e influence of a known gradient on the velocity dispersion. The experimental situation requires solution of the inverse problem: obtaining the profile from measured velocity data. In this paper the inverse problem is solved exactly, and the gradient is related simply to the dispersion through Laplace transforms. This approach allows rapid determination of either the gradient or the dispersion, once the other is known. Several gradients are discussed and excellent agreement is obtained with published experimental dispersion data." (Author) 8 refs.

1974-212+

Chen, J. N. C., Papadakis, E. P., Carnevale, E. H. and Carey, C. A.,
    "High Temperature Attenuation and Modulus Measurements," Proc. 1974
    Ultrason. Symp., 530-533 (1974).

"An ultrasonic traveling wave method is reported for studying
attenuation in metals approaching their melting point. Broadband
torsional and extensional waves centered on 80 kHz and 120 kHz are
generated in thin wires. Simultaneous measurements of torsional and
extensional velocity are used to calculate modulus vs. temperature.
The attenuation and modulus are presented for temperatures from 25°C to
900°C. Fourier transforms of broadband pulses are used to determine the
attenuation as a function of frequency at fixed temperature. Results
principally in copper and silver are reported. High temperature
attenuation peaks found in the present work correlate well with the
(low frequency) low temperature grain boundary peaks from earlier
independent torsional pendulum work." (Author) 10 refs.

1968-213+

Davey, C. N., "The Ultrasonic Interference Micrometer," Ultrasonics,
    103-106 (April 1968).

"In nuclear reactors, thickness of the walls of tubes are measured
by two ultrasonic instruments: the continuous wave micrometer and the
pulsed wave micrometer. The principles and recording techniques of the
instruments are described. The continuous wave micrometer is used for
simultaneous determination of thicknesses during defect-testing operations;
while the pulsed wave instrument is preferred when better surface discrimination is needed." (Author) 4 refs.

1961-214+

Redwood, M., "Piezoelectric Generation of an Electrical Impulse,"
    J. Acoust. Soc. Am. 33 (10), 1386-1390.

"The following problem is examined theoretically: A pressure is
applied to an open-circuited bar of piezoelectric ceramic; after steady-
state conditions have been achieved, an electrical resistance R is
connected between the electrodes and simultaneously the pressure is
removed. General equations describing the subsequent current and energy
dissipation in the resistor are developed, and examined in detail for a
specific transducer of lead zirconate-titanate ceramic. The general
equation for current consists of a series of time-delayed functions,
signifying physically that a mechanical wave propagates through the
transducer with successive reflections at the end faces, releasing
strain energy originally stored in the material; this energy is then
dissipated in electrical form in the resistance. The initial current
has two components, an exponential function representing the discharge
of energy stored in electrical form in the transducer capacitance C_0
(about 10% of the total stored energy), and a step function representing
the release of energy stored in mechanical form (remaining 90%). If R

is small the exponential function predominates, and has a time constant which is approximately RC_0; the stored electrical energy is released very rapidly, while the mechanical energy is released much more slowly. If R is large (of the order of 10^4 ohms), exponential function and step function are of comparable magnitudes, and the time constant of the exponential is modified by the mechanical impedance of the transducer." (Author) 3 refs.

1977-215

Scott, W. R., "Ultrasonic Reflection Spectrum Analysis for NDT of Periodic Layered Media," NADC Report #77093-30 (may also appear in Ultrasonics).

No abstract.

1976-216+

Johnson, D. M., "Model for Predicting the Reflection of Ultrasonic Pulses from a Body of Known Shape," J. Acoust. Soc. Am. 59 (6), 1319-1323.

"An approximate model for determining the response functions of reflectors of known shape has been used to provide theoretical support for experimental investigations of flaw characterization. The reflection of broad-band ultrasonic pulses from an object can result in modifications to both the frequency spectrum and phase information in the pulses. These changes are largely controlled by the parameters of the reflector, and relating these factors has been the subject of many experimental investigations. Within the fundamental approximations, the model can readily be applied to any reflector of known shape and provides a simple interpretation of what is physically occurring. It has the advantage that additional experimental factors can be included to determine their effect on the results." (Author) 11 refs.

1975-217+

Greguss, P., "Methods and Apparatus for Analyzing Scattered Ultrasonic Fields," Ultrasonics International 1975, IPC Science and Technology Press (London).

"It was recently reported that, acoustically, human soft tissues behave as semi-ordered three-dimensional matrices in which the matrix spacing and arrangement are characteristic of the particular tissue. This phenomenon, however, constitutes an acoustic analogue to light scattering in crystalline-like polymer films and so the Debye-Bucche theory could be applied, naturally, with a slight modification, if the optical replica of the scattered ultrasonic field could be displayed. The purpose of this presentation is to show whether there is any change in the coherence of the ultrasonic radiation scattered in tissues, and whether the scattering has any effect on the beam diameter of the interrogating ultrasonic radiation. The concept of a parameter of relative importance of scattering and free space diffraction is introduced. The possibility of using holographic Fourier spectrometry to investigate the spectral intensity distribution of the scattered ultrasonic field will be discussed." (Author) 23 refs.

1978-218+

Couchman, J. C., Chang, F. H., Yee, B. G. and Bell, J. R., "Resonance
     Splitting in Ultrasonic Spectroscopy," IEEE Trans. Son. Ultrason.,
     SU-25 (5), 1978.

     "Ultrasonic spectroscopy is currently being developed as a method
for determining physical properties and dimensions of discontinuities in
materials. These determinations are made from the Fourier spectra of
transmitted or reflected broad-band ultrasound and are a manifestation
of spectral energy addition or removal when normal modes of vibrations
in media cause constructive or destructive interference. It has been
observed experimentally in laminates that when different layers in a
test specimen resonate at the same frequency, there is a curious
splitting of resonances and sometimes triplet resonances in Fourier
transform spectra where only one resonance might be expected. A
theoretical basis for the resonance splitting is established by a
classical acoustics treatment of the laminate transmission problem.
Triple peaking is affirmed and parametrically investigated. Theoretical
analysis, and parametric evaluation of resonance splitting are presented
in this paper. Experimental verification of the resonance triplet
effect is shown. The potential usefulness of resonance splitting in
process control is discussed." (Author) 2 refs.

1977-219+

Miki, T., Yamaguchi, H. and Nagaki, Y., "An Accurate Wide-Band Automatic
     Waveform Analyzer," IEEE Trans. Instrum.,Msmt., IM-26 (4).

     "This paper presents accurate and quantitative measurements of wave-
form parameters and eye pattern which is important as an evaluation
parameter of digital transmission characteristics in the development of
wide-band automatic waveform analyzer from dc to 10 GHz. The measurements
and error of transfer characteristics by using FFT is also described.
New techniques are adopted in averaging, calibrating, and compensating
the time axis for the accurate measurements. Relevant software offers a
method for quantitative measurements. The methods have made it possible
to measure the waveform parameters which are about ten times more accurate
than can be obtained through conventional oscilloscopes.
     In the quantitative evaluation of eye pattern, the measuring error
has been improved from 10 down to 2 or 3 percent as compared with
measurements by sight using a conventional oscilloscope.
     An equation which gives the approximate value of the error in the
measurement of transfer function in the frequency domain from the time-
domain pulse response is shown and confirmed to correspond with the actual
case." (Author) 6 refs.

1977-220+

Jerri, A. J., "The Shannon Sampling Theorem—Its Various Extensions and
     Applications: A Tutorial Review," Proc. IEEE, 65 (11).

     "It has been almost thirty years since Shannon introduced the sampling
theorem to communications theory. In this review paper we will attempt to

present the various contributions made for the sampling theorems with the necessary mathematical details to make it self-contained.

We will begin by a clear statement of Shannon's sampling theorem followed by its applied interpretation for time-invariant systems. Then we will review its origin as Whittaker's interpolation series. The extensions will include sampling for functions of more than one variable, random processes, nonuniform sampling, nonband-limited functions, implicit sampling, generalized functions (distributions), sampling with the function and its derivatives as suggested by Shannon in his original paper, and sampling for general integral transforms. Also the conditions on the functions to be sampled will be summarized. The error analysis of the various sampling expansions, including specific error bounds for the truncation, aliasing, jitter and parts of various other errors will be discussed and summarized. This paper will be concluded by searching the different recent applications of the sampling theorems in other fields, besides communications theory. These include optics, crystallography, time-varying systems, boundary value problems, spline approximation, special functions, and the Fourier and other discrete transforms." (Author) 248 refs.

1977-221+

Pao, Y. H. and Sachse, W., "Ultrasonic Phase Spectroscopy and the Dispersion of Elastic Waves in Solids," Proc. 9th Int. Cong. Acoust., Madrid, 1977, Session N-26.

"We show how the dispersion relation and the phase and group velocites of elastic waves in wave guides or solids with microstructures can be determined from the phase spectra of spectral-analyzed broadband pulses." (Author) 2 refs.

1977-222+

Loew, M. H., Shankar, R. and Mucciardi, A. N., "Experiments with Echo Detection in the Presence of Noise Using the Power Cepstrum and a Modification," Proc. 1977 Int. Conf. Acoust. Speech & Sig. Processing, May 1977.

"The power cepstrum of a function is found by computing the power spectrum of the logarithm of the power spectrum of that function. In addition, some applications of the cepstrum employ a smoothing (windowing) function immediately before or after the logarithm operation.

"This paper describes some experiments in echo-time detection and frequency-band power measurements performed on ultrasound waveforms, both by the standard cepstral method and by the deletion of either or both of the window and the logarithm from that method. It is shown that echo-time detection in the presence of noise is achieved more reliably without the logarithm step for the waveforms of interest. Our experimental findings compare favorably with theoretical results reported recently by Hassab and Boucher." (Author) 7 refs.

1975-223+

Lakestani, F., Baboux, J. C. and Fleischmann, P., "Broadening the
    Bandwidth of Piezoelectric Transducers by Means of Transmission
    Lines," Ultrasonics 13 (4), 176-180.

    "The transfer functions of a transducer at emission and reception
of ultrasonic waves are recalled and then verified. In order to broaden
its bandwidth, an original method based on transmission line properties
is described." (Author) 6 refs.

1958-224+

Lutsch, A., "Ultrasonic Reflectoscope with an Indicator of the Degree
    of Coupling Between Transducer and Object," J. Acoust. Soc. Am. 30
    (6), 1958.

    "A further development of Firestone's Reflectoscope is described,
indicating the degree of coupling between transducer and the test object.
It is shown that the distortion of a radio-frequency pulse—produced by
the transducer—is a function of its coupling to the test object.
Therefore, the distortion can be used to measure the coupling. The
trailing edge of the radio-frequency pulse is amplified by a time-
selective amplifier, rectified and connected to a galvanometer. Different
conditions of coupling are analyzed. Figures are given for a barium
titanate transducer working at 500 kcps. References such as back
boundaries are no longer necessary in the field of nondestructive testing."
(Author) 6 refs.

1973-225+

Obarz, J. and Svuss, B., "A New Method and Equipment for Measuring the
    Ultrasonic Attenuation Caused by Scattering," Proc. 7th Int. Conf.
    on NDT, Warsaw, June, 1973, Session J-03.

    "Attenuation of ultrasonic waves in polycrystalline materials is
caused in part by thermal losses - the so-called attenuation by absorption,
in part by scattering on the structure of grains. Which of the two kinds
of attenuation will predominate and what form the scattering attenuation -
frequency relationship will assume, depends on the ratio between the
dimensions of the average grain size $\overline{D}$ and the wave length λ. The
attenuation by absorption always grows linearly with frequency. As the
attenuation due to scattering depends not only on frequency but also on
the elastic constants, their anisotropy, the density and the size of
grains, the coefficient of this attenuation makes it possible to form
opinion about the examined structure." (Author) 3 refs.

1977-226+

Yanagida, M. and Kakusko, O., "Logarithmic-Frequency-Scale Representation
    of Discrete Frequency Spectrum of a Band-Pass Signal," Proc. 9th Int.
    Cong. on Acoust., Madrid, 1977, Session R-18.

"Here presented is an efficient approximation method to obtain the logarithmic-frequency-scale representation of discrete frequency spectrum of a band-pass signal using the double sampling technique (1) and an approximate interpolation formula with the sampling function." (Author) 1 ref.

1974-227+

Posakony, G. J., "Challenges for Electrical Engineering in Ultrasonic Nondestructive Testing," IEEE Trans. Son. Ultrason., SU-21(4), 305-311 (1974).

"Nondestructive testing remains a most challenging field for research and development requiring state-of-the-art electrical engineering technologies. With the rapid growth in material sciences, fundamental advancements are needed to provide a better means for characterizing material defects to give metallurgists the detailed information necessary to assess the failure pottential of known anomalies. Ultrasonic technologies have the greatest potential for providing advanced means of describing the size, population, geometry and location of material anomalies. Current ultrasonic test methods must be advanced. The solutions involve interdisciplinary efforts between individuals involved in physical acoustics, electrical engineering, metallurgy, fracture mechanics, and structural design." (Author)

1975-228+

Preston, K., "Use of Pattern Recognition for Signal Processing in Ultrasonic Histopathology," Proc. Sem. Ultrason. Tissue Char., NBS Special Pub. 453, 51-59.

"New developments in high-speed electronics have made possible real-time digitization of the ultrasonic A-scan as a block of binary words. This makes computer processing of such blocks feasible using techniques previously developed in time-series analysis for communications, speech processing, word recognition, etc. The research reported here is concentrating on: (1) advanced digitization techniques having variable block length and skip-block capability; (2) application of speech spectrogram representation and analysis to A-scan processing; and (3) use of human audition (of the recorded A-scan, properly frequency-translated) to assist in defining the features most useful for A-scan pattern recognition." (Author) 13 refs.

1975-229

LeCroissette, D. H. and Heyser, R. C., "Attenuation and Velocity Measurements in Tissue," Proc. Sem. Ultrason. Tissue Char. NBS Special Publication 453.

"A practical ultrasound system is described which is capable of making measurements of attenuation and velocity in tissue as a function of frequency. This method is based upon a technique known as Time Delay Spectrometry which employs a swept frequency output signal and has the ability to provide anechoic ultrasonic measurements. A system operating between 2 and 3 MHz has already been shown to be capable of producing images in soft tissue with a resolution of less than 2 mm. Preliminary measurements on excised tissue using this system have indicated a frequency dependence of attenuation in pathological tissue that is substantially different from that of normal tissue. The method generates the time domain response simultaneously with the received frequency sweep. It is shown that the time response in the apparatus is available as a displayed phasor quantity and that the arrival time of the directly received signal can therefore be measured to within a few nanoseconds." (Author) 12 refs.

1978-230+

Foster, F. S. and Hunt, J. W., "The Design and Characterization of Short Ultrasound Transducers," Ultrasonics, 116-122.

"The construction and testing of short pulse (bipolar and unipolar) ultrasound transducers are discussed and several examples are analyzed. Equations predicting the radiation field of the short pulse transducer are derived using a new approach. Agreement was obtained between the theoretical predictions and the experimentally measured field for one short pulse transducer. Some discrepancy was noted between the predicted pulse shapes and those observed experimentally." (Author) 21 refs.

1971-231+

Cross, B. T., "Sound Beam Directivity: A Frequency-Dependent Variable," Non-Destr. Test. (London), 119-125.

"This report is part of a study of high frequency sound beam propagation and the information concerns the frequency-dependence observed in ultrasonic test results. The author has mathematically predicted and empirically verified that ultrasonic response curves from area-amplitude references are influenced by frequency. The frequency influence is identified as the beam directivity associated with the d/λ ratio. Schlieren photography provides supporting data by showing the reflected sound beam patterns for various reflector sizes. These findings are significant accurate prediction and may be made for a given ultrasonic inspection by considering the frequency range of its operation. Also insight is given into the reasons why two different instruments can yield noncorrelating test data at the same apparent frequency." (Author) 6 refs.

1978-232+

Nabel, E. and Mundry, E., "Evaluation of Echoes in Ultrasonic Testing by Deconvolution," Mat. Eval., 59-61, 77, Jan., 1978.

"This paper is based on what has turned out to be a good tool for
ultrasonic echo analysis, namely, linear system theory, and will report
some effects of fundamental importance for ultrasonic spectroscopy and
deconvolution. In many cases the importance of these effects is not
recognized, although they are responsible for most of the differences
between experimental results and theory. The paper deals with the
improvements in the analysis of echo indications by the calculation of the
impulse response of a reflector (deconvolution of ultrasonic echoes).
The importance of phase information is emphasized, and the influence of
the low frequencies (i.e., the low diffraction orders) on the experimental
results is shown. As an example, the result of a simulated convolution
will be demonstrated with respect to the theory of replica pulses published
by Freedman." (Author) 11 refs.

1977-233

Tamburelli, C., "Use of Ultrasound in Assessing the Susceptibility of
    Steel to Lamellar Tearing," NDT International, 3-8, Feb., 1977.

"Increasingly, developments in welded constructions means that steel
plate with high resistance to lamellar tearing has to be used. An experimental programme has been carried out on 17 C/Mn Fe52D steels, to indicate
what features of longitudinal wave transmission ultrasonics may be used
in assessing susceptibility to lamellar tearing. A combination of two
kinds of evaluation proved promising: the contribution of matrix toughness,
measured from the anisotropy of ultrasonic velocity; and the contribution
of the amount of inclusions, assessed by ultrasonic attenuation measurements and inclusion counting by ultrasonic reflection. No useful results
have been obtained from ultrasonic spectroscopy." (Author) 7 refs.

1978-234+

Budiansky, B. and Rice, J. R., "On the Estimation of a Crack Fracture
    Parameter by Long-Wavelength Scattering," J. App. Mech. 45(2).

"Attention is focussed herein on the possibility of estimating the
fracture-mechanics parameter $k_I = (K_I)_{max}/\sigma$ associated with a flat crack
of initially unknown dimensions and orientation by using long-wavelength
NDE measurements. Here K_I is the mode I stress-intensity factor
associated with tension σ normal to the plane of the crack, and "max"
denotes the largest value along the crack perimeter. The estimates will
be made on the basis of the long wavelength studies by Gubernatis, et al.
[1][3], and certain properties of elliptic cracks that are nearly shape
invariant." (Author) 2 refs.

1975-235+

Canella, G., "The Ultrasonic Field in Water and Steel," NDT (London),
    38-42 (Feb., 1975).

"The investigations of the ultrasonic field of transducers used in
ndt, had two objects. The first was to compare directly propagation in
steel and water. The second was to study the contribution of the beam

spread to the attenuation of the echo. The beam spread is considered as a function of the area of the reflector and its distance from the transducer." (Author) 9 refs.

1977-236+

Baboux, J. C., Lakestani, F., Fleischmann, P., and Perdrix, M., "Calibration of Ultrasonic Transmitters," NDT International.1977, 135-138 (June, 1977).

"A simple experimental method is described which enables the absolute measurement of the acoustic pressure transmitted by a transducer. After a theoretical study of the method used, some experimental results on industrial transducers are given." (Author) 2 refs.

1974-237+

Highmore, P. J., "Nondestructive Testing of Bonded Joints—The Depth Location of Non-Bonds in Multi-Layered Media," NDT (London), 327-330, Dec., 1974.

"In some types of multi-layered structure it is important to detect not only the presence of non-bonds but also to determine their depth locations. This paper describes a simple ndt method, designed especially to meet this requirement, based on the well-known ultrasonic resonance technique." (Author) 1 ref.

1977-238+

Thompson, L. A., "Method of Response Equalization for a Piezoelectric Transducer," J. Acoust. Soc. Am. 62(6).

"A method of using the skirt of a low-pass filter to flatten the projecting response of a piezoelectric transducer is described. An example is given in which the 30 dB response variation of an F-27 standard transducer over the 20-80-kHz frequency range is reduced to 9 dB." (Author) 1 ref.

1972-239+

Myers, G. H., Thumin, A., Feldman, S., deSantis, G. and Lupo, F. J., "A Miniature Pulser-Preamplifier for Ultrasonic Transducers," Ultrasonics, 87-89 (March, 1972).

"In order to avoid difficulties experienced with conventional ultrasonic systems using several feet of shielded cable to connect transducer and electronics, a miniature pulser-preamplifier was developed to fit the transducer holder. Its characteristics proved superior to most devices in common use." (Author) 3 refs.

1977-240+

Hill, J. J., "A Simple Digital Pulse-Shaping Circuit," Proc. IEEE., 1517
     (1977)

     "A method of generating pulses of prescribed shape is described
where binary-coded samples of the pulse are stored in a READ-ONLY memory.
     The proposed circuit is used to generate pulses having a Gaussian
shape." (Author) 2 refs.

1976-241

Cousins, R. R., "The Mathematical Theory for Ultrasonic Spectroscopy,"
     Seminar on Ultrasonic Spectroscopy, City University (London),
     27 October 1976.

     "A viscoelastic model is proposed to describe the loss of signal
of an ultrasonic pulse due to both viscoelastic dissipation and scattering
from defects. The material specimen is divided into a number of
theoretical layers normal to the direction of travel of pulse, and an
analysis of the frequency spectrum of the reflected pulse enables the
properties of successive layers to be determined. In the case of
laminated material the amplitude of the spectrum is sufficient to
determine the depth of delaminated regions, and an experimental example
is given." (Author)

1976-242

Adler, L., "Scattering of Broad-Band Ultrasound from Geometrical Shapes
     Embedded in Metal," Seminar on Ultrasonic Spectroscopy, City
     University (London), 27 October 1976.

     "In order to develop realistic models for flaw characterization in
NDE scattering of broadband ultrasonic pulses from various shaped
cavities embedded in diffusion-bonded titanium was measured. The frequency
and angular distribution of the scattered energy were analyzed and compared
with two existing theories: (1) Keller's geometrical theory of diffraction
for the region $ka \geq 1$, and (2) Born's approximation for the region
$ka \leq 1$." (Author)

1976-243

Haines, N. F., "Deconvolution in Ultrasonic Spectroscopy," Seminar on
     Ultrasonic Spectroscopy, City University (London), 27 October 1976.

     "Ultrasonic spectroscopy techniques have been used in CEGB Nuclear
reactors since 1974 to make measurements of corrosion layers on
inaccessible steel surfaces. As from 1977 two of the boards regions will
have their own spectroscopy systems with personnel trained to make these
measurements on an operational basis.
     Research work at Berkeley Nuclear Laboratories is continuing into
possible other areas of application. During the past 18 months theoretical

work and more recently experimental work has demonstrated where spectroscopy techniques may be of use in understanding the reflection of pulses from real defects. The theoretical model developed can predict the reflected waveform and hence peak amplitude as a function of size, shape, orientation and surface roughness of a reflector.

The underlying theme of the talk will be the relationship between the time domain and frequency domain. In some cases one domain may give greater physical insight into the system being considered than the other. In particular the reflection from surfaces are perhaps better understood in the time domain and hence the mathematical techniques of convolution and deconvolution become important." (Author)

1976-244

Markham, M. F., "Polymeric and Composite Materials," Seminar on
    Ultrasonic Spectroscopy, City University (London), 27 October 1976.

"An ultrasonic pulse transmission technique using a spectrum analyzer is described for studying secondary relaxation in polymers. Application to epoxide resins show that the technique yields the dynamic mechanical properties over a wide ultrasonic frequency band. A γ relaxation is located, and its behaviour under different cure conditions is investigated. A long term aging effect is also noticed, the study of which could lead to useful information regarding the molecular processes involved at various stages of cure." (Author)

1976-245

Lloyd, E. A., "Predictive Modeling of Ultrasonic Responses," Seminar
    on Ultrasonic Spectroscopy, City University (London), 27 October 1976.

"In testing of structures fabricated from welded plate sections it is required to distinguish between signals returned from a weld bead of uncertain geometry and an equally variable defect echo. In order to understand better the factors governing the visibility of defects in this situation, a computer based modelling technique has been developed with which it is possible to vary the skip distance, probe angle, aperture and frequency response of a synthetic transducer.

Whereas in the time-domain, the visibility of a defect will depend largely on the relative size and position of the defect response to the signal generated by the associated weld bead, no such obvious restriction need apply when the dual signal is examined in the frequency domain. There, the choice exists for examining the relative contributions of the defect and weld-bead over the whole or any part of the spectral "window" available for testing. It is for this reason that facilities have been built into the model so that both time-domain and frequency domain representations of a weld-bead associated defect signal can be synthesized. Results will be discussed.

As a further test of the validity of the model, a short pulse shear wave transducer was used to interrogate slots of various depths milled over a flat plate. The features predicted are readily discernible." (Author)

1976-246

Quentin, G., "Progress Towards an Ultrasonic Characterization of Rough
     Surfaces," Seminar on Ultrasonic Spectroscopy, City University
     (London), 27 October 1976.

"New non-contact methods are described which lead to quite precise
estimates of the r.m.s. roughness k of randomly rough surfaces as well
as to a separation of periodic surface defects from random roughness.
For k < 25 µm accuracy is of the order of ± 1µm while for very rough
surfaces with h > 50µm accuracy is ± 3µm. Some process has also been
made in the measurement of the autocorrelation length." (Author)

1976-247

Nabel, E., "The Importance of Phase for Ultrasonic Spectroscopy and
     Deconvolution," Seminar on Ultrasonic Spectroscopy, City University
     (London), 27 October 1976.

"Much work has been done during recent years in the field of
ultrasonic spectroscopy. Due to the fact that in most cases only swept
frequency receivers were used for spectrum analysis which do not yield
any phase information, phase spectrum was often regarded as being of no
value.
     During the research work described in this paper spectrum analysis
by means of a swept frequency receiver was replaced by calculating the
complex spectrum that means amplitude and phase spectrum with a minicomputer.
     It became evident that the phase spectrum cannot be regarded
separately from the amplitude spectrum. Amplitude and phase spectrum
must be considered to be one unique set of data. Only that allows
correct deconvolution, and makes it possible to calculate the impulse
response of a reflector by inverse Fourier transform.
     The paper will describe some model experiments in an immersion
tank and on a steel specimen, and that it is possible to describe the
used reflectors easily by using a deconvolution technique and calculating
the impulse response.
     The effects caused by neglecting phase information will be
demonstrated with simulated and natural data." (Author)

1976-248

Lepper, R. D., Decker, D., Trier, H. G., and Reuter, R., "Attempts to
     Characterize Tissues Using Spectroscopy," Seminar on Ultrasonic
     Spectroscopy, City University (London), 27 October 1976.

"The paper deals mainly with the practical problems of implementing
ultrasonic spectroscopy in the field of tissue differentiation.
     Firstly, the limitations of the authors' experimental set-up of
1970 - 74 are discussed. A new set-up now gives considerable improvements, notably on-line digitization and less RF noise.
     Results are given of in-vitro experiments which reveal the
difficulties of using the transfer function of tissue as well as experi-

mental difficulties with highly damped transducers. A calibration method is proposed to overcome these difficulties." (Author)

1976-249

Gore, J. C., and Leeman, S., "Cardiac Tissue Characterization by Ultrasonic Spectroscopy," Seminar on Ultrasonic Spectroscopy, City University (London), 27 October 1976.

"An analysis of ultrasound scattering from tissues shows that the spectral content of the backscattered sound contains useful information about tissue density fluctuations. Such an analysis can be used to characterize tissues but it is shown that the information is limited by the nature of the interrogating sound pulse.
When the echoes from heart wall are analyzed it is demonstrated that their spectra reveal information about the contractile state of the cardiac muscle. The extension of this method to the recognition of myocardial disease is presently being attempted and the utility of such information is discussed." (Author)

1976-250

Nicholas, D. and Hill, C. R., "A Spectral Evaluation of Ultrasonic Backscattering from Human Tissues," Seminar on Ultrasonic Spectroscopy, City University (London), 27 October 1976.

"A number of human soft tissues exhibit structural features, and corresponding patterns of acoustic impedance variation, with characteristic spacings of the order of a millimeter. Bulk scattering from such tissues in the megahertz frequency region is thus diffractive in nature, exhibiting marked orientation dependence and a complex dependence on frequency. A series of investigations that have been carried out on these phenomena will be reported and their potential interest for in vivo tissue characterization will be discussed." (Author)

1969-251+

Truell, R., Elbaum, C., and Chick, B. B., Ultrasonic Methods in Solid State Physics, Academic Press, New York (1969), 464 pp, 335 refs.

This book contains chapters dealing with (1) propagation of stress waves in solids, (2) measurement of attenuation and velocity by pulse methods and (3) causes of losses and associated velocity changes. A section is included on the "specific application of the spectrum analyzer: factors affecting the spectrum." The appendices contain diagrams of automated velocity and attenuation measurement systems.

1978-252

Davis, M. C., "Coal Slurry Diagnostics by Ultrasound Transmission," J. Acoust. Soc. Am. 64 (2), 406-410.

"Several application of ultrasonic detectors are suggested for monitoring coal water slurries in coal conversion processes. These include mass flow, particle size, and temperature. Modeling of transmission losses include viscous and thermal transport processes as well as multiscattering effects. Simple monitoring of sound attenuation versus frequency yields a unique dependence from which the value of characteristic parameters may be deduced, all from a single transmitter-receiver pair." (Author) 8 refs.

1978-253

Ting, C. S. and Sachse, W., "Measurement of Ultrasonic Dispersion by Phase Comparison of Continuous Harmonic Waves," J. Acoust. Soc. Am. 64 (2), 852-857.

"The method of phase comparison of continuous waves is applied to determine the dispersion relation, phase, and group velocities as a function of dispersive materials. A combination of the variable frequency method and the variable path-length method is found necessary to eliminate any uncertainty in the dispersion relation determination. Experiments are performed on specimens of various thickness. A constraint equation can be derived since the dispersion relation is unique and independent of the specimen thickness. This equation provides a procedure for determining the absolute number of wavelengths in the specimen. Measurements in unidirectional, fiber-reinforced boron-epoxy specimens show good agreement with results reported previously." (Author) 14 refs.

1978-254

Bray, D. E., Egle, D. M. and Reiter, L., "Rayleigh Wave Dispersion in the Cold-Worked Layer of Used Railroad Rail," J. Acoust. Soc. Am. 64 (2), 845-851.

"The first shear mode (Sezawa mode) and the fundamental Rayleigh mode have been identified as they propagated through the work-hardened layer on the top surface of used, steel railroad rail. Longitudinal wave velocities and densities are very nearly equal for the work-hardened layer and the subadjacent layer. The ratio of shear wave velocity in the upper layer to that in the subadjacent layer is near to 0.95. Experimental data obtained at several frequencies (0.5-2.0 MHz) showed good agreement with expected velocities for a layer thickness ranging from 3 to 5 mm." (Author) 13 refs.

1970-255

Meyer, H. J., "Inspection of Grey Iron Castings by Ultrasonic Attenuation," Non-Destruc. Test. (London), 99-104, April 1970.

"Ultrasonic pulses of definite frequency and wavelength undergo a varying degree of scatter depending upon the size and quantity of graphite flakes in grey cast iron. The amount of sound energy left after a sound beam has passed a given cross-section provides, therefore, a measure of the structure and content of the graphite and consequently, the physical strength of the cross-section." (Author) 10 refs.

1973-256+

Chang, F. H., Couchman, J. C. and Yee, B. G. W., "Transmission Frequency
    Spectra of Ultrasonic Waves through Multi-Layer Media," Proc. 1973
    IEEE Ultrasonic Symp.

    "The frequency spectrum of ultrasonic plane waves transmitted through
a multi-layered laminate structure at normal incidence was analyzed to study
the amplitude distribution of the frequency components. In supporting the
theoretical calculations, the wave equation was solved to evaluate the
displacement field with the appropriate boundary conditions in a six-
region laminate. Cavity resonation of the plane waves in the layers
produced peaks in the transmission frequency spectrum. Experiments were
conducted using a pair of broad-band acoustical transducers transmitting
a pulsed ultrasound centered at 5 MHz through multi-layer adhesive-bonded
aluminum plates with different plate thicknesses. Resonant peaks in the
experimental frequency spectra were compared with those theoretically
calculated for regions of good bond and regions of disbond. Applications
of this technique to nondestructive testing of bonded structures are
described." (Authors) 9 refs.

1978-257+

Hill, J. J., "Digital Generation of a Nonlinear Time-Base," IEEE Trans.
    Instrum. Msmt., IM-27 (3), 298-300.

    "A Method of generating nonlinear time-bases is described. The
approximation may be either step or piecewise linear and involves using
a READ-ONLY memory as a look-up table to store digitally the shape of
the required function.
    The case of a logarithmic time-base is considered in detail."
(Author) 4 refs.

1978-258+

Khuri-Yakub, B. T. and Kino, G. S., "A New Technique for Excitation of
    Surface and Shear Acoustic Waves on Nonpiezoelectric Materials,"
    Appl. Phys. Lett., 32 (9), 513-514 (May 1978).

    "An interdigital transducer deposited on a piezoelectric substrate
has been used to excite SAW on nonpiezoelectric materials by using a fluid
couplant. The piezoelectric substrate is held at an angle to the
nonpiezoelectric material so as to match the tangential k vectors of the
surface waves. Experiments have been carried out with a $LiNbO_3$ piezoelectric
substrate and a ceramic such as SiC or Si_3N_4 with a fluid couplant. At a
center frequency of 100 MHz, the estimated conversion efficiency of the
surface wave from the piezoelectric to the nonpiezoelectric material is
-3.5 dB. The results compare favorably with a normal mode coupling theory
we have developed which predicts -2.7 dB efficiency." (Authors) 8 refs.

1976-259+

Lakin, K. M. and Fedotowsky, A., "Characterization of NDE Transducers and
      Scattering Surfaces Using Phase and Amplitude Measurements of Ultra-
      sonic Field Patterns," IEEE Trans. Son. Ultrason., SU-23 (5), 317-322.

"The characterization of transducers for quantitative NDE applications
requires that the radiation pattern, conversion efficiency, and bandwidth
be accurately determined. These quantities may, in principle, be
determined if the transducer's construction and constituent parts are
independently known. However, most often the internal details of the
transducer are unknown and subject to statistical variations and aging.
A measurement technique and system for characterizing transducers based
upon external measurements is described, which does not rely upon knowledge
of the transducer's construction." (Author) 11 refs.

1978-260+

Fraser, J., Khuri-Yakub, B. T. and Kino, G. S., "The Design of Efficient
      Broadband Wedge Transducers," Appl. Phys. Lett. 32 (11), 698-700
      (June 1978).

A simple coupled-mode theory has been developed for acoustic-surface-
wave wedge transducers. Surface-wave transducers have been fabricated
to operate on aluminum using water as the wedge material. The measured
efficiency was 68% at 2.75 MHz, the theoretical value being 81%. Transducers have also been fabricated to operate on glass with a rubbery
solid, RTV 615, as the wedge material. The experimental and theoretical
efficiencies of this transducers at 3.2 MHz were 35 and 50%, respectively.
The surface-wave leakage coefficient of RTV 615 on glass has been measured
and found to be in excellent agreement with theory." (Author) 5 refs.

1966-261

vander Pauw, L. J., "The Planar Transducers - A New Type of Transducer
      for Exciting Longitudinal Acoustic Waves," Appl. Phys. Lett. 9 (3),
      129-131 (August 1966).

"We have constructed a new type of transducer for exciting longitudinal
acoustic waves. It has only one interface, for which reason we shall call
it a "planar" transducer. The planar transducer has comb-shaped electrodes
which can be applied with a standard photo-mask technique. The frequency
characteristics of the planar transducer turns out to be favorable
compared with the frequency characteristic of the conventional transducer.
      We shall first mention the essential characteristics of the conventional
transducer and then compare these with the characteristics of the planar
transducer." (Author).
      It is shown that the planar transducer has a wide-band frequency
response, although the overall response seems to be about 10 dB lower than
that of a plate transducer. 2 refs.

1978-262+

Rhyne, T. L., "An Improved Interpretation of Mason's Model for Piezoelectric
    Plate Transducers," IEEE Trans. Son. Ultrason. SU-25 (2), 98-103
    (March 1978).

"A new interpretation of Mason's model for a piezoelectric plate
transducer is presented. The network model emphasizes a series connection
for the two acoustic loads while utilizing lumped impedance elements
expressed as functions of the delay operator $Z = \exp sT$. An exact analysis
of the plate dynamics permits a simplified resistive model for conditions
of light asymmetry in acoustic loading. A simplified resistive structure
provides transmission (reception) loss near the half-wave resonance. Air
backed front loading is modeled. Finally, RLC lumped component models
are provided for evaluation of transducers as lumped element filters."
(Author) 18 refs.

1973-263+

Wright, H., "Impulse-Response Function Corresponding to Reflection from
    a Region of Continuous Impedance Change," J. Acoust. Soc. Am. 53
    (5), 1356-1359 (1973).

"Acoustic reflection from a region of continuously varying specific
acoustic impedance is characterized, in the linear circuit sense, by a
unit reflection impulse-response function $R(t)$. It is shown that in the
absence of attenuation and for modest impedance excursions, the impulse-
response function corresponding to reflection from a region which has
continuously variable impedance along the incident axis is given by
$R(2t) \approx (dz/dt)/4z$, where $z(t)$ is the impedance profile and t is acoustic
travel time." (Author).

1973-264+

Sigelmann, R. A. and Reid, J. M., "Analysis and Measurement of Ultrasound
    Backscattering from an Ensemble of Scatterers Excited by Sine-Wave
    Bursts," J. Acoust. Soc. Am. 53 (5), 1351-1355 (1973).

"This paper develops a practical approximation for the backscattering
of periodic bursts of sine waves by a volume of randomly distributed
scatterers. The approximation is applied to the measurement of a
'volumetric backscattering cross section,' using a substitution method
in which the rms value of the gated backscattered signal is compared with
the rms value of a wave reflected from a target of known coefficient of
reflection. It is shown that the signal backscattered from the ensemble
depends on the attenuation of the wave in the volume and upon the burst
and gate lengths. An equation to obtain the volumetric backscattering
cross section from experimental data is derived." (Author) 6 refs.

1977-265+

Rhyne, T. L., "Radiation Coupling of a Disk to a Plane and Back or a
     Disk to Disk: An exact Solution," J. Acoust. Soc. Am. 61 (2), 318-324
     (February 1977).

     "The radiation coupling or coupling by propagating waves is solved
for a disk in an infinite baffle to a plane and back or equivalently a disk
to a disk both in infinite baffles. The radiation coupling is defined as
a linear filter operating between lumped mechanical components which may
be incorporated into transducer models. The impulse response of the
radiation-coupling filter and the Fourier transfer function for the
radiation-coupling filter are solved in closed form. The radiation-
coupling gain (loss) is applicable to the correction of experimental
data and to the absolute calibration of circular transducers by self-
reciprocity measurements." (Author) 14 refs.

1977-266+

Weight, J. P. and Ravenhall, F. W., "An Inexpensive Wideband Recording
     Facility," J. Phys. E. (Sci. Instr.) 10 (4), 424-428 (1977).

     "A series of modifications is described whereby a commercial video
tape recorder was adapted for non-TV-format signals. The recorder used
had a typical twin-head helical scanning system, with slow-motion and
stop-playback capabilities. The signals to be recorded in this case
were developed during crack propagation of metal specimens under tension,
i.e., stress-wave emissions. These occur at random intervals and are in
the frequency range 20 kHz to 2 MHz." (Author) 3 refs.

1977-267+

Harnik, E., "A Broadband Probe for Studies of Acoustic Surface Waves,"
     J. Phys. E. (Sci. Instr.) 10 (12), 1217-1218 (1977).

     "An acoustic surface wave probe has been developed for use in
broadband non-destructive testing and in seismological modelling. The
probe has a bandwidth of about 0.4-4 MHz but the design is suitable for
use up to centre frequencies of about 5 MHz. It takes a negligible
amount of energy from the ultrasonic beam and appears to reproduce
accurately the shape of an ultrasonic pulse. The probe is characterized
by simplicity of construction and operation." (Author) 6 refs.

1970-268+

Gericke, O. R., "Theory and NDT Applications of Ultrasonic Pulse-Echo
     Spectroscopy," Proc. Symp. on the Future of Ultrasonic Spectroscopy,
     London, October 1970, Paper #1.

     The author discusses the use of a pulsed swept-frequency system for
ultrasonic spectroscopy. The main advantage is that one is able to produce

a flat frequency response with such a device. Unfortunately the time resolution and signal-to-noise ratio are both poor.

The use of ultrasonic spectroscopy for examining material micro-structure, assessing severity of defects and investigating the frequency-dependence of ultrasonic beam spreading, is reviewed.

Many spectra, produced by the author's apparatus, illustrate the productive uses of spectroscopy.

1970-269+

Lloyd, E. A., "Wide-Band Ultrasonic Techniques," Proc. Symp. on the Future of Ultrasonic Spectroscopy, London, October 1970, Paper #2.

Wide-band ultrasonic techniques are categorized as either measurements of distributed phenomena (grain size, etc.) or detection and classification of discrete targets. Furthermore, the target may be non-penetrable (a void) or penetrable. The signals from the penetrable target, being in general the more complex.

Response of the defect is approximated by considering only its main scattering features. Discrete scattering centers result in modulations in the magnitude spectrum. Cepstral processing is discussed as a method for extracting defect-size information from the modulated spectrum.

The possibility of compact data presentation as the coefficients in a power series expansion (in frequency) of the material's transfer function is mentioned.

Several novel designs for wide-bandwidth transducers are presented.

1970-270+

Aldridge, E. E., "Ultrasonic Spectroscopy at the NDT Centre, Harwell: Progress Report," Proc. Symp. on the Future of Ultrasonic Spectroscopy, London, October 1970, Paper #3.

Work to date at Harwell has centered about the design and construction of spectroscopic instrumentation. The importance of gating circuits, with minimal switching transients, for use with analog spectrum analyzers is noted. Problems of signal pick-up from fast digital circuitry is mentioned.

The author notes that if the flaw affects frequency components, which are at the same time attenuated by the material surrounding the defect, information concerning the defect may be difficult to extract.

1970-271+

Clipson, W. R., "Ultrasonic Spectroscopy Development for Inclusion Cloud Assessment," Proc. Symp. on the Future of Ultrasonic Spectroscopy, London, October 1970, Paper #4.

Frequently, defects occur as "clouds" of small inclusions. Whereas the size of individual defects may be such as to preclude detection, the cloud is detectable. Although no theoretical work is given, experimental spectral measurements (utilizing a wave analyzer) is presented.

Spectral differences in the signals from flat-bottom and side-drilled holes were observed, indicating hope for flaw characterization.

1970-272+

Mitchell, R. F., "Wide-Band Acoustic Bulk Wave Transducers," Proc. Symp. on the Future of Ultrasonic Spectroscopy, London, October 1970, Paper #5.

Observation that a CdS thin-film transducer gave a flat frequency response led the author to investigate the properties of a transducer with a piezoelectric constant which varies through its thickness. For certain functional relationships of piezoelectric constant versus distance, the transducer's response will be broadbanded.
An interdigital bulk-wave device and a transducer with a shaped back surface also hold possibilities for wide-band response.

1942-273+

Mason, W. P., Electromechanical Transducers and Wave Filters, Van Nostrand Co., Inc., New York (1942).

1964-274+

Berlincourt, D. A., Curran, D. R., and Jaffee, H., "Piezoelectric and Piezomagnetic Materials and Their Function in Transducers," Ch. 3 in Physical Acoustics, Vol. I, Part A (W. P. Mason, ed.), Academic Press, New York (1964).

1978-275+

Fox, M. D. and Donnelly, J. F., "Simplified Method for Determining Piezoelectric Constants for Thickness Mode Transducers," J. Acoust. Soc. Am. 64 (5), 1261-1265.

"A procedure is described for obtaining the stress constant e_{33} for an arbitrary piezoelectric transducer operating in the thickness mode. A closed-form solution is developed which uses measurements of the electrical impedance of the transducer as input. Verification of the calculated parameters is accomplished by incorporating them into a computer model of the transducer. A discussion of various methods of obtaining the clamped capacitance C_0 is included as well as a calculation of an equivalent resistance to represent losses in the ceramic. Numerical examples are presented.

1972-276+

Papadakis, E. P., "Ultrasonic Diffraction Loss and Phase Change for Broad-Band Pulses," J. Acoust. Soc. Am. 52 (3), 847-849.

"The effective diffraction loss and phase change in the field of broad-band transducers is computed in terms of the normalized distance $S_c = z\lambda_c/a^2$ at the center frequency of the pulse. Diffraction corrections for attenuation and velocity are explained, and their limitations stated for broad-band pulses. It is shown that the corrections are functions of bandwidth as well as of S_c." (Author)

1964-277

Carome, E. F., Parks, P. E., and Mraz, S. J., "Propagation of Acoustic Transients in Water," J. Acoust. Soc. Am. 36, 946-952.

"A technique is described for investigating the propagation of acoustic transients in liquids. Thick piezoelectric plates are employed as acoustic sources and detectors. The results of recent theoretical works on the transient response of such elements are extended to determine the relationships between the time profiles of the voltage applied to the source, the stress wave in the liquid, and the output voltage of the detector. Effects of transient processes in the field of a piston source also are considered. Results are presented of an experimental study of the propagation in water of low-amplitude pressure steps and impulses as narrow as 0.05 µsec. The data are strongly dependent on the parameters of the source-detector configuration. This limits the range of applicability of the technique, but its usefulness for studies of absorbing liquids is indicated."

1970-278

Freedman, A., "Sound Field of Plane or Gently Curved Pulsed Radiators," J. Acoust. Soc. Am. 48, 221-227.

"When a single pulse is applied to a plane radiator in a large rigid baffle or to a convexly curved baffled radiator having dimensions and radii of curvature large compared with the relevant wavelengths, the pressure at a field point is shown theoretically to consist, generally, of a sequence of pulses, each of which is, approximately, a scaled replica of the applied pulse. The number of pulses and their relative size and spacing are functions of position of the field point. In the direction of the main beam, if the radiating surface is plane, these pulses are not resolved and a single nearly undistorted pulse is obtained. A form of reciprocity is shown to exist between the structure of the acoustic signal at a point in the field of a pulsed transducer when transmitting and the structure of the electrical signal when the same transducer receives an acoustic pulse. Simple relationships are presented between the formulas for pulsed radiation, reception, and backscattering from a plane surface." (Author)

1976-279+

Kazhis, R. I. and Lukoshevichyus, A. I., "Wideband Piezoelectric Transducers with an Inhomogeneous Electric Field," Sov. Phys. Acoust. 22, 167-168.

"The bandwidth of piezoelectric transducers is mainly limited by
the presence of two sources of ultrasound near the faces of the piezoelectric element and multiple reflections of the generated ultrasonic
waves in the transducer. . . . We have investigated piezoceramic transducers with an inhomogeneous electric field, for which the distribution
of the force lines is determined by a special placement of the working
electrodes relative to the crystallographic axes of the piezoelectric."
(Author)

1970-280

Stephanishen, P. R., "Transient Radiation from Pistons in an Infinite
    Baffle," J. Acoust. Soc. Am. 49, 1629-1638.

"An approach is presented to compute the near- and farfield transient
radiation resulting from a specified velocity motion of a piston or array
in a rigid infinite baffle. The approach, which is based on a Green's
function development, utilizes a transformation of coordinates to simplify
the evaluation of the resultant surface integrals. A simple expression
is developed for an impulse response function, which is the time-
dependent velocity potential at a spatial point resulting from an impulse
velocity of a piston of any shape. The time-dependent velocity potential
and pressure for any piston velocity motion may then be computed by a
convolution of the piston velocity with the appropriate impulse response.
The response of an array may be computed using superposition. Several
examples illustrating the usefulness of the approach are presented. The
farfield time-dependent radiation from a rectangular piston is discussed
for both continuous and pulsed velocity conditions. For a pulsed velocity
of time duration T it is shown that the pressure at several of the field
points can consist of two separate pulses of the same duration, when T is
less than the travel time across the piston." (Author)

1974-281

Robinson, D. E., Lees, S., and Bess, L., "Near Field Transient Radiation
    Patterns for Circular Pistons," IEEE Trans. Acoust. Speech Sig. Proc.
    22 (6), 395-403.

"The exact impulse response of field parameters for any field point
on or off axis for the case where a circular disc radiator face is
subjected to a displacement step corresponding to a velocity impulse is
reviewed. By convolution, the transient field pattern for any arbitrary
motion of the disc can be obtained. The exact response for a half-sine
monopulse is computed. An approximate representation of the transient
pressure response to the velocity impulse input at the disc is derived,
and it is shown to correspond to the replica pulses described previously.
The regions of validity of the approximation are quite limited and the
replica pulses are displaced in time from the positions formerly attributable
to them. The displaced replica approximation is applied to an examination
of the structure of the near field for continuous sinusoidal excitation
and a plot of positions of extrema is produced. It is shown that this
approximation gives good agreement with the exact values and is superior
to the previous published approach in this regard. For short sinusoidal
pulses the effect of pulse length on the field pattern, and of field point

on the time history of a transient wave are shown. When the excitation is a short sinusoidal pulse the effect of the pulse length and field point position on the field pattern and wave shape are demonstrated." (Author).

1966-282

Kossoff, G., "The Effects of Backing and Matching on the Performance of Piezoelectric Ceramic Transducers," IEEE Trans. Son. Ultrason. SU-13 (1), 20-30.

"The effects of backing and matching on the performance of transmitting and receiving PZT7A transducers working into a water load are analyzed. Although backing widens the bandwidth, it also increases the transmission loss, and more efficient and wider bandwidth transducers are obtained by quarter-wave matching the transducer to the water load. By quarter-wave matching the transducer to low impedance absorbing backings, reflected high impedance absorbing backings may be obtained; and very wide bandwidth and efficient transducers are obtained by quarter-wave matching both to the backing and to the load. In pulse detection applications, the pulse width of these transducers has been found to be nearly independent of such increases in bandwidth. The explanation for this effect and a procedure for determining the approximate echo pulse waveform is presented." (Author)

1972-283

Meeker, T. R., "Thickness Mode Piezoelectric Transducers," Ultrasonics, 26-36 (January 1972).

"This paper is a tutorial review of the theory of the simple thickness mode piezoelectric transducer. The usual differential equations and constitutive relations are used to obtain general impedance equations for the transducer with arbitrary boundary conditions. In the derivations, special attention is given to showing what basic assumptions are made, and which material constants must be used in the equations. As usual, the thickness mode theory is only valid if no quantities depend on the lateral co-ordinates of the plate. It is shown that certain elastic and piezo-electric constants must be zero in the plate co-ordinate system for the simple thickness mode theory to be valid for the transducer. Four geometrical and material variables, and three boundary conditions completely determine the transfer and impedance properties of the transducer. Exact expressions are given for the electrical impedance of the simple thickness mode resonator with free surfaces, and for three electrical properties of a bonded and backed thickness mode transducer (namely, the electrical impedance at low frequency, and the electrical impedance and transfer loss at the halfwave frequency). The transfer loss is 3.3 dB for an unbacked, untuned, and acoustically matched transducer with no series electrical resistance and a piezoelectric coupling factor of 0.5." (Author)

1969-284

Sittig, K. E., "Effects of Bonding and Electrode Layers on the Transmission Parameters of Piezoelectric Transducers Used in Ultrasonic Digital Delay Lines," IEEE Trans. Son. Ultrason. SU-16 (1), 2-10.

"In ultrasonic delay lines with thickness-driven piezoelectric
transducers, it is necessary to have electrode and, possibly, bonding
layers in the sound transmission path. If these layers have characteristic
impedances that are substantially different from those of the piezoelectric
transducer and the delay medium, they act as mismatched transmission line
sections between the transducer and its load, and transform the normally
real load impedance into a complex one.

The resulting shifted and deformed response curves are computed for a
large number of layer parameters by means of Mason's equivalent circuit.
From these plots, information as to permissible layer thickness, etc., may
be obtained and used in the design procedure of ultrasonic delay lines.

In digital delay lines, where linear phase response is a design
requirement, any intermediate layers should be as thin as possible or be
closely matched to the delay medium in order to avoid fast ripples in the
frequency response, which would give rise to side lobes far away from the
main signal in the time domain." (Author)

1967-285

Sittig, K. E., "Transmission Parameters of Thickness Driven Piezoelectric
Transducers Arranged in Multilayer Configurations," IEEE Trans. Son.
Ultrason. SU-14, 167-174.

"The individual transducers of an ultrasonic delay line may consist
of a multiplicity of piezoelectrically active layers electrically connected
in series, parallel, or grouped in series-parallel combinations interspersed
with electrically conductive or nonconductive layers of different
characteristic acoustic impedances. The stack of transducer layers may be
loaded by an absorptive or reactive backing and coupled to the delay medium
through bonding and matching layers.

The transmission parameters for such configurations are written in a
form well suited to digital computation. Inspection of numerical results
reveals effects which may be qualitatively understood by visualizing
separately the effects due to the mechanical resonances of the layer
assembly and those due to the arrangement of piezoelectric material with
respect to the stress distribution within the stack.

The examples given indicate that transducers consisting of
alternately poled stacked $\lambda/2$ layers of a low coupling factor material such
as CdS give an insertion loss improvement at the cost of bandwidth reduction
little different from that obtained with narrow-band tuned terminations.
For high coupling factor layers, no significant improvement is obtainable."
(Author)

1971-286

Mattiat, O. E. (editor), Ultrasonic Transducer Materials, Plenum, New York.

1975-287

Martin, R. W. and Sigelmann, R. A., "Force and Electrical Thevenin
Equivalent Circuits and Simulations for Thickness Mode Piezoelectric
Transducers," J. Acoust. Soc. Am. 58 (2), 475-489.

"A simple model is reported for thickness-mode piezoelectric elements used as ultrasonic transducers in measurement systems. The model represents the excitation system and transducer as a Thevenin mechanical equivalent for the transmitting mode and a Thevenin electrical equivalent for the receiving mode. Computer programs based on the model have been developed, and computer simulations to study the effects of backing materials, element areas, and excitation sources are reported. The nature in which the source impedance alters the Thevenin mechanical output impedance and its importance in determining peak transmission frequency and in computing acoustic coating layers for matching are found. A total transfer improvement of 28 dB was shown for epoxy-backed elements radiating into fluid with the transducer used as both transmitter and receiver by using high values of source and load impedances in contrast to low values. The model was found to agree closely with experimental data of a 2.7-MHz transducer." (Author)

1973-288+

Legros, D. and Lewiner, J., "Electrostatic Ultrasonic Transducers and Their Utilization with Foil Electrets," J. Acoust. Soc. Am. 53 (6), 1663-1672.

"Electrostatic ultrasonic transducers are very attractive when considered from the point of view of simplicity. They are constituted by a condenser, the ultrasonic wave being directly excited on the electrodes. These transducers are currently used at low frequencies (microphones) and sometimes at higher frequencies (up to a few megahertz). At higher frequencies the bias voltage applied across the condenser has to be quite large and electrification of the central dielectric layer can appear. This paper describes such effects and presents the experimental conditions allowing the transducer to operate. The electrification of the dielectric layer is studied and the problems related to the conservation of the deposited charges are considered for Mylar and polypropylene foils of about 10-μ thickness. In the present work the ultrasonic waves generated or received by these transducers have frequencies ranging from 10 to 200 MHz." (Author)

1973-289+

Sessler, G. M. and West, J. E., "Electret Transducers: A Review," J. Acoust. Soc. Am. 53 (6), 1589-1600.

"A review of the history, design, performance, and application of electret transducers is presented. Particular emphasis is placed on foil-electret transducers incorporating a thin-film electret made of Teflon or related materials. Such transducers have excellent frequency response, low distortion, small vibration sensitivity, and have been used over a frequency range extending from 10^{-3} to 2×10^{8} Hz. They can be made in a variety of shapes over a large range of sizes and are generally not affected by adverse environmental conditions. More than 10 million electret transducers are being manufactured annually as microphones with various directivity patterns for use in amateur and studio applications, tape recorders, sound-measuring instruments, telephone-operators' headsets, hearing aids, and acoustic-graphic tablets, and as transducers in earphones

and phonograph cartridges. Electret transducers are also used for experimental and research applications in such widely different fields as gas
analysis, opto-acoustic spectroscopy, aeronautics, atmospheric studies,
telephony, ultrasonics, acoustic holography, data transmission, and leak
detection in space stations." (Author)

1962-290+

Freedman, A., "A Mechanism of Acoustic Echo Formation," Acustica 12, 10-21.

"Using a method of analysis analogous to that of physical optics, the
direct backscattering of small amplitude acoustic waves from a rigid body,
immersed in an ideal fluid medium is re-examined. The incident radiation is
a pulse of general type, and at long ranges, no restrictions are imposed on
the directivity patterns of the transmitting and receiving transducers.
Clarification of the echo-formation mechanism applicable at small wavelengths is obtained.
   The echo is shown to be composed of a number of discrete pulses, each
a replica of the transmission pulse, and hence termed an 'image pulse.'
An image pulse is generated whenever there is a discontinuity with respect
to range, r, in $d^nW(r)/dr^n$, where $W(r)$ is the solid angle subtended at the
transducers by that part of the scattering body within range r. n may be
zero or any positive integer. It is shown that four types of echo envelope
arise from varying degrees of overlap of these image pulses.
   The combination of 'image pulse' and 'creeping wave' mechanisms is
believed to account for the main scattering phenomena from rigid convex
bodies, the former mechanism being paramount outside the shadow region at
small wavelengths, the latter mechanism predominating at large wavelengths."
(Author) 7 refs.

1960-291

Filipczynski, L., "Transients and the Equivalent Electrical Circuit of a
   Piezoelectric Transducer," Acustica 10, 149.

"The subject of the present paper is an X-cut quartz transducer, in
which one-dimensional mechanical vibrations are discussed. Starting with
piezoelectric equations in terms of the electrical enthalpy, transients
of the transducer, frequency-response characteristics and the input
impedance are analyzed. The results of experiments confirm the described
mechanism of the vibrations in the transducer, as well as the frequency-
response characteristics. On the basis of the results obtained, an equivalent electrical circuit of the transducer has been constructed in terms of
a transmission line. The given circuit is valid for steady states and
for transients as well." (Author)

1974-292

Beaver, W. L., "Sonic Nearfields of Pulsed Piston Radiators," J. Acoust.
   Soc. Am. 56, 1043-1048.

"The formation of sonic pulses in the nearfield region of a pulsed piston radiator has been investigated by similation on a digital computer. The results give insight into the sonic radiation process, showing the formation of pulses that replicate the piston motion, with trailing disturbances which originate from the rim. A comparison is made between pulsed and CW beam profiles, showing that there is little difference for moderate pulse lengths." (Author)

1975-293+

Tabuchi, D., Inoue, N., Okuwa, T., and Ohno, K., "Ultrasonic Spectroscopy and Automatic Ultrasonic Spectrometer," Acustica 32, 236-243.

"An automatic recording ultrasonic spectrometer has been constructed in order to measure and record the frequency spectra of ultrasonic absorption and velocity. For the measurement of absorption an ultrasonic pulse echo method is used, and an ultrasonic pulse circulation method is applied for the measurement of velocity. The data of absorption and velocity are punched on a tape, and typed. If the tape is put into a digital computer, the ultrasonic spectra and the characteristic values are computed, punched on a tape, and typed. When the tape is put into the ultrasonic spectrometer, the ultrasonic spectra of absorption and velocity are recorded by a digital plotter. The process is completely automated by a method of sequence control." (Author) 6 refs.

1971-294+

Papadakis, E. P. and Fowler, K. A., "Broadband Transducers: Radiation Field and Selected Applications," J. Acoust. Soc. Am. 50 (3), 729-745.

"In many applications, broad-band ultrasonic transducers capable of producing short video pulses are required. Previously, plane-wave analysis with equivalent circuits has proven successful in predicting pulse shape in the time and frequency domains. The present approach is to recognize that piston sources radiate nonplanar waves, and that the frequency spectrum of a broad-band piston source can be measured experimentally. With the spectrum as a weighing function for the field profiles of a monofrequency piston source, a superposition is performed to find the pressure and phase profiles in the radiation field of a broad-band transducer. Experimental measurements are presented that take advantage of the broad-band pulse technique combined with spectrum analysis. These include thickness gauging of thin materials and interface layers, and relative viscosity measurements." (Author)

1971-295+

Papadakis, E. P., "Effects of Input Amplitude Profile Upon Diffraction Loss and Phase Change in a Pulse-Echo System," J. Acoust. Soc. Am. 49 (1), 166-168.

"The particle velocity profile V(p) across the face of a transmitting transducer is shown to have large effects upon the diffraction loss and

phase change in the ultrasonic field of the transducer. Various functions
V(p), monotonic decreasing from the center to the rim of a circular
transducer, were employed. The pulse-echo response of the transducer was
calculated by numerical integration on an electronic computer. It was
found that the functions V(p) chosen caused the diffraction-loss and phase-
change curves to be smoother than in the piston case and caused the
respective peaks and plateaus to shift with distance in the Fresnel
region." (Author)

1960-296+

Brekhovskikh, L. M., Waves in Layered Media, Academic Press, New York
    (translated from the Russian by D. Lieberman and edited by R. T.
    Beyer).

1963-297+

Redwood, M., "A Study of Waveforms in the Generation and Detection of
    Short Ultrasonic Pulses," Appl. Mat. Res. 2, 76-84.

    "Investigations of the properties of materials frequently make use
of short ultrasonic pulses consisting of from one to about ten oscillations
which are roughly sinusoidal in shape. The ultrasonic signal is usually
generated by applying an electrical signal to a piezoelectric transducer.
After passing through the material under investigation it is detected by
using a second piezoelectric transducer. The nature of the electrical
waveforms observed in such experimental systems and their relation to the
ultrasonic signal is frequently not well understood, as they are dependent
on a complex combination of circumstances involving (1) the thickness of
the transducers, (2) the acoustic impedances of the materials in contact
with both faces of each transducer, and (3) the nature of the electrical
resistance into which the receiving transducer feeds. Sometimes the pulse
shape is also considerably affected by the nature of the material under
investigation, particularly if this material is highly absorbent. Lack
of understanding of the change in shape of the waveform which can be
produced, particularly by the receiving transducer, frequently leads to
misconceptions concerning the actual shape of the ultrasonic pulse and
its frequency spectrum. This may also lead to considerable errors in
estimates of its velocity and attenuation.
    The generation and detection of ultrasonic pulses by using
piezoelectric transducers are treated here in some detail. Methods of
predicting the various ultrasonic and electrical waveforms are developed
and illustrated by application to a particular system designed for the
measurement of velocity in small samples of material and hence using as
short an ultrasonic pulse as possible." (Author) 3 refs.

1978-298+

Simpson, W. A., Jr., "A Microcomputer-Controlled Ultrasonic Data
    Acquisition System," Oak Ridge National Laboratory Tech. Memo
    ORNL/TM6531.

"The large volume of ultrasonic data generated by computer-aided test procedures has necessitated the development of a mobile, high-speed data acquisition and storage system. This approach offers the decided advantage of on-site data collection and remote data processing. It also utilizes standard, commercially available ultrasonic instrumentation. This system is controlled by an Intel 8080A microprocessor. The MCS80-SDK microcomputer board was chosen, and magnetic tape is used as the storage medium. A detailed description is provided of both the hardware and software developed to interface the magnetic tape storage subsystem to Biomation 8100 and Biomation 805 waveform recorders. A boxcar integrator acquisition system is also described for use when signal averaging becomes necessary. Both assembly language and machine language listings are provided for the software." (Author)

1976-299

Robinson, D. E. and Williams, B. G., "Digital Acquisition and Interactive Processing of Ultrasonic Echoes," Ultrasound in Med. and Biol. 2, 199-212.

"The requirements for an interactive digital signal processing system for ultrasonic pulse-echo signals are discussed. A system based on an Interdata Model 80 mini-computer and micro-processor interface is described. The system is capable of acquiring ultrasonic data at a sampling rate of 6 MHz. Ultrasonic B-mode data may be acquired in Line Mode, when echo waveform data and transducer position and orientation are stored, or in Section Mode when the data is converted directly into picture form in memory in the same way that a standard echogram is formed on the screen of an oscilloscope. In each case the data for single complete high resolution echogram may be acquired in less than 15 sec. It is shown that the 6 MHz sampling rate is sufficient to faithfully preserve the echo waveshape of a 2 MHz system independently of the relation to the phase of the sampling. Also shown is a cross-sectional echogram of the pregnant uterus, and its digital representation with a raster density of 80 x 100 and 160 x 200 picture elements.
The computer is programmed with an interactive program to allow ultrasonic signals to be acquired, stored, processed and examined with the convenience of a desk calculator. Sample operations are illustrated including data interpolation, spectrum analysis, filtering and complex signal deconvolution. The ability of deconvolution techniques to resolve targets separated by less than one wavelength in depth is demonstrated. Possibilities of further processing techniques are outlined." (Author)
20 refs.

1978-300

Elsley, R. K., "Accurate Ultrasonic Measurements with the Biomation 8100 Transient Recorder," Proc. First Intl. Symp. on Ultrason. Mat. Characterization, June 1978.

"The Biomation 8100 Transient Recorder performs 8-bit analog-to-digital (A/D) conversions at a 100 MHz sample rate and is widely used for data acquisition of high frequency ultrasonic signals. Due to the nature of the A/D method used, the accuracy is substantially less than 8-bits under some conditions, particularly at high frequencies. The errors which occur are found to be partially random and partially systematic (called "preferred states" by the manufacturer). The accuracy which can be obtained depends not only on the signal which is being acquired, but also on what features of that signal the experimenter is interested in measuring. By using signal averaging and offset variation, dynamic ranges in excess of 70 dB (12-bits) have been obtained, and subtle but important features in the signals being analyzed have been thereby measured." (Author) 1 ref.

1956-301

Ying, C. F. and Truell, Rohn, "Scattering of a Plane Longitudinal Wave by a Spherical Obstacle in an Isotropically Elastic Solid," J. Appl. Phys. 27, 1086-1097 (1956).

The first consideration of the scattering of an acoustic wave propagating in a solid.

1960-302

Einspruch, Norman G., Witterholt, E. J., Truell, Rohn, "Scattering of a Plane Transverse Wave by a Spherical Obstacle in an Elastic Medium," J. Appl. Phys. 31, 806-818 (1960).

A consideration of the scattering of a shear wave by a spherical discontinuity.

1965-303

Johnson, Gregert and Truell, Rohn, "Numerical Computations of Elastic Scattering Cross Sections," J. Appl. Phys. 36, 3466-3475 (1965).

A brief review of the calculation of cross section expressions for the scattering of an elastic wave in an elastic medium, with numerical calculations.

1975-304

Gubernatis, J. E., Domany, E., Huberman, M. and Krumhansl, J. A., "Theory of the Scattering of Ultrasound by Flaws," Proc. 1975 IEEE Ultrason. Symp., 107-110.

"An integral equation governing the scattering of ultrasound by an arbitrarily shaped flaw is presented, and features of the scattered displacement and stress fields are discussed for the case of a flaw embedded in an isotropic medium. Also discussed are differential cross sections for the scattered power. These cross sections for a spherical flaw (cavity and inclusion) are evaluated by an approximation analogous to the first Born approximation in quantum mechanical scattering. The results of the calculations are compared with exact results for scattering of ultrasound by spheres. The relevance of this comparison to NDE, i.e., flaw identification, is discussed." (Author) 3 refs.

1958-305

White, R. W., "Elastic Wave Scattering at a Cylindrical Discontinuity in a Solid," J. Acoust. Soc. Am. 30, 771-785 (1958).

Deals with the scattering of plane compressional and shear waves at oblique incidence on an infinite elastic rod embedded in another isotropic elastic medium. It pays particular attention to mode conversion. It also reports some measurements.

1948-306

Fridman, M. M., "The Diffraction of a Plane Elastic Wave by a Semi-Infinite Rigid Plane," Dokl. Akad. Nauk. USSR 60, 1145-1148 (1948), (in Russian).

The first solution for scattering from a two-dimensional flaw.

1953-307

Maue, A.-W., "Die Bengung Elasticher Wellen an der Halbebene," Z. F. Ang. Math. and Mech. 33, 1-10 (1953).

An exact reduction to quadratures of the two-dimensional problem of elastic scattering of an infinite half-plane weak crack. The differential wave equations and boundary conditions are combined into integral equations for the potentials, which are represented as plane-wave superpositions. The integral equations are solved by the method of Clemmow. (In German.)

1964-308

Ang, D. D. and Knopoff, L., "Diffraction of Scalar Elastic Waves by a Clamped Finite Strip," Proc. N.A.S. 51, 471-476 (1964).

The long-wavelength limit to the solution of the far field for the title problem.

1964-309

Ang, D. D. and Knopoff, "Diffraction of Scalar Elastic Waves by a Finite Crack," Proc. N.A.S. 51, 593-598 (1964).

Similar to (308) except that the strip has weak boundary conditions (weak crack), which is a more important problem in applications.

1964-310

Ang, D. D. and Knopoff, L., "Diffraction of Vector Elastic Waves by a Clamped Finite Strip," Proc. N. A. S. 52, 201-207 (1964).

An extension of the calculation of (308).

1964-311

Ang, D. D. and Knopoff, L., "Diffraction of Vector Elastic Waves by a Finite Crack," Proc. N.A.S. 52, 1075-1081 (1964).

This paper considers the problem of the diffraction of an incident plane longitudinal wave by a finite crack, and evaluates the far fields by the method of steepest descents.

1976-312

Tan, T. H., "Theorem on the Scattering and the Absorption Cross Section for Scattering of Plane, Time-Harmonic, Elastic Waves," J. Acoust. Soc. Am. 59, 1265-1267 (1976).

A method, due to de Hoop, is extended to the scattering of elasto-dynamic waves to derive the "cross-section theorem" (the optical theorem).

1977-313

Varatharajulu, V.,* "Reciprocity Relations and Forward Amplitude Theorems for Elastic Waves," J. Math. Phys. 18, 537-543 (1977).

This paper derives the forward scattering (optical) theorem and receiprocity relations (including polarization change on scattering) for plane, monochromatic elastic waves scattered by obstacles of arbitrary shape.

*Now Varadan.

1977-314

Gubernatis, J. E., Domany, E., and Krumhansl, J. A., "Formal Aspects of the Theory of the Scattering of Ultrasound by Flaws in Elastic Materials," J. Appl. Phys. 48, 2804-2811 (1977).

This paper considers the general theory of the scattering of ultrasound by flaws. It considers an incident plane wave scattering from a single homogeneous flaw in an isotropic elastic medium, and obtains an integral equation to describe the problem. The integration is over a volume, and this appears to be the first report to present the volume formulation for elasticity in a reasonably complete form. It derives expressions for scattered amplitudes and differential cross sections, and an optical theorem.

1977-315

Tan, T. H., "Reciprocity Relations for Scattering of Plane, Elastic Waves," J. Acoust. Soc. Am. 61, 928-931 (1977).

Reciprocity relations for scattering of plane, elastic waves incident upon a finite, linear, reciprocal obstacle in a homogeneous, isotropic, perfectly elastic medium are investigated.

1976-316

Pao, Yih-Hsing and Mow, C. C., "Theory of Normal Modes and Ultrasonic Spectral Analysis of the Scattering of Waves in Solids," J. Acoust. Soc. Am. 59, 1046-1056 (1976).

A theory of the spectral analysis of the scattering of elastic waves is presented and illustrated with numerical results for the scattering by a circular cylindrical fluid inclusion in a solid. From overtone frequencies the ratio of the wave speed to the radius of the inclusion can be determined. The application of this technique to nondestructive testing is discussed.

1976-317

Pao, Yih-Hsing and Varatharajulu,* Vasundara, "Huygens' Principle, Radiation Conditions, and Integral Formulas for the Scattering of Elastic Waves," J. Acoust. Soc. Am. 59, 1361-1371 (1976).

By using the divergence theorem this paper shows how Helmholtz- and Kirchhoff-type integral formulas can be derived. Both "interior" and "exterior" formulas are obtained; these formulas are necessary for investigating the scattering of elastic waves by bounded objects. The results illustrate Huygens' principle for the two wave fronts of the elastic wave field.

*Now Varadan.

1977-318

Gubernatis, J. E., Domany, E., Krumhansl, J. A., and Huberman, M., "The Born Approximation in the Theory of Scattering of Elastic Waves by Flaws," J. Appl. Phys. 48, 2812-2819 (1977).

The integral equation formulation obtained in 77-1 is used to derive an approximation scheme, which may be applied relatively easily to scatterers of complicated shapes. The approximation works best for backscattered long waves, but in certain cases is surprisingly good even for short wavelengths and all angles.

1959-319

Karal, Frank C., Jr., and Keller, Joseph B., "Elastic Wave Propagation in Homogeneous and Inhomogeneous Media," J. Acoust. Soc. Am. 31, 694-705 (1959).

This is the extension of Keller's geometrical theory of diffraction to elastic waves. It gives a general method for solving linearized elastic-wave problems which does not depend on the possibility of separation of variables. The method should work well for short wavelengths, but experience had shown that it was still useful for wavelengths of the same order of magnitude as other dimensions in the problem.

1978-320

Weight, J. P. and Hayman, A. J., "Observations of the Propagation of Very Short Ultrasonic Pulses and Their Reflection by Small Targets," J. Acoust. Soc. Am. 63 (2), 396-404.

"The field of a circular ultrasonic transducer emitting a single-cycle pulse into water has been observed using a specially constructed small (150 μm) wide-band receiving probe and a compact stroboscopic schlieren system. The theoretically predicted plane-wave and diffracted edge-wave components of the field have been resolved. Good agreement with the theory for a pistonlike source is obtained, except in a region less than 1.5 transducer radii from the transducer. The output of the transducer used in the transmit-receive mode to detect small targets has been measured and the results are in accord with a time-domain principle of reciprocity between transmission and reception. Implications of the results for field plotting and for the location and characterization of small targets are considered." (Author) 27 refs.

1975-321

Richardson, J. M. and Tittmann, B. R., "Deducing Subsurface Property Gradients from Surface Wave Dispersion Data," Proc. 1975 IEEE Ultrason. Symp., 488-491.

"Because of the ill-posed nature of the problem, special mathematical techniques must be used to convert surface wave dispersion data into subsurface property measurement. The solution is approached here within the framework of estimation theory. This approach starts with a mathematical model giving a probabilistic description of the possible results of measurement and then the optimal estimate is obtained as the most probable value within the constraints imposed by the actual measurements. Estimation theory

also yields auxiliary measures pertaining to bias, data vs. model
dominance, resolution and a posteriori variance. The theory is applied
to actual experimental data consisting of the phase velocities of Rayleigh
surface waves in surface-hardened steel at a set of four wavelengths.
The estimated profile of hardening is compared with independent destructive
measurements. As a test, the theory is also applied at the same set of
wavelengths to a set of synthetic data calculated from an assumed profile.
The above auxiliary measures giving properties of the estimator are also
discussed." (Authors) 5 refs.

1978-322

Lewis, D. K., Szilas, P., Fitting, D. W., and Adler, L., "Spectrum Analysis
    of Elastic Wave Scattering from Cracks in Metals," J. Acoust. Soc. Am.
    63, Suppl. No. 1, 974.

Here experiments are compared to Keller's theory for elastic wave
diffraction, with Maue's solution serving as the canonical problem.

1977-323

Achenbach, J. D. and Gautesen, A. K., "Geometrical Theory of Diffraction
    for Three-D Elastodynamics," J. Acoust. Soc. Am. 61, 413-421.

Here Keller's geometrical diffraction theory is applied to three-
dimensional elastodynamics, particularly to the diffraction of longitudinal
waves by a crack. This yields approximations useful for large frequencies
and/or large distances from the crack edge. As an example the diffraction
of a point-source field by a semi-infinite crack is worked out in detail.

1978-324

Gautesen, A. K., Achenbach, J. D., and McMaken, H., "Surface-Wave Rays in
    Elastodynamic Diffraction by Cracks," J. Acoust. Soc. Am. 63, 1824-
    1831.

This is the first study of the contributions to the diffracted fields
which come, not from diffracted rays of longitudinal and transverse motion,
but from rays of surface waves. These provide the main contributions on
the faces of the crack. As an example the problem of a plane longitudinal
wave normally incident on a penny-shaped crack is worked out in some detail.

1977-325

Datta, S. K., "Diffraction of Plane Elastic Waves by Ellipsoidal
    Inclusions," J. Acoust. Soc. Am. 61, 1432-1437.

The method of matched asymptotic expansions is used to get a low-
frequency solution for the diffraction of a plane wave by an elastic

ellipsoidal inclusion. Numerical results are given, and applicability
to NDE is discussed.

1976-326

Waterman, P. C., "Matrix Theory of Elastic Wave Scattering," J. Acoust. Soc.
    Am. 60, 567-580.

    Earlier developments of a matrix theory for acoustic and EM scattering
are here extended by their developer to elastic waves. If certain matrix
elements which express mode conversion are set to zero, the elastic matrix
equations reduce to a superposition of acoustic and EM equations, providing
a unified theory of scattering of acoustic, EM and elastic waves by an
obstacle of arbitrary geometry and making available the entire body of
acoustic and EM results to compare the elastic theory with. The matrices
are symmetric and unitary.

1976-327

Varatharajulu, V.,* and Pao, Y.-H., "Scattering Matrix for Elastic Waves.
    1. Theory," J. Acoust. Soc. Am. 60, 556-566.

    This paper extends the already existing scattering matrix approach
of Waterman to the scattering of elastic waves. The method is applicable
to obstacles of arbitrary shape so one does not have to calculate a special
set of wave functions for each geometry. The matrices are symmetric and
unitary, which is very nice because these properties are essential for
checking the numerical accuracy.

*Now Varadan.

1978-328

Waterman, P. C., "Matrix Theory of Elastic Wave Scattering. II. A New
    Conservation Law," J. Acoust. Soc. Am. 63, 1320-1325.

    A new conserved elastodynamic field quantity is found; this new
conservation requirement may lead to deeper physical understanding and
to simpler computational methods using the scattering-matrix theory.

1978-329

Varadan, Vasundara V., "Scattering Matrix for Elastic Waves. II. Appli-
    cation to Elliptic Cylinders," J. Acoust. Soc. Am. 63, 1014-1024.

    The scattering-matrix approach to elastic wave scattering is here
employed to give numerical results for scattering of obliquely-incident
plane waves from a cylinder of elliptic cross section. It is much more
useful for short wavelengths than for long.

1978-330

Varadan, Vijay K., Varadan, Vasundara V., and Pao, Yih-Hsing, "Multiple
    Scattering of Elastic Waves by Cylinders of Arbitrary Cross Section.
    I. SH Waves., J. Acoust. Soc. Am. 63, 1310-1319.

The problem here is many identical, long, parallel randomly
distributed cylinders of arbitrary cross section, scattering time-harmonic
polarized plane shear waves. The method combines the scattering-matrix
approach and a statistical averaging technique. Numerical results are
presented.

1972-331

Boore, David M., "Finite Difference Methods for Seismic Wave Propagation
    in Heterogeneous Materials," Methods in Computational Physics 11, 1-37.

A review article on the computation of elastic wave propagation in
media whose properties change with position, by finite difference methods.

1972-332

Lysmer, John and Drake, Lawrence A., "A Finite Element Method for
    Seismology," Methods in Computational Physics 11, 181-216.

A presentation of a finite element method for surface waves which
pays attention to computational feasibility.

1976-333

Datta, S. K., "Scattering of Elastic Waves by a Distribution of Inclusions,"
    Arch. Mech. Stos. 28, 317-324.

The problem of scattering of plane P-waves off a uniform distribution
of rigid spheroids is treated by combining the method of matched asymptotic
expansions and a suitable configurational averaging method.

1975-334

Vary, A., "Feasibility of Ranking Fracture Toughness by Ultrasonic Measure-
    ments," Proc. 1975 IEEE Ultrason. Symp., 588-590.

"Preliminary experimental verification was made of the expected
correlation between ultrasonic attenuation parameters and fracture
toughness measurements on a set of maraging steel specimens. An empirical
equation is proposed for relating the fracture toughness property K_c to the
ultrasonic properties of a polycrystalline solid. The pertinent ultrasonic
factors in this case involve the attenuation coefficient α, frequency f, and
β, the slope of the α vs. f curve. The proposed relation has the form

$K_c = \phi\beta_f$. It predicts that the fracture toughness property K_c will be
proportional to the attenuation slope β evaluated over an appropriate
frequency range. The results of this feasibility study with maraging
steel specimens indicate that if various specimens of a given metal
possess different fracture toughness, it is possible to rank them in order
of toughness by ultrasonic testing." (Author) 9 refs.

1976-335

Sobczyk, K., "Elastic Wave Propagation in a Discrete Random Medium," Acta
    Mechanica 25, 13-28.

    This paper considers propagation of elastic waves in an infinite
solid containing a random configuration of identical finite scatterers. The
work was stimulated by practical questions in geophysics and ultrasonic
spectroscopy. It gives a general formulation for scatterers of arbitrary
shape, and solutions for specific cases of spherical scatterers. The
English has a Polish flavor.

1974-336

Keer, L. M. and Luong, W. C., "Diffraction of Waves and Stress Intensity
    Factors in a Cracked Layered Composite," J. Acoust. Soc. Am. 56, 1681-
    1686.

    Layers in composite materials act somewhat as waveguides; this study
considers the effect of a crack perpendicular to the layer, and shows that
it gives rise to scattered waves in the layer which could be detected at
a large distance from the flaw.

1975-337

Keer, L. M., Luong, W. C. and Achenbach, J. D., "Elastodynamic Stress
    Intensity Factors for a Crack in a Layered Composite," J. Acoust. Soc.
    Am. 58, 1204-1210.

    This studies the effect of a crack parallel to the layer (cf. 74-2);
the ease of detection of the flaw depends on the stiffness of the layer,
with flaws in relatively stiff layers being harder to detect.

1974-338

Christensen, R. M., "Wave Propagation in Elastic Media with a Periodic
    Array of Discrete Inclusions," J. Acoust. Soc. Am. 55, 700-707.

    This studies the propagation of waves in a homogeneous, isotropic
medium containing an array of discrete inclusions of another material.
Full account is taken of multiple scatterings. The direction of
propagation is restricted to be one of the symmetry directions of the
material (which has cubic symmetry). A perturbation method is used.

1978-339

Simons, Donald A., "Reflection of Rayleigh Waves by Strips, Grooves and
    Periodic Arrays of Strips or Grooves," J. Acoust. Soc. 63, 1292-1301.

Devices incorporating grooves and strips are used to perform certain
microwave signal-processing applications, and this is the most recent
paper on the title problem, with references to earlier work. Integral
equations are solved by perturbation techniques.

1978-340

Gaunard, G. C. and H. Uberall, "Theory of Resonant Scattering from
    Spherical Cavities in Elastic and Viscoelastic Media," J. Acoust. Soc.
    Am. 63, 1699-1712.

This paper studies theoretically the scattering of a plane p-wave by a
fluid-filled spherical cavity in elastic and viscoelastic (hence absorbing)
media. The approach, new to elastodynamics and acoustics, is familiar in
nuclear scattering theory. Numerical computations are presented.

1977-341

Tan, T. H., "Scattering of Plane, Elastic Waves by a Plane Crack of Finite
    Width," Appl. Sci. Res. 33, 75-100.

This paper considers the diffraction of time-harmonic, vertically
polarized (the problem involving horizontally polarized waves has already
been dealt with extensively in the literature), plane elastic waves by a
crack of finite width using the integral-equation method. Numerical
solutions are presented.

1976-342

Tan, T. H., "Diffraction of Time-Harmonic Elastic Waves by a Cylindrical
    Obstacle," Appl. Sci. Res. 32, 97-144.

An integral equation formulation of the diffraction of two-dimensional
elastic waves by a cylindrical obstacle is presented. For a number of
configurations the integral equations are solved numerically. Also numerical
results on power scattering and extinction cross sections are given.

1975-343

Rose, Joseph L. and Paul A. Meyer, "Model for Ultrasonic Field Analysis in
    Solids," J. Acoust. Soc. Am. 57, 598-605.

"This paper concentrates on one of the most basic ultrasonic problems
in NDT: that of evaluating the ultrasonic field characteristics in a solid

material resulting from a pulsed piezoelectric crystal" (authors). It presents a theoretical model that can be used to evaluate analytically ultrasonic transducer longitudinal wave-generation characteristics in homogeneous isotropic solids. Results depend on the spectrum of the input pulse.

1974-344

Chow, T. S., "Scattering of Elastic Waves in an Inhomogeneous Solid," J. Acoust. Soc. Am. 56, 1049-1051.

Plane harmonic elastic waves are propagating in an isotropic material containing randomly distributed inhomogeneities; results are expressed in terms of correlation functions.

1976-345

Israilov, M. Sh., "Certain Exact Solutions to Problems of Diffraction of Elastic Waves at a Segment," Sov. Phys. Dokl. 21, 756-757 (from DAN SSSR 231, 1074-1076).

Exact solutions for particular cases of diffraction of longitudinal and transverse waves in elastic media are given; one problem corresponds to diffraction at a rigid plate, another to diffraction at a slit. These are for transient plane waves.

1971-346

Kraft, David W., and Michael C. Franzblau, "Scattering of Elastic Waves from a Spherical Cavity in a Solid Medium," J. Appl. Phys. 42, 3019-3024.

This is extending the work of Truell and his collaborators; it gives the first numerical computations of the scattering cross section for an incident transverse wave.

1963-347+

Bogert, B. P., Healy, M. J. R., and Tukey, J. W., "The Frequency Analysis of Time Series for Echoes: Cepstrum, Pseudo-Autocovariance, Cross-Cepstrum and Saphe Cracking," Chapter 15 in Time Series Analysis, Rosenblatt (editor), Wiley (1963).

The authors introduce the technique of cepstral processing. The use of this method for detecting time separation of "echoes" in the presence of various sources of noise is explored.

1969-348+

Cooley, J. W., Lewis, P. A. W., and Welch, P. D., "The Fast Fourier
   Transform and its Applications," IEEE Trans. Educ. E-12 (1), 27.

   "The advent of the fast Fourier transform method has greatly extended
our ability to implement Fourier methods on digital computers. A description
of the algorothm and its programming is given here and followed by a theorem
relating its operands, the finite sample sequences, to the continuous
functions they often are intended to approximate. An analysis of the error
due to discrete sampling over finite ranges is given in terms of aliasing.
Procedures for computing Fourier integrals, convolutions and lagged
products are outlined" (author). A FORTRAN subroutine is given for
computing the discrete Fourier transform by the FFT method.

1967-349+

Singleton, R. C., "A Method for Computing the Fast Fourier Transform
   with Auxiliary Memory and Limited High-Speed Storage," IEEE Trans.
   Audio Electroacoust. AU-15, 91-97.

   "A method is given for computing the fast Fourier transform of
arbitrarily large size using auxiliary memory files, such as magnetic
tape or disk, for data storage. Four data files are used, two in and two
out. A multivariate complex Fourier transform of $n = 2^m$ data points is
computed in m passes of the data, and the transformed result is permuted
to normal order by m-1 additional passes. With buffered input-output,
computing can be overlapped with reading and writing of data. Computing
time is proportional to $n \log_2 n$. The method can be used with as few as
three files, but file passing for permutation is reduced by using six or
eight files. With eight files, the optimum number for a radix 2 transform,
the transform is computed in m passes without need for additional
permutation passes.
   An ALGOL procedure for computing the complex Fourier transform with
four, six, or eight files is listed, and timing and accuracy test results
are given. This procedure allows an arbitrary number of variables, each
dimension a power of 2" (author).

1969-350+

Singleton, R. C., "An Algorithm for Computing the Mixed Radix Fast Fourier
   Transform," IEEE Trans. Audio Electroacoust. AU-17, 93-103.

   "This paper presents an algorithm for computing the fast Fourier
transform, based on a method proposed by Cooley and Tukey. As in their
algorithm, the dimension n of the transform is factored (if possible),
and n/p elementary transforms of dimension p are computed for each factor
p of n. An improved method of computing a transform step corresponding to
an odd factor of n is given; with this method, the number of complex multiplications for an elementary transform of dimension p is reduced from $(p-1)^2$ to
$(p-1)^2/4$ for odd p. The fast Fourier transform, when computed in place,
requires a final permutation step to arrange the results in normal order.
This algorithm includes an efficient method for permuting the results in
place. The algorithm is described mathematically and illustrated by a
FORTRAN subroutine" (Author).

1975-351

Burgess, J. C., "On Digital Spectrum Analysis of Periodic Signals,"
    J. Acoust. Soc. Am. 58 (3), 556-567.

    "Digital spectrum analysis of harmonic signals can result in
amplitude estimates in error as much as 3.92 dB. The corresponding frequency estimates are not exact. The paper presents a general method for
obtaining significantly improved estimates of amplitude and frequency.
Criteria are given which allow specification of an error limit. Specific
equations are given for sample signals that are unmodified (open window)
and for signals modified by a Hamming data window. The phenomenon called
"leakage" is shown to result from discontinuities imposed by the computation process at the periodically extended "ends" of a sample signal
and not, as is often supposed, by discontinuities presumed to exist (but
do not) at the "ends" of the open window. Criteria for window selection
to reduce leakage are discussed. Calibration is specifically treated.
When any data window other than the open window is used, a different
calibration must be applied to periodic and random components in a signal.
Although discussion is limited to a single harmonic signal, the method can
be applied in a straightforward way to signals with multiple harmonics"
(Author) 32 refs.

1966-352+

Gauster, W. B. and Breazeale, M. A., "Detector for Measurement of Ultra-
    sonic Strain Amplitudes in Solids," Rev. Sci. Instrum. 37 (11), 1544-
    1548.

    "A capacitive detector has been developed for strain amplitude
measurements of longitudinal ultrasonic waves in the frequency range from
5 to 100 MHz. The sensitivity of the device is such that displacement
amplitudes of the order of 10^{-10} cm can be detected. As a check of the
technique, quantitative measurements of the harmonic distortion of ultrasonic waves in a single crystal of germanium were made. From the results,
some combinations of third-order elastic constants are calculated and are
compared with values obtained with the same sample by another method"
(authors) 11 refs.

1977-353+

Cantrell, J. H., Jr. and Breazeale, M. A., "Elimination of Transducer
    Bond Corrections in Accurate Ultrasonic Wave Velocity Measurements
    by Use of Capacitive Transducers," J. Acoust. Soc. Am. 61 (2), 403-
    406

    "A capacitive-driver-capacitive detector system for generation and
detection of ultrasonic waves has been developed. This eliminates the
necessity of bonding piezoelectric transducers to solid samples. With the
capacitive-driver-capacitive-detector system, free-free boundary conditions
exist at the sample surfaces and longitudinal ultrasonic-wave velocities
in solids can be measured accurately without correcting for ultrasonic-wave
phase shifts due to sample-bonded transducer interfaces. The capacitive

driver has a mica dielectric which increases the breakdown potential, but
maintains the free-free boundary conditions at the solid specimen surfaces.
This allows for a larger-amplitude ultrasonic signal to be generated in the
sample than is possible with an air-gap capacitive driver. This improves
the precision of the measurement. The accuracy of the method is comparable
with that of bonded-transducer methods, after bond corrections are made"
(author) 14 refs.

1971-354

Kazys, R. J. and Domarkas, V., "The Frequency and Transient Response of
    Piezotransducers with Intermediate Layers and Electrical Matching
    Circuits," _Proc. 7th Int. Congress on Acoustics_, Session 25U3
    (Budapest, 1971).

    "This investigation includes the action of intermediate layers, backing
and electrical matching circuits on frequency and transient responses of
thickness vibrating piezotransducers operating in liquids . . . Consideration
is given to frequency and phase response of the system piezotransmitter-
piezoreceiver . . ." (author) 5 refs.

1977-355

Kazys, R. and Lukosevicius, A., "Optimization of the Piezoelectric Trans-
    ducer Response by Means of Electrical Correcting Circuits," _Ultrasonics_,
    111-116 (May 1977).

    "A method of shortening the transient response of a piezoelectric
transducer is described. It can be applied to thickness mode piezoelectric
transducers of arbitrary electromechanical coupling. The system incorporates
electrical correcting circuits and can produce a transient response with a
duration much shorter than the transit time of an ultrasonic wave traveling
through the piezoelectric plate" (authors) 6 refs.

1976-356

Lizzi, F., Katz, L., St. Louis, L. and Coleman, D. J., "Applications of
    Spectral Analysis in Medical Ultrasonography," _Ultrasonics_, 77-80
    (March 1976).

    "Spectral analysis of ultrasonic reflections from biological tissues
can be used to determine basic tissue parameters for use in differential
diagnosis. This paper describes the use of the technique under circumstances encountered in several types of clinical examinations. The
applications are illustrated with results obtained from laboratory measurements with a system now being employed in a clinical evaluation programme.
The test objects studied simulate tissues with planar boundaries, tissues
with heterogeneous interior structure, and tissues causing acoustic
'shadowing' of posterior regions" (authors) 6 refs.

1975-357

Lele, P. P., Mansfield, A. B., Murphy, A. I., Namery, J., and Senapati,
N., "Tissue Characterization by Ultrasonic Frequency-Dependent
Attenuation and Scattering," Proc. Sem. Ultrason. Tissue Charac.,
NBS Special Pub. 453.

"Studies conducted in this laboratory to explore the feasibility of
utilizing acoustic impedance, attenuation, and scattering characteristics
of tissues for enhancing the diagnostic capabilities of ultrasound are
described. Frequency-dependent ultrasonic attenuation is found to be
sufficiently greater in infarcted or otherwise necrotized tissues than in
normal controls to permit their positive identification. Superficial
and internal scattering properties of tissues hold the promise of being
significant for diagnostic applications. The difficulties that will have
to be overcome to successfully utilize these properties are discussed"
(authors) 24 refs.

1978-358

Serabian, S. and Williams, R. S., "Experimental Determination of Ultrasonic Attenuating Characteristics Using the Roney Generalized
Theory," Mat. Eval. 55-62 (July 1978).

"To date, little use has been made of the generalized theory of
ultrasonic attenuation in polycrystalline materials proposed by Roney.
It is the only generalized theory which appears to run the gamut of grain
size and frequency dependency of attenuation from the hysteresis loss
mechanism through the complete scattering losses, i.e., Rayleigh, phase
and diffusion. The theory requires only two constants—a hysteresis
constant for the hysteresis losses and a scattering coefficient to describe
those losses due to scattering. In the frequency range normally associated
with the ultrasonic interrogation method the hysteresis losses are essentially
negligible, thus, the scattering coefficient can fully describe the ability
of a given material to propagate ultrasound. Moreover, this assessment of
the material can be made without necessitating direct inferences to the
grain size or frequency involved" (authors) 30 refs.

1973-359

Kesler, N. A., Merkulov, L. G., Shmurun, Y. A., and Tokarev, V. A.,
"Ultrasonic Spectral Method for Attenuation Measurement and Device
for Automatic Testing of Microstructure of Materials," Proc. 7th
Int. Conf. on NDT, Session J-34 (Warszawa, 1973).

This paper presents the mathematics and an experimental technique for
determining the attenuation in a plane parallel plate, over a band of
ultrasonic frequencies. 3 refs.

1974-360

Heyser, R. C. and Le Croissette, D. H., "A New Ultrasonic Imaging System Using Time Delay Spectrometry," Ultrasound in Med. and Biol. 1, 119-131.

"A new method of forming a visual image by ultrasound is described. A shadowgraphic transmission image similar to an x-ray radiograph is produced by the application of a technique known as Time Delay Spectrometry. The system uses a repetitive frequency sweep with a linear relationship between frequency and time and the transmitting and receiving crystal are scanned in raster fashion about the subject. By electronic processing, an image may be built up which represents the energy transmitted through the specimen with a given time delay. An intensity modulated picture encompassing the full shades-of-gray capability of the recording system can be produced. A second type of image showing transmission time through the specimen may also be formed. Brightness changes in the displayed image in this case correspond to changes in the ultrasonic transmission time through the specimen. There is no analog for this type of image in current x-ray or ultrasonic practice. Examples of both types of images of specimens both in vitro and in vivo are shown. The advantages and potentials of this method for biomedical ultrasonic imaging and analysis are discussed" (author) 6 refs.

1975-361

Alers, G. and Graham, L. J., "Reflection of Ultrasonic Waves by Thin Interfaces," Proc. 1975 IEEE Ultrason. Symp., 579-582.

"In order to measure the quality of an adhesive bond using ultrasonic waves, it is important to recognize those features in a reflected echo that carry information about the structure of the thin, chemically different interface between the adhesive and the adherend. We have studied the frequency dependence of the phase and amplitude of ultrasonic pulses reflected from very thin bonds formed between identical blocks of Lucite so that the reflection process is dominated by the nature of the interface and not by the impedance mismatch that occurs in practical adhesive to metal joints. The results show a frequency independent reflection coefficient over the range of 2.5 to 10 MHz which is very difficult to fit with currently available models of reflection from thin layers" (authors) 5 refs.

1971-362

Lees, S., "Ultrasonic Measurement of Thin Layers," IEEE Trans. Son. Ultrason., SU-18 (2), 81-86.

"The shape of a pulse echo from a thin layer embedded between two thicker media is changed because the successive echoes from the two close interfaces overlap. A simple computer algorithm is developed for real time computation of the change in shape as a function of the film thickness. It is only necessary to know the specific acoustic impedances of the three media. In one experiment castor oil was embedded between glass

and steel. The calculated echoes closely resembled the experimental
results for films between 1- and 38-µ thick. A curve was devised for
estimating the film thickness from peak ratios in the echo. A second
experimental situation appeared in testing acoustical transmission across
an amalgam-tooth dentin boundary with water as the film medium. Numerical
calculations produced the same echo patterns as were observed indicating
that there is a gap in the interface between 1 and 10 µ in the samples"
(author).

1978-363

Heyman, J. S., "Phase Insensitive Acoustoelectric Transducer," J. Acoust.
    Soc. Am. 64 (1), 243.

    "Conventional ultrasonic transducers transform acoustic waves into
electrical signals preserving phase and amplitude information. When the
acoustic wavelength is significantly smaller than the transducer diameter,
severe phase modulation of the electrical signal can occur. This results
in anomalous attenuation measurements, background noise in Non-Destructive
Evaluation (NDE), and in general complicates data interpretation. In this
article, we describe and evaluate a phase insensitive transducer based on
the acoustoelectric effect. Theory of operation of the Acousto-Electric
Transducer (AET) is discussed and some optimization procedures outlined
for its use. Directivity data for the AET is contrasted with a conventional
piezoelectric transducer. In addition, transmission scanning data of
phantom flaws in metal plates is presented for both transducers and
demonstrates a significant improvement in resolution with the AET" (author).

1966-364

Carome, E. F., Moeller, C. E. and Clark, N. A., "Intense Ruby-Laser-
    Induced Acoustic Impulses in Liquids," J. Acoust. Soc. Am. 40 (6),
    1462.

    "An experimental study has been made of the acoustic signals induced
in liquids by the focused beam from a Q-spoiled ruby laser. Very intense
acoustic impulses have been produced with laser pulses of less than 0.05 J
total energy. These appear to be generated by dielectric breakdown and
not associated with the hypersonic waves that may be produced simultaneously
by stimulated Brillouin scattering. The observed impulses have peak
pressures of approximately 500 atm and frequency components in excess of
2400 Mc/sec." (Authors), 8 refs.

1977-365

von Gutfield, R. J. and Melcher, R. L., "20 MHz Acoustic Waves from Pulsed
    Thermoelastic Expansions of Constrained Surfaces," J. Acoust. Soc. Am.
    30 (6), 257-259.

    "Repetitive pulses from lasers with pulse widths 5-10 nsec or a
current generator with 10-25-nsec widths have been used to launch acoustic

waves by thermoelastic expansions. For the laser case, when transparent
media such as quartz plates are used to acoustically constrain the energy
absorbing surface, an increase of up to 46 dB at 20 MHz was observed over
that generated from a free surface. An experiment using a scannable laser
to generate elastic waves for flaw detection in a metallic sample is
described." (Author), 5 refs.

1973-366

Thompson, D. O., editor of Proceedings of the Interdisciplinary Workshop
    on Nondestructive Testing - Materials Characterization,
    AFML-TR-73-69, April 1973.

"The field of nondestructive testing and materials characterization
is examined with emphasis on new approaches that may lead to significantly
improved future capabilities. The presentations range from examples of
present capabilities and limitations to field of basic research. The
recommendations of four panels are presented for future research and
development to advance the present state-of-the-art." (Editor)

1975-367

Thompson, D. O., "Interdisciplinary Program for Quantitative Flaw
    Definition - Special Report First Year Effort," ARPA/AFML Contract
    F33615-74-C-5180

This report contains summaries of work performed in:
(1) Quantitative Flaw Definition
    - piezoelectric and electromagnetic transducers
    - data processing
    - theoretical and experimental work on scattering of ultrasound
      from defects
    - system integration
    - sample preparation
(2) Bond Strength
    - acoustical interactions at thin interfaces
    - nature of bonded interface degradation in composites
(3) Failure Prediction
    - determination of residual stresses in structural material
    - acoustic emission

1975-368

Lakin, K. M., "Piezoelectric Transducers," Project I, Unit I, Task 1, in
    Interdis. Pgm. for Quant. Flaw Def. - Spec. Rpt. 1st Yr. Effort,
    8-16.

Work is presented on transducer construction, transducer circuit
modeling and the construction of a data acquisition system for use in
radiation field pattern analysis. 2 refs.

1975-369

Maxfield, B. W., "Optimization and Application of Electrodynamic Acoustic
    Wave Transducers," Project I, Unit I, Task 2, in Interdis. Pgm. for
    Quant. Flaw Def. - Spec. Rpt. 1st Yr. Effort, 17-32.

    Electromagnetic acoustic wave transducers (EMATS) are analyzed
theoretically. The acoustic field predicted is compared to that produced
in an experimental system. 1 ref.

1975-370

White, R. M, and Kerber, G. L., "Analog Data Processing," Project I, Unit
    II, Task 2, in Interdis. Pgm. for Quant. Flaw Def. - Spec. Rpt. 1st
    Yr. Effort, 51-56.

    The rationale for deconvolution filtering is discussed. A system,
utilizing a surface acoustic wave (SAW) filter, is presented. 1 ref.

1975-371

Yee, B. G. W., Couchman, J. C. and Bell, Jr., "Digital Techniques for
    Ultrasonic Flaw Characterization," Project I, Unit II, Task 3, in
    Interdis. Pgm. for Quant. Flaw Def. - Spec. Rpt. 1st Yr. Effort,
    57-85.

    The hardware and software for these investigator's data collection and
signal processing system are described. Numerous timed and frequency
domain signatures for spheroids and flat-bottomed holes were acquired—
many are presented. Preliminary comparisons with theory are discussed.

1975-372

Adler, L., "Models for the Frequency Dependence of Ultrasonic Scattering
    from Real Flaws," Project I, Unit III, Task 1, in Interdis. Pgm. for
    Quant. Flaw Def. - Spec. Rpt. 1st Yr. Effort, 86-111.

    The author discusses a geometrical theory of diffraction for acoustic
waves (based on the electromagnetic theory of Keller). The author's
experimental system for studying the angular and frequency dependence of
acoustic wave scattering from defects is presented. Theory and experiment
agree reasonably well, with excellent agreement as to the spacing of nulls
in the spectra. 4 refs.

1975-373

Packman, P. F. and Coyne, E. J., "Defect Characterization by Spatial
    Distribution of Ultrasonic Scattered Energy," Project I, Unit III,
    Task 2, in Interdis. Pgm. for Quant. Flaw Def. - Spec. Rpt. 1st Yr.
    Effort, 112-127.

"The ability of the ultrasonic indicia to characterize the shape and size of imbedded defects has been developed and examined." (Author), 23 refs.

1975-374

Tittmann, B. R., "Comparison of Theory and Experiment for Ultrasonic Scattering from Spherical and Flat-Bottom Cavities," Project I, Unit III, Task 3, in Interdis. Pgm. for Quant. Flaw Def. - Spec. Rpt. 1st Yr. Effort, 128-139.

The author compares his experimental measurements to the theory of Ermolov (solution of a scalar potential equation for normal incidence of longitudinal waves on a rigid, motionless disk in a fluid). Agreement is good considering the simplicity of Ermolov's solution. 5 refs.

1975-375

Krumhansl, J. A., Gubernatis, J. E., Huberman, M., and Domany, E., "Theoretical Studies of Flaws and NDE," Project I, Unit III, Task 4, in Interdis. Pgm. for Quant. Flaw Def. - Spec. Rpt. 1st Yr. Effort, 140-144.

"1. The general information of integral equation scattering theory for vector elastic waves has been reviewed and summarized in a report just being completed.

While there is much literature (acoustic) for scalar wave problems, we believe this to be the first time that the details for elastic wave systems have been documented. Our writeup can serve as a source for this theory.

2. The first Born approximation to the general integral equation has been obtained both in analytic form, and programmed for computations.

3. For spherical flaws, the exact partial wave solutions in the literature (Truell et al.) have been checked [some algebraic corrections], programmed, and evaluated.

4. Thus, we have computed Born approximation and exact scattering, of incident longitudinal or transverse waves by spherical scatterers - as a function of scattering angle (0 - 180°) and for kr_s from 0 to about 6.

The cases considered are (a) spherical holes in Al, Ti, and stainless steel, and (b) Al and stainless steel spheres in aluminum.

5. The practically useful conclusion is that there are many useful regimes of the first Born approximation - which because of its relative simplicity does not require extensive computing effort for use. This shows promise as a first approximation to explore scattering pattern features (signatures)." (Authors)

1975-376

Kraut, E. A., "Review of Theories of Scattering of Elastic Waves by
    Cracks," Project I, Unit IV, Task 1, in Interdis. Pgm. for Quant.
    Flaw Def. - Spec. Rpt. 1st Yr. Effort, 145-162.

    Scattering of elastic waves by a 2-dimensional crack of arbitrary
shape in an unbounded elastic medium is considered. The Kirchhoff approximation is discussed, and the scattering from a penny-shaped crack is
investigated. 44 refs.

1975-377

Alers, G. A. and Graham, L. J., "Ultrasonic Wave Interaction with
    Interfaces," Project II, Unit I, Task 1, in Interdis. Pgm. for Quant.
    Flaw Def. - Spec. Rpt. 1st Yr. Effort, 183-196.

    "Adhesive bonds of different mechanical strength, between identical
materials, were fabricated by both chemical and thermal means." (Author)
The frequency dependence of the amplitude of ultrasonic signals from the
bonds is experimentally determined and compared to theoretical models
and to the bond strength. 5 refs.

1975-378

Rose, J. L. and Meyer, P. A., "Ultrasonic Signal Processing Methods for
    Adhesive Bond Strength Measurements," Project II, Unit I, Task 2, in
    Interdis. Pgm. for Quant. Flaw Def. - Spec. Rpt. 1st Yr. Effort,
    197-231.

    "The purpose of this work is to examine the effects of selected
attenuation functions in adhesive bond modeling problems so that the
attenuation in signal processing and interpretation can be treated
adequately. Bond models are presently being used to study such problems
as improper substrate surface preparation, improper adhesive cure, or
chemical segregation of the adhesive. .... . Results indicate that
attenuation effects can substantially alter the ultrasonic reflection
even though the bondline is relatively thin. . . ." (Authors)

1974-379

Tittmann, B. R. and Cohen, E. R., "Acoustic Wave Scattering from a
    Sphere," Proc. Interdis. Workshop for Quant. Flaw Def., Tech. Rpt.
    AFML-TR-74-238, 173-186.

    "The objective of this program is to calculate the frequency and
angular dependence of the ultrasonic energy scattered from a solid
ellipsoid of revolution embedded in another solid. This calculation
must take into account the mode conversion that takes place at the
boundaries of the ellipsoid. The first phase will concentrate on the
spherical case so that a simple ellipsoid can then be treated by
perturbation methods. The vector and scalar potential problem for the

sphere has been solved and is programmed onto the computer at the Science
Center. In order to check this program, the case of a rigid, motionless
sphere is being calculated because it can be compared to the results of a
published calculation by Morse.

An integral part of this program is an experimental check on the
calculations performed by making accurate measurements of the angle and
frequency dependence of the scattering of ultrasonic waves in the 1 to
15 MHz range. A 2-1/2 inch diameter by 2-1/2 inch thick sample containing
a single spherical void 400 microns in diameter is being prepared by
diffusion bonding techniques. Pure titanium has been chosen for the host
material because it showed a minimum amount of attenuation and background
scattering. Spheres of tungsten carbide and magnesium will also be
embedded in other titanium samples so that a detailed study of mode
conversion effects can be made." (Author) 3 refs.

1974-380

Mucciardi, A. N., "Adaptive Nonlinear Modeling for Ultrasonic Signal
　　　Processing," Proc. Interdis. Workshop for Quant. Flaw Def., Tech.
　　　Rpt. AFML-TR-74-238, 194-212.

". . . The main objectives of this project are to evaluate the
efficacy of adaptive nonlinear signal processing techniques to model
material flaw descriptors with high accuracy. This modeling synthesis
procedure has its expression in a nonlinear adaptive trainable network.
The procedure is unique because a detailed knowledge of the underlying
physical phenomena is not required. Indeed, it is believed that such
information is contained implicitly in the experimental data, and it is
the purpose of the methodology to extract this information and to generate
models accordingly. . . . In this current project, the feasibility of
adaptive nonlinear signal processing techniques for UNDT will be
demonstrated. In particular, adaptive trainable networks will be
synthesized for characterization of UNDT waveforms for accurate inferences
of: (1) flat-bottom-hole sizes, and (2) the length of fatigue cracks. . . .
These results will provide important information to metallurgical
investigators regarding the relationships between the best-found UNDT
waveform parameter subsets and the underlying physical phenomena."
(Author)

1974-381

Moran, T. J., "Studies of Electromagnetic Sound Generation for NDE,"
　　　Proc. Interdis. Workshop for Quant. Flaw Def., Tech. Rpt.
　　　AFML-TR-74-238, 213-223.

"The technique of electromagnetic sound generation has been known
since 1967. Since it provides a contactless means of generating ultrasonic
waves in metals, application of the technique to NDE would eliminate the
problem of coupling the transducer to the sample under evaluation. It
is also an extremely flexible technique since it can be used to generate
bulk and surface waves of all polarizations. At its present state of
development, the technique is relatively inefficient in converting RF
energy to sound energy in comparison to standard transducer techniques and

it is also material dependent since the generation of the sound occurs
inside the material near the surface. The goal of the present work is
to first optimize the efficiency of the generation process and secondly,
to perform a systematic study of the generation efficiency in many materials
of present interest in manufacturing. We will use both unflawed samples
and those with well-characterized flaws to determine the detection
capabilities." (Author), 4 refs.

1974-382

Felix, M. P., "Laser-Generated Ultrasonic Beams," Proc. Interdis. Workshop
    for Quant. Flaw Def., Tech. Rpt. AFML-TR-74-238, 224-240.

    "A device has been developed which uses a Q-switched laser pulse to
produce a plane compressive stress pulse or a slowly decaying sinusoidal
stress wave train in any solid or liquid material. The device utilizes
a thin liquid layer to totally absorb the laser pulse and generate a stress
pulse by rapid thermal expansion. Compressive stress pulses of 200 nano-
second duration and up to 5 kilobars amplitude have been obtained. Wave
trains of about 30 cycle duration and 1/4 kilobar amplitude (peak-to-peak
in typical solids) have been obtained at frequencies between 1-25 MHz.
Stress amplitudes may be varied by filtering the incident laser radiation.
This device should prove useful wherever large amplitude stress pulses
or large amplitude sinusoidal wave trains are required—such as in
nondestructive testing." (Author)

1974-383

Meyer, P., "Ultrasonic Procedures for Predicting Adhesive Bond Strength,"
    Proc. Interdis. Workshop for Quant. Flaw Def., Tech. Rpt. AFML-TR-
    74-238; 340-351.

    "Theoretical wave propagation models that treat ultrasonic wave
interaction with an adhesively bonded system have been developed. These
models allow the selection of appropriate ultrasonic transducers for bond
inspection analysis. Such problems as a variation in bondline thickness,
the presence of density gradients in the adhesive bond and improper surface
preparation are treated in detail. Preliminary results indicate that the
potential for success appears quite high for obtaining a correlation
between a bond performance parameter and some specific ultrasonic test
parameter." (Author)

1974-384

Yee, B. G. W., "Applicability of Ultrasonic Resonance Spectroscopy to NDE
    of Adhesive Bonds," Proc. Interdis. Workshop for Quant. Flaw Def.,
    Tech. Rpt. AFML-TR-74-238, 352-371.

    "Work being done at General Dynamics involving computerized signal
processing of ultrasonic wave forms from metals, laminates and composites

is discussed. The method has application for materials characterization and defect detection.

A Hewlett-Packard 2100A digital computer was included into a laboratory tool for the signal processing described. The signal processing system includes an ultrasonic pulser, broad-band piezoelectric transducer, stepless gate oscilloscope, display scanner, and computer interface converter channels. Digitized wave forms are filtered and Fourier transformed by computer sof-ware and then displayed on an X-Y grid.

A detailed description of wave-form digitization Fourier transforms, signal convolution and interpretation of results will be presented. The applicability of the computerized scheme to crack width detection, acoustic impedance determination, partial bond characterization and sound velocity measurements will be discussed. The prospective use for measuring the strength of bonded materials will also be discussed." (Author)

1975-385

Cohen, E. R., "Analysis of Ultrasonic Scattering from Simply Shaped Objects," Proc. of the ARPA/AFML Review of Quant. NDE, AFML-TR-75-212, 47-55.

The mathematics of acoustic wave scattering from spheres and spheroids is developed.

1975-386

Krumhansl, J. A., "Basic Theory of Ultrasonic Scattering by Defects: Numerical Studies and Features for Experimental Application," Proc. ARPA/AFML Rev. of Quant. NDE, AFML-TR-75-212, 57-66.

The author summarizes his theoretical work on scattering of ultrasound by defects. An integral equation governing the scattering by an arbitrary shaped flaw is used. 3 refs.

1975-387

Packman, P. F. and Coyne, E. J., "Defect Characterization by Spatial Distribution of Ultrasonic Scattered Energy," Proc. ARPA/AFML Rev. of Quant. NDE, AFML-TR-75-212, 129-146.

Essentially the same as (1975-373).

1975-388

Sachse, W., "Scattering of Ultrasonic Pulses from Cylindrical Inclusions in Elastic Solids," Proc. ARPA/AFML Rev. of Quant. NDE, AFML-TR-75-212, 147-168.

A data collection and analysis system is presented. The angular and
frequency-dependent scattering from imbedded cylinders is studied. The
author performs the important task of identifying the received signals
with the probable ray paths of the ultrasound. 8 refs.

1975-389

Couchman, J., "Digital Measurements of Scattering from Spheroids and
      Flat-Bottom Holes," Proc. ARPA/AFML Rev. of Quant. NDE, AFML-TR-
      75-212, 169-194.

      Essentially the same as (1975-371).

1975-390

Tittmann, B., "Comparison of Theory and Experiment for Ultrasonic
      Scattering from Spherical and Flat Bottom Cavities," Proc. ARPA/AFML
      Rev. of Quant. NDE, AFML-TR-75-212, 195-217.

      Essentially the same as (1975-374).

1975-391

Adler, L., "Angular Dependence of Ultrasonic Waves Scattered from Flat
      Bottom Holes," Proc. ARPA/AFML Rev. of Quant. NDE, AFML-TR-75-212,
      219-245.

      Essentially the same as (1975-372).

1975-392

White, R., "Surface Acoustic Wave Filters for Real Time Processing of
      Ultrasonic Signals," Proc. ARPA/AFML Rev. of Quant. NDE, AFML-TR-75-
      .212, 321-341.

      An expanded version of (1975-370).

      A clearly written and detailed explanation of how a SAW filter may
be used for deconvolution.

1975-393

Mucciardi, A. N., "Adaptive Learning Network Approach to Defect
      Characterization," Proc. ARPA/AFML Rev. of Quant. NDE, AFML-TR-75-
      212, 363-383.

      The feasibility of employing pattern recognition techniques to
ultrasonic NDE is assessed. A data collection system is described and
an adaptive learning network (ALN) is presented. The ALN flat-bottom hole
classifier is found to be extremely accurate. 20 refs.

1975-394

Forsen, G., "Interactive Pattern Analysis and Recognition," Proc. ARPA/AFML
Rev. of Quant. NDE, AFML-TR-75-212, 385-398.

The general approach of applying pattern analysis and recognition
to NDE is discussed. 2 refs.

1975-395

Maxfield, B., "Optimization and Application of Electrodynamic Ultrasonic
Wave Transducers," Proc. ARPA/AFML Rev. of Quant. NDE, AFML-TR-75-212,
399-412.

Essentially the same as (1975-369).

1975-396

Thomas, R., "Acoustic Surface Wave Generation with Electromagnetic
Transducers," Proc. ARPA/AFML Rev. of Quant. NDE, AFML-TR-75-212,
413-428.

The author has extended the technique of EMAT-generated surface
waves to frequencies in the MHz range (typically 4.5-10 MHz). The
possibility of extending the range to 40 MHz seems to be good if the
coil can be placed near enough to the sample surface.

1975-397

Frost, H. M. and Szabo, T. L., "Transducers Applied to Measurements of
Velocity Dispersion of Acoustic Surface Waves," Proc. ARPA/AFML Rev.
of Quant. NDE, AFML-TR-75-212, 429-450.

Wedge transducers, comb transducers, and flat cable and hand wound
EMATS were used to measure surface wave dispersion. Results of the
experiments are presented. 2 refs.

1975-398

Lakin, K., "Piezoelectric Transducers for Quantitative NDE," Proc. ARPA/
AFML Rev. of Quant. NDE," AFML-TR-75-212, 463-478.

A somewhat expanded version of (1975-368).

1975-399

Alers, G. and Graham, L., "Ultrasonic Wave Interactions with Interfaces,"
Proc. ARPA/AFML Rev. Quant. NDE, AFML-TR-75-212, 579-593.

Essentially the same as (1975-377).

1975-400

Rose, J., "Attenuation Influences in Adhesive Bond Modeling," Proc. ARPA/
    AFML Rev. of Quant. NDE, AFML-TR-75-212, 595-611.

    Essentially the same as (1975-378).

1975-401

Seydel, J. A., "Methods Development for Nondestructive Measurement of
    Bond Strength in Adhesively Bonded Structures," Proc. ARPA/AFML
    Rev. of Quant. NDE, AFML-TR-75-212, 613-630.

    The author uses an equivalent-time sampling and digitization system
to acquire pulse-echo data from adhesive bonds. Attempts are made to
characterize adhesive bond strength by a measurement of ultrasonic
reflectivity as a function of frequency. 8 refs.

1975-402

Szabo, T., "Residual Stress Measurements from Surface Wave Velocity
    Dispersion," Proc. ARPA/AFML Rev. of Quant. NDE, AFML-TR-75-212,
    749-767.

    A method of inferring a subsurface residual stress gradient from
surface wave dispersion data is presented. The techniques involve an
inverse Laplace transformation of normalized dispersion data. 5 refs.

1975-403

Richardson, J., "Deducing Subsurface Property Gradients from Surface Wave
    Dispersion Data," Proc. ARPA/AFML Rev. of Quant. NDE, AFML-TR-75-212,
    769-790.

    Two situations are analyzed: (1) the dense data case, in which
dispersion data are assumed to be available for all wavelengths, and
(2) the sparse data case. An estimation theory approach is used to give
the most probable subsurface gradient.

1976-404

Ulrych, T. J. and Clayton, R. W., "Time Series Modelling and Maximum
    Entropy," Physics of the Earth and Planetary Interiors 12, (2/3),
    188-200.

"This paper briefly reviews the principles of maximum entropy spectral
analysis and the closely related problem of autoregressive time series modelling. The important aspect of model identification is discussed with
particular emphasis on the representation of harmonic processes with noise
in terms of autoregressive moving-average models. It is shown that this
representation leads to a spectral estimator proposed by Pisarenko in
1973." (Author), 35 refs.

1976-405

Thompson, D. O., editor of Interdisciplinary Program for Quantitative
    Flaw Definition-Special Report Second Year Effort, Report for ARPA/
    AFML under Contract F33615-74-C-5180.

"The technical results of the second year of effort sponsored by
the ARPA/AFML Center for Advanced NDE . . . are assembled in this report.
They are grouped into three projects . . .
(1) Flaw Characterization by Ultrasonic Techniques
    - electromagnetic transducers
    - characterization of NDE transducers
    - signal processing with SAW devices
    - high frequency ultrasonics
    - adaptive learning
    - sample preparation
    - fundamental scattering studies
    - standards
    - flaw detection in ceramics
(2) Measurement of Strength Related Properties
    - adhesive bond strength
    - strength of composites
(3) Nondestructive Measurement of Residual Stress in Metals
    - inference from harmonic generation
    - inference from efficiency of the electromagnetic generation of
      ultrasound . . ." (Editor).

1976-406

Thompson, R. B. and Fortunko, C. M., "Optimization of Electromagnetic
    Transducer Systems," Interdis. Pgm. for Quant. Flaw Def.-Spec. Rpt.
    2nd Yr. Effort, 1-19.

The electronics and coil designs are optimized to provide the maximum
signal to noise ratio. 13 refs.

1976-407

Lakin, K. M., "Characterization of NDE Transducers," Interdis. Pgm. for
    Quant. Flaw Def.-Spec. Rpt. 2nd Yr. Effort, 30-43.

Field pattern measurements are made utilizing the system described
in previous publications. S-parameters are introduced as a "convenient
means of describing devices involving transmission line type behavior."
11 refs.

1976-408

White, R. M., "Signal Processing with SAW Devices," Interdis. Pgm. for
    Quant. Flaw Def.-Spec. Rpt. 2nd Yr. Effort, 44-99.

An update on the author's work with SAW inverse filters.

1976-409

Elsley, R. K., "Quantitative Estimation of Properties of Ultrasonic
    Scatterers," Interdis. Pgm. for Quant. Flaw Def.-Spec. Rpt. 2nd Yr.
    Effort, 63-86.

The author describes how a data base of theoretical and ultrasonic
scattering results was assembled. The data are to be used to train on
adaptive learning network. "Some efforts were also directed toward
investigating simple, non-adaptive learning techniques for inferring at
least the size of the scattering object from the scattering data."
2 refs.

1976-410

Mucciardi, A. N., "Application of Adaptive Learning Networks to NDE
    Methods," Interdis. Pgm. for Quant. Flaw Def.-Spec. Rpt. 2nd Yr,
    Effort, 87-88.

Describes the objectives of a project to model flaw characteristics
obtained from theoretical ultrasonic scattering waveforms via adaptive
learning decision algorithms.

1976-411

Krumhansl, J. A., "Theoretical Studies of Ultrasonic Scattering and
    Defects," Interdis. Pgm. for Quant. Flaw Def.-Spec. Rpt. 2nd Yr.
    Effort, 102-122.

The regions of applicability and validity of the Born approximation
are determined. Other approximations (static, quasistatic) are evaluated.
Work on scattering from flat cracks is detailed. 6 refs.

1976-412

Tittmann, B. R., "Measurements of Scattering of Ultrasound by Ellipsoidal
    Cavities," Interdis. Pgm. for Quant. Flaw Def.-Spec. Rpt. 2nd Yr.
    Effort, 123-139.

Scattering from ellipsoidal cavities is experimentally investigated.
A contact method is utilized with the cavity at the center of a "doorknob" shaped sample. Incident longitudinal and shear waves are used.
1 ref.

1976-413

Adler, L. and Lewis, D. K., "Models for the Frequency Dependence of
  Ultrasonic Scattering from Real Flaws," Interdis. Pgm. for Quant.
  Flaw Def.-Spec. Rpt. 2nd Yr. Effort, 140-166.

   "Scattering of elastic waves at flaws embedded in titanium was
analyzed by measuring frequency and angular dependence of the scattered
intensity pattern. This scattered intensity pattern was also calculated
from two existing theories: (1) Keller's geometrical theory of
diffraction, which was solved for two-dimensional, crack-like flaws of
circular and elliptical symmetries; (2) "Born approximation," a scattering
theory (introduced by Krumhansl et al., Cornell) for the spherical oblate
and prolate spheroidal cavities. The experimental result was favorable
compared to theory." (Authors)

1976-414

Evans, A. G., Tittmann, B. R., Kino, G. S., and Khuri-Yakub, P. T.,
  "Ultrasonic Flaw Detection in Ceramics," Interdis. Pgm. for Quant.
  Flaw Def.-Spec. Rpt. 2nd Yr. Effort, 177-209.

   A high frequency (200 MHz) ultrasonic system is described. Techniques
for accurate attenuation measurements have been made. Microstructure and
scattering from defects have been studied. 7 refs.

1976-415

Alers, G. A. and Thompson, R. B., "Trapped Acoustic Modes for Adhesive
  Strength Determination," Interdis. Pgm. for Quant. Flaw Def.-Spec.
  Rpt. 2nd Yr. Effort, 215-237.

   "The experiments discussed in this report were designed to consider
the case in which the acoustic energy propagates parallel to the metal
adhesive interface (of an adhesive bond) so that small differences in the
boundary conditions could accumulate over a large interaction distance,"
(Author), 6 refs.

1976-416

Flynn, P. L., "Cohesive Strength Prediction of Adhesive Joints," Interdis.
  Pgm. for Quant. Flaw Def.-Spec. Rpt. 2nd Yr. Effort, 238-263.

   "An analytical study has been carried out to derive the acoustic
spectral response of an attenuating adhesive bondline in terms of the
physical properties of the adhesive . . . Experimental verification of
the derived correlations was provided by systematically varying the
properties of Chemlok 304, . . . correlated well with the ultrasonic amplitude ratio, sound velocity, attenuation coefficient and resonance depth.
Correlation was not evident between resonance quality and strength because
the sound velocity and attenuation of the adhesive were inversely
related." (Author), 6 refs.

1976-417

Rose, J. L. and Thomas, G. H., "Ultrasonic Attenuation Effects Associated
    with the Metal to Composite Adhesive Bond Problem," Interdis. Pgm. for
    Quant. Flaw Def.-Spec. Rpt. 2nd Yr. Effort, 321-342.

"Two different composite modeling approaches were used in this study.
The first considered a 5-layer composite with the interfacial reflection
caused by a very thin epoxy layer between composite layers. The second
model consisted of an area discontinuity factor between each composite
layer that accounted for the reflection factor at the interface. . . . The
effect of composite masking was not significant in the 0-7 MHz range.
After 7 MHz, however, the composite masking begins to show some significant effects, the effects however still being separable from the surface
preparation or bond quality information." (Author), 5 refs.

1977-418

Thompson, D. O., editor of Interdisciplinary Program for Quantitative Flaw
    Definition-Special Report Third Year Effort, Report for ARPA/AFML
    under Contract F33615-74-C-5180.

"This report presents technical summaries of the various research
tasks that have been pursued in the third year of effort by the ARPA/AFML
Center for Advanced NDE. . . . They are grouped into two projects:
    (1) Flaw Characterization by Ultrasonic Techniques
        - electromagnetic and piezoelectric transducers
        - signal processing (SAW and CCD)
        - sample preparation
        - fundamental scattering studies (experimental and theoretical)
        - imaging
        - adaptive learning
        - inversion techniques
        - failure prediction in ceramics
        - detection and characterization of surface flaws
    (2) Measurement of Strength Related Properties
        - adhesively bonded materials
        - composite materials
        - residual stress"

1977-419

Lakin, K. M. and Strand, T., "Characterization of NDE Transducers," Interdis.
    Pgm. for Quant. Flaw Def.-Spec. Rpt. 3rd Yr. Effort, 21-41.

"The problem of characterizing NDE transducers has been approached
from two directions. First, the radiation pattern of the transducer has
been analyzed in terms of Fourier transform reconstructions that yield
information about the magnitude and phase of the fields anywhere in the
region beyond the very near field. . . . The program (also) resulted in
a simple but concise method for modelling the transducers as two-part
hybrid networks . . ." (Authors), 10 refs.

1977-420

White, R. M., "Signal Processing Research in Connection with Ultrasonics in Non-Destructive Testing," Interdis. Pgm. for Quant. Flaw Def.-Spec. Rpt. 3rd Yr. Effort, 43-58.

An analog data acquisition is described using a change coupled device (CCD) video delay line. A fast clock is used during acquisition; a slow clock for readout. Also a CCD transversal filter is used for matched filtering.

1977-421

Krumhansl, J. A., "Development and Application of Ultrasonic Scattering Theory to Non-Destructive Evaluation—Three Year Summary," Interdis. Pgm. for Quant. Flaw Def.-Spec. Rpt. 3rd Yr. Effort, 64-69.

A summary of the following work is presented.
"(a) Careful examination of the general long wave limit in order to determine the maximum number of independent defect parameters which can be determined from scattering data.
 (b) Long wave and high frequency limits of scattering by cracks. (The Born approximation is not well defined for cracks.)
 (c) Addressing the 'inverse' problem.
 (d) Attempts to obtain an 'exact' (calibration) scattering solution for a few spheroidal geometries, to complete evaluation of Born approximation errors." (Author), 19 refs.

1977-422

Domany, E., "Utilization of Physical Features of Scattered Power for Defect Characterization," Interdis. Pgm. for Quant. Flaw Def.-Spec. Rpt. 3rd Yr. Effort, 70-81.

"The first Born approximation provides a useful means to study scattering of ultrasound by various defects. In particular, it seems to yield qualitatively good results for the scattered power when averaged over a range of frequencies. Features of the scattered power that have been discovered by this method are reviewed. A convenient way to summarize the scattering data, by numerical projections, was used to assemble a library of scattered power from various defects. Addressing the particular problem of an oblate spheroidal cavity, a step-by-step procedure to determine its orientation and shape is suggested. Areas of future development are indicated." (Author), 10 refs.

1977-423

Tittmann, B. R., Elsley, R. K., Nadler, H., and Cohen, E. R., "Experimental Measurements and Interpretation of Ultrasonic Scattering by Flaws," Interdis. Pgm. for Quant. Flaw Def.-Spec. Rpt. 3rd Yr. Effort, 82-121.

"The objective of this investigation was to develop procedures for deducing key geometric features of flaws from the details of the ultrasonic scattered fields, and, in particular, those features necessary for the evaluation against quantitative accept/reject criteria derived from fracture mechanics. To accomplish this objective, the investigation sought to correlate flaw characteristics such as size, shape, orientation and content of the flaw with the absolute value of scattered power and its variation with scattering angle and ultrasonic frequency, to verify theoretical scattering models developed by Krumhansl et al. (this report), and to lay the basis for inversion procedures developed by Mucciardi (see this report) and Bleistein (see this report)." (Author), 14 refs. ("This report" refers to Ref. 418.)

1977-424

Adler, L., "Identification of Flaws from Scattered Ultrasonic Fields as Measured at a Planar Surface," Interdis. Pgm. for Quant. Flaw Def.-Spec. Rpt. 3rd Yr. Effort, 122-158.

"The objective of this investigation was to correlate ultrasonic scattering data—such as the variation of scattered power with angle and frequency, mode conversion, etc.—to characteristics of a flaw in solids such as size, shape, and orientation by using flat samples and an immersed system." (Author), 10 refs.

1977-425

Kino, G. S., "New Techniques for Acoustic Nondestructive Testing," Interdis. Pgm. for Quant. Flaw Def.-Spec. Rpt. 3rd Yr. Effort, 159-175.

A phased array imaging system is described. The circuitry and inverse filter system are useful to anyone constructing an ultrasonic system (quantitative or imaging).

1977-426

Mucciardi, A. N., Shankar, R., Shaley, M. F., and Johnson, M. D., "Application of Adaptive Learning Networks to NDE Methods," Interdis. Pgm. for Quant. Flaw Def.-Spec. Rpt. 3rd Yr. Effort, 176-231.

Adaptive learning networks to the following problems.
- measurement of the size and acoustic impedance of spherical defects from the analysis of theoretically scattered waveforms.
- actual scattering from spherical defects.
- estimate the size and orientation of spheroidal defects from analysis of the Born approximation model. 6 refs.

1977-427

Bleistein, N. and Cohen, J., "Application of a New Inverse Method for
    Nondestructive Evaluation," Interdis. Pgm. for Quant. Flaw Def.-Spec.
    Rpt. 3rd Yr. Effort, 232-246.

"The application of a new inverse method to nondestructive evaluation
is described. In particular, detection of a small hole in an otherwise
homogeneous solid is discussed. The scattering of an acoustic probe by
the hole is considered. It is shown that the scattered wave is proportional
to the Fourier transform of the characteristic function of the domain
occupied by the hole. The characteristic function is equal to unity in that
domain and zero outside. Thus, knowledge of this function characterizes
the domain. The basic result is derived under the assumption that the
scatterer is small - allowing use of the Born approximation - and "far"
from the surface of the solid. Some features of aperture limited - band
limited and aspect angle limited - observations are discussed. The
applicability of this inverse method to non-destructive evaluation is
demonstrated by this preliminary analysis." (Authors), 12 refs.

1977-428

Kino, G. S., Khuri-Yakub, B. T., Tittmann, B. R., Ahlberg, L., Evans, A. G.,
    Biswas, R., and Fulrath, R., "Ultrasonic Failure Prediction in
    Ceramics," Interdis. Pgm. for Quant. Flaw Def.-Spec. Rpt. 3rd Yr.
    Effort, 247-266.

Defect characterization in the size range (10-100 micrometer) in
structural ceramics was performed. High frequency techniques (200 MHz)
were employed. A new ultrasonic technology, based on ZnO was developed
as part of the problem. 8 refs.

1977-429

Alers, G. A. and Elsley, R. K., "NDE Techniques for Measuring the Strength
    of Adhesion," Interdis. Pgm. for Quant. Flaw Def.-Spec. Rpt. 3rd Yr.
    Effort," 271-285.

"It has been the objective of this study to find ultrasonic techniques
that can give a quantitative measure of the status of the metal to
adhesive interface so that the adhesion strength of an adhesive bond could
be predicted. . . . it is possible to predict the approximate strength
of the adhesive bond from a measurement of the splitting of the lowest
standing wave resonance in the adherends." (Authors), 7 refs.

1977-430

Flynn, P. L. and Henslee, S. P., "Cohesive Bond Strength Prediction,
    FM-400 a Realistic Adhesive System," Interdis. Pgm. for Quant. Flaw
    Def.-Spec. Rpt. 3rd Yr. Effort, 286-307.

". . . Scattering analysis was applied to an adhesive layer and provided a basis for choosing measureable ultrasonic parameters that characterized the acoustic properties of the layer. This method was applied to simple adhesive systems with good results, but found some problems in general application. The largest problem in the scrimmed adhesive was entrapment of small voids in the scrim pattern. The small voids affected the attenuation measurements, but did not affect the cohesive strengths." (Authors), 8 refs.

1977-431

Thompson, D. O., Proceedings of the ARPA/AFML Review of Progress in Quantitative NDE, Tech. Rpt. AFML-TR-77-44.

These edited transcripts contain information relating to quantitative NDE. Included are summaries of work on:
Adhesives and composites
New materials and techniques
Measurement of internal stress
Fundamentals of acoustic emission
Signal acquisition and processing
Defect characterization
    - fundamentals (experimental and theoretical) and techniques.

"In addition a Mini-Symposium is presented related to Advances in Electromagnetic Transducers." (Editor)

1977-432

Alers, G. A., "Trapped Acoustic Modes for Adhesive Strength Determination," Proc. ARPA/AFML Rev. Prog. Quant. NDE, AFML-TR-77-44, 52-58.

"In order to extend the time and distance over which an ultrasonic wave can interact with the strength determining layer at a metal to adhesive interface, we have considered the acoustic wave modes that propagate parallel to the interface in the plane of a sandwich type adhesive bond between two metal plates. A detailed mathematical analysis of these bound modes was made with special attention to the role played by the boundary conditions at the adhesive to metal interface so that the frequencies and modes that are most sensitive to the bounding conditions could be predicted. Experiments to verify these predictions were carried out using surface waves launched into the adhesive layer from the external metal plates and by current carrying wires embedded in the adhesive subjected to an external, static magnetic field. The theoretical analysis also predicted that the fundamental thickness mode of vibration of the entire sandwich structure should also exhibit a sensitivity to the boundary conditions. This was studied experimentally by taking the Fourier transform of low frequency echos reflected from the structure." (Author)

1977-433

Flynn, P. L., "Cohesive Strength Prediction of Adhesive Joints," Proc. ARPA/AFML Rev. Prog. Quant. NDE, AFML-TR-77-44, 59-65.

"An analytical study was carried out to determine the influence of changes in bondline properties on measurable ultrasonic parameters in both the time and frequency domains. An experimental study was conducted in which the adhesive properties were varied by mixing a paste adhesive in different ratios of resin and hardener. The properties of the adhesive bondlines were measured in-situ with high frequency, broad-band ultrasonics. Physical properties extracted from the ultrasonic data included the sound velocity, acoustic impedance and the attenuation of the adhesive layer. The expected correlations were seen between the NDE parameters identified by the analytical study and the strength and stiffness of the bonded specimens." (Author), 3 refs.

1977-434

Buckley, M. J. and Raney, J. M., "The Use of Continuous Wave Ultrasonic Spectroscopy for Adhesive-Bond Evaluation," Proc. ARPA/AFML Rev. Prog. Quant. NDE, AFML-TR-77-44, 66-73.

"For certain NDE applications, the use of CW ultrasonic spectroscopy to acquire ultrasonic transmission and reflection data has several advantages over pulse techniques. The specific system currently used to record amplitude and phase as a function of frequency over the range of 20 KHz to 20 MHz will be discussed. In addition, theoretical calculations of the ultrasonic spectra for adhesively bonded structures will be presented along with initial results obtained in fitting the theoretical calculations to the experimental data in order to determine the acoustic properties of the adhesive layer." (Author), 2 refs.

1977-435

Evans, A. G., "Ultrasonic Flaw Detection in Ceramics," Proc. ARPA/AFML Rev. Prog. Quant. NDE, AFML-TR-77-44, 74-77.

"A high frequency ultrasonic approach for determining defects in ceramic materials (in the size range required for failure prediction) has been outlined. A 200 MHz A-scan device pulsed with a short (2 ns) pulse has been constructed and shown to have a good dynamic range (70 dB) and a depth resolution of at least 25 microm. A B-scan system for defect detection studies has also been developed and is ready for use.

Techniques for accurate attenuation measurements in ceramics have been developed and automated. Preliminary data have also been obtained on a range of ceramic polycrystals. Calculations of the scattering from defects in ceramics, and of bond losses in thin gold foils, have been used in cylinders with the attenuation data to predict typical defect detectabilities. These calculations predict that defects in the size range 20-100 microm. should be detectable (with the present transducers) in fully-dense, fine-grained ceramics. Preliminary defect detection studies have confirmed that defects at least as small as 100 microm. are readily detectable in these materials." (Authors)

1977-436

Lakin, K. M., "Characterization of NDE Transducers," Proc. ARPA/AFML
    Rev. Prog. Quant. NDE, AFML-TR-77-44, 116-122.

"A system for characterizing NDE transducers has been implemented
which involves both electrical circuit modeling and measurements of the
amplitude and phase of the radiation patterns. The field pattern
measurements allow a determination of the field at the transducer surface
as well as at locations distant from the transducer. The technique is
also adaptable to characterizing scattering surfaces treated as apparent
sources. The electrical characterization centers around a network model
involving a hybrid set of S-parameters. Using simple and readily available references, the four parameters of the transducer may be determined
and then used to quantitively predict the transducer performance in
scattering experiments. In its simplest form the technique uses water
bath experiments and scattering off the transducer surface." (Author)

1977-437

Szabo, T. L., "Surface Acoustic Wave Electromagnetic Transducer Modeling
    and Design for NDE Applications," Proc. ARPA/AFML Rev. Prog. Quant.
    NDE, AFML-TR-77-44, 128-132.

"Recent progress in SAW electromagnetic transducer (EMT) modeling
and fabrication techniques have greatly increased EMT versatility for NDE
applications. Unlike other types of SAW NDE transducers that suffer from
variability in manufacture and coupling conditions, virtually identical
EMT's can be made that have predictable characteristics. During the past
year we have developed a new equivalent circuit model for the noncontact
EMT that describes both its acoustical and electrical characteristics.
This model is useful for design and for assessing the effects of electrical
matching and transduction on different materials. Perhaps the most striking
result of the model is the similarity in frequency response between the
meander line SAW EMT and the SAW interdigital transducer. For conventional
EMT's, design is simple and in excellent agreement with experiment.
By modification of transducer geometry, as with IDT's, more advanced fre-
quency response shapes can be realized. This result implies that for SAW
EMT's signal processing functions can be combined with transduction for
NDE applications." (Author), 10 refs.

1977-438

Maxfield, B. W., "Optimization and Application of Electrodynamic Acoustic
    Wave Transducers," Proc. ARPA/AFML Rev. Prog. Quant. NDE, AFML-TR-
    77-44, 133-135.

"We have used electromagnetic acoustic-wave transducers (EMATs) to
measure the intensity distribution of shear ultrasonic waves scattered
by a cylindrical flaw and a conventional flat-bottomed hole. Results are
in reasonable qualitative agreement with the behavior expected for these
scattering arrangements. Quantitative measurements, however, have proven
very difficult to obtain because of fundamental problems in providing
adequate shielding for the receiver coil while avoiding distortion of the

drive field. Our experimental work has defined the problem areas and in most cases suggested solutions. It now seems quite probable that other work, both in this program and outside, may yield definitive solutions to the problems that we have identified so that in the near future it may be possible to have quantitative shear wave scattering information from scanned EMAT measurements." (Author)

1977-439

Moran, J. J., "Characteristics and Applications of Electromagnetic Surface Wave Transducers," Proc. ARPA/AFML Rev. Prog. Quant. NDE, AFML-TR-77-44, 136-141.

"The steerability and relative generation efficiency of bulk acoustic waves generated by a meander line EMAT in aluminum have been determined. The frequency dependence of the propagation angle relative to the surface was found to agree well with theory. Efficiency for shear-wave generation (frequency range, 5-9 MHz) was about 4 dB per conversion less than the Rayleigh-wave generation efficiency at 4.6 MHz, and for longitudinal waves (frequency range, 10-24 MHz) about 10 dB less.

A second effort to achieve piezoelectric SAW device signal processing capabilities with EMAT designs has shown that a pulse compression device is completely feasible. We have demonstrated that it is possible to compress a 3.5 μsec (2-6 MHz) chirp signal to a 0.25 μsec pulse. Possible applications will be discussed." (Author)

1977-440

Thompson, R. B. and Fortunko, C. M., "Optimization of Electromagnetic Transducer Systems," Proc. ARPA/AFML Rev. Prog. Quant. NDE, AFML-TR-44, 142-147.

"The results of a program to realize maximum dynamic ranges with lightweight electromagnetic transducers suitable for hand-held use is described. Transducers were constructed using compact samarium-cobalt permanent magnets and: (1) spiral coils for generating axially polarized shear waves, (2) masked coils for generating plane polarized shear waves, and (3) meander coils for generating surface waves. A high power, line generator was constructed which uses a spark-gap to switch currents of several hundred amperes in either single pulse or tone burst mode of operation. A new transformer coupling network and a broadband, low noise preamplifier have been demonstrated for an overall 30 dB increase in sensitivity. Dynamic ranges as high as 80 dB were obtained in the tone burst mode ($\Delta f/f \sim 10\%$). Applications of these transducers to problems of the Army and EPRI will be briefly discussed." (Author), 7 refs.

1977-441

White, R. M., "Signal Processing with Surface Acoustic Wave Devices," Proc. ARPA/AFML Rev. Prog. Quant. NDE, AFML-TR-77-44, 148-153.

"A surface acoustic wave (SAW) filter was designed and constructed having a response inverse to that of a simulated NDT system. Further testing of this filter was undertaken with the aim of characterizing the filter more accurately. A redesign of the filter to include an input transducer having less loss was then carried out. It appears that the use of a two-pair of three-pair input IDT is advantageous, and a filter incorporating such a transducer is now being fabricated for use with the simulated NDT system (5 MHz PZT transducer cemented onto an aluminum block). With this filter, determination is to be made of all the relevant electrical characteristics, as well as determination of the minimum spatial resolution obtainable at the filter output when two closely-spaced impedance discontinuities produce reflection." (Author), 3 refs.

1978-442

Haines, N. F., Bell, J. C., and McIntyre, P. J., "The Application of Broadband Ultrasonic Spectroscopy to the Study of Layered Media," J. Acoust. Soc. Am. 64 (6), 1645-1663.

"An investigation has been made of the frequency dependence of amplitude and phase information when broadband ultrasonic pulses, in the region 1-30 MHz, are reflected from layered targets. An on line computer performing Fourier analysis of sampled ultrasonic pulses allowed both amplitude and phase information to be studied. Layers of various acoustic impedances, velocities, and attenuation have been investigated, and in particular, layers of magnetite grown on mild steel. In all cases excellent agreement between experiment and theory has been achieved. The possible use of the techniques of deconvolution has also been considered for the measurement of the thickness of layers. The methods developed have found application in the problem of determining the thickness of a corrosion layer on the inside surface of a component where access can only be gained through the outer surface." (Authors), 64 refs.

1977-443

Heyman, J., "A Non-Phase Sensitive Transducer for Ultrasonics," Proc. ARPA/AFML Rev. Prog. Quant. NDE, AFML-TR-77-44, 154-156.

The phase-insensitive nature of a CsS acoustoelectric converter is discussed for applications in NDE.

1977-444

Krumhansl, J. A., "Interpretation of Ultrasonic Scattering Measurements by Various Flaws from Theoretical Studies," Proc. ARPA.AFML Rev. Prog. Quant. NDE, AFML-TR-77-44, 164-172.

"From a review and extension of the general theory of ultrasonic scattering by defects in elastic media we have developed computer programs and approximation methods for analyzing experimental scattering data. Few exact theoretical expressions are available, but significant information can be obtained from the "Born approximation," the quasi-static approximation, and the exact long wave limit. Computed results for spheres and spheroids (prolate and oblate), for both longitudinal and transverse incident and scattered waves will be presented. In addition, we explore several methods for visual and graphical presentation of the analytical results for most convenient use in test situations." (Author), 6 refs.

1977-445

Tittmann, B. R., "Scattering of Ultrasound by Ellipsoidal Cavities," Proc. ARPA/AFML Rev. Prog. Quant. NDE, AFML-TR-77-44, 173-179.

"Experimental results have been compared with theory for ellipsoidal and spherical cavities embedded in titanium alloy by the diffusion bonding process. The measurements comprised the cases of incident longitudinal and shear waves including mode conversion. Whenever possible, comparisons were performed with the results of exact theory and those of the Born approximation. The Born approximation was found useful in the back scattering directions for low ka values (k is the wave vector of the sound wave and a is the radius of the scatterer). In the experiments, a reciprocity relation was discovered which should prove very useful in further studies: The same angular dependence is obtained in mode conversion when the mode of the incident and scattered wave is interchanged. This result has now been corroborated by both the exact theory and the Born approximation. The results are discussed in the context of failure prediction." (Author), 4 refs.

1977-446

Adler, L. and Lewis, K., "Models for the Frequency Dependence of Ultrasonic Scattering from Real Flaws," Proc. ARPA/AFML Rev. Prog. Quant. NDE, AFML-TR-77-44, 180-186.

"Scattering of elastic waves at flaws embedded in titanium was analyzed by measuring frequency and angular dependence of the scattered intensity pattern. This scattered intensity pattern was also calculated from two existing theories: (1) Keller's geometrical theory of diffraction, which was solved for two-dimensional crack-like flaws of circular and elliptical symmetries; (2) "Born approximation," a scattering theory (introduced by Krumhansl et al., Cornell) for the spherical oblate and prolate spheroidal cavities. The experimental result was favorably compared to theory." (Authors)

1977-447

Mucciardi, A. N., "Measurement of Subsurface Fatigue Crack Size Using
      Nonlinear Adaptive Learning," Proc. ARPA/AFML Rev. Prog. Quant. NDE,
      AFML-TR-77-44, 194-199.

      "A new NDE nonlinear signal processing system has been developed to
detect and measure small, subsurface fatigue cracks. The system synthesized
from nondestructive evaluation (NDE) waveform parameter inputs is capable
of detecting and measuring quantitatively subsurface fatigue cracks in the
size range of 0 to 279 mils to within 70 percent of their nominally
characterized lengths. Previous investigations had achieved a 50 percent
detection rate for cracks larger than 30 mils. However, the fatigue
crack measurement system reported herein is the first known fatigue crack
NDE system capable of detection and measurement for this wide range."
(Author)

1977-448

Rose, J. L., Eisenstein, B., Fehlauer, J., and Avioli, M., "Defect
      Characterization—Fundamental Flaw Classification Solution
      Potential," Proc. ARPA/AFML Rev. Prog. Quant. NDE, AFML-TR-77-44,
      200-207.

      "Emphasis in the paper will be placed on a work description and
analysis associated with a flaw classification problem of discriminating
between ultrasonic signals that have been reflected from elliptical
and circular side drilled electro discharge machined slots in a steel
block. The flaw types used in this experiment are several elliptical
holes with eccentricities, from .15 to 1.0. The signals are sampled at
a 100 MHz rate and quantized with an 8 bit word length. The signal
processing is performed on a PDP 11/05 mini-computer. . . . Results
obtained thus far indicate that for minor diameter to major diameter
ratios e in excess of 0.7, discrimination between elliptical and
circular flaws is very difficult. For e less than 0.3, discrimination
is easy. Consequently, the feature extraction and pattern classification
techniques have been concentrated on e in the range 0.3 to 0.7 in order
to establish the efficacy of the research protocol." (Authors)

1978-449

Thompson, D. O., Proceedings of the ARPA/AFML Review of Progress in
      Quantitative NDE, Tech. Rpt. AFML-TR-78-55.

      "The edited transcripts of the ARPA/AFML Review of Quantitative
NDE . . . are presented in this document." (Editor)
      Topics covered in the review include
      Flaw Characterization by Quantitative Ultrasonics
      - long wave scattering
      - experimental measurements
      - use of physical features of scattered power for defect
        characterization
      - inversion methods

Minisymposium on Scattering Theories
New Techniques and Phenomena
- prediction of remaining life of fatigue damaged samples
- exoelectron emission
- measurement of stress profiles
- ultrasonic tomography
- imaging
- ultrasonic transducers
- acoustic emission
- penetrants
NDE for Advanced Materials
- fracture mechanics
- bond strength
Reliability of Structural Ceramics
- high frequency ultrasonics
- ultrasonic microscopy
New Technology
- electronic equipment
- transducers
- standards
- imaging
- adhesive bonds

1978-450

Gubernatis, J. E., "Long Wave Scattering of Elastic Waves from Volumetric
   and Crack-Like Defects of Simple Shapes," Proc. ARPA/AFML Rev. Prog.
   Quant. NDE, AFML-TR-78-55, 21-25.

"The development of several approximations appears to permit accurate
and practical calculations of the scattering of elastic waves from
volumetric and crack-like defects of simple shapes if the wavelength of
the incident wave is larger than the characteristic length of the shape.
These approximations, which I call the quasi-static and extended quasi-
static, use static solutions of defects in uniform strains to predict
scattered (dynamic) fields. Since static solutions for several simple
defect shapes (oblate and prolate spheroid, ellipsoid, and circular and
elliptical cracks) are available, scattering predictions are possible, and
the results of such calculations are presented." (Author), 10 refs.

1978-451

Tittmann, B. R. and Elsley, R. K., "Experimental Measurements and Interpretation of Ultrasonic Scattering by Flaws," Proc. ARPA/AFML Rev.
   Prog. Quant. NDE, AFML-TR-78-55, 26-35.

"Experimental measurements have been carried out on the scattering
of elastic waves from ellipsoidal and spherical cavities embedded in
titanium alloy by the diffusion bonding process. The scattering data
have been compared with calculations from exact theory and those from
the Born approximation. The results allow the following conclusions:
(1) the new concept of sample fabrication by the diffusion bonding process
is proven to be successful, (2) the scattering data have been found to be

a valuable test for the evaluation and refinement of scattering theories which treat scattering from ellipsoidal cavities on an approximate basis, (3) the scattering data provide a useful data base for use in developing scattering inversion techniques (i.e., deterministic, probabilistic, or adaptive schemes) from which quantitative properties of the scattering center can be rapidly extracted; (4) the work has provided a preliminary definition of the minimum quantity and type of data acquisition needed for a "smart" NDE system." (Authors), 3 refs.

1978-452

Adler, L., Lewis, K., Szilas, P., and Fitting, D., "Identification of Flaws from Scattered Ultrasonic Fields as Measured at a Planar Surface," Proc. ARPA/AFML Rev. Prog. Quant. NDE, AFML-TR-78-55, 36-43.

"Ultrasonic wave scattering from ellipsoidal and cylindrical cavities embedded in titanium was measured and analyzed with a newly designed signal processing system. Using an immersion system and samples with flat faces, the range of waves incident, at certain polar and azimuthal angles, was determined for both L-L and L-S scattering. Attempts were made to define key parameters from both amplitude and phase spectra for characterizing cavities. Results are compared to predictions of Born approximations (developed by Krumhansl et al. at Cornell) and to experimental results taken by a contact system (Tittmann et al. at Rockwell). A new (Keller type) theory for crack-like defects which includes mode conversion will also be presented." (Authors), 2 refs.

1978-453

Domany, E., "Utilization of Physical Features of Scattered Power for Defect Characterization," Proc. ARPA/AFML Rev. Prog. Quant. NDE, AFML-TR-78-55, 44-49.

"The first Born approximation provides a useful means to study scattering of ultrasound by various defects. In particular, it seems to yield qualitatively good results for the scattered power, when averaged over a range of frequencies. Features of the scattered power that have been discovered by this method will be reviewed. Closer study of some of these features leads to a step procedure to characterize an oblate spheroidal defect; to determine its orientation and shape. Procedures for extension to other shapes can also be given. Areas of future development will be indicated." (Author), 10 refs.

1978-454

Shankar, R., Mucciardi, A. N., Whaley, M. F., and Johnson, M. D., "Inversion of Ultrasonic Scattering Data to Measure Defect Size, Orientation and Acoustic Properties," Proc. ARPA/AFML Rev. Prog. Quant. NDE, AFML-TR-78-55, 50-72.

"Empirical solutions via the adaptive learning network methodology have been obtained to measure characteristics of three-dimensional defects (spherical and spheroidal) from the analysis of theoretically-modeled scattered waveforms. The solutions have been successfully applied to measure defects from actually observed ultrasonic scattering data.

Spherical voids and inclusion in Ti-6-4, varying in diameter from 0.02 cm to 0.12 cm, and varying in acoustic impedance ratio (with respect to the host alloy (Ti-6-4) from zero for air cavities to four for tungsten-carbide inclusions, can be directly measured via:

(i) The phase cepstrum - which yields an unambiguous measurement of defect diameter and is independent of its acoustic impedance ratio.

(ii) Adaptive Learning Networks (ALN) - synthesized from the amplitude spectrum and which yield accurate measurements of defect diameter and the acoustic impedance ratio of the included material.

The two empirical solutions, synthesized from the scattering data from an exact model for spheres, yield similar accurate results when applied to actual scattering observed from the defects." (Authors), 6 refs.

1978-455

Bleistein, N. and Cohen, J. K., "Application of Geometrical Diffraction Theory to Scattering by Cracks," Proc. ARPA/AFML Rev. Prog. Quant. NDE, AFML-TR-78-55, 102-107.

"We formulate the inverse problem as an equation or system of equations in which one of the unknowns is a function which directly characterizes the irregularity to be determined. Under the assumption of small sized anomalies or small changes in media properties, our system reduces to a single linear integral equation for this "characteristic" function. In many cases of practical interest, this equation admits closed form solutions. Even under the constraints of practical limitations on the data, information about the irregularity can be deduced.

As an example, we consider the case of a void in a solid probed by acoustic waves. We show how high frequency data can be directly processed to yield the actual shape of the anomaly in a region of the surface covered by specular reflection of the probe. In the low frequency case, we show how to directly process the data to yield the volume, centroid, and "products of inertia" of the void." (Authors), 10 refs.

1978-456

Sachse, W., "Measurement of Phase and Group Velocities of Dispersive Waves in Solids," Proc. ARPA/AFML Rev. Prog. Quant. NDE, AFML-TR-78-55, 122-126.

"The dispersion relation and the propagational speeds of waves in dispersive solids are determined by a newly developed technique in which the phase function of spectral analyzed broad band pulses is determined. The method is simpler and in agreement with the continuous wave-resonance technique. Application is made to ultrasonic stress waves and pulses propagating in fiber-reinforced composite materials." (Author)

1978-457

Heyman, J. S., "A Phase Insensitive Ultrasonic Receiver," Proc. ARPA/AFML
   Rev. Prog. Quant. NDE, AFML-TR-78-55, 151-156.

"A phase insensitive receiver transducer called an acousto-electric converter (AEC) is described and several applications to NDE are presented. Although phase information in ultrasonics is used for many sensitive material measurements, the phase signal may effectively result in noise under certain circumstances. When the propagating phase front is spacially variant and the acoustic receiver is larger than the acoustic wavelength, severe modulation of the resulting electrical signal occurs. Under these conditions, the electrical output of the transducer is a complicated superposition of incident waves of different phase and amplitude.

These effects are particularly troublesome in applied studies such as NDE where non-flat and parallel inhomogeneous materials are investigated. To examine the application of the new receiver to NDE, comparison data is presented obtained under identical conditions for ultrasonic transmission through materials using conventional as well as AEC transducers. The materials investigated contain phantom flaws in both metals and composites. The results indicate that an AEC may be a useful device for material characterization." (Authors), 11 refs.

1978-458

Kwan, S. H., White, R. M., and Muller, R. S., "Integrated Ultrasonic
   Transducer," Proc. ARPA/AFML Rev. Prog. Quant. NDE, AFML-TR-78-55,
   157-160.

"Ultrasonic transducers composed of integrated assemblies of double-diffused MOS transistors (DMOST) and thin-film piezoelectric transducing elements are described. The entire transducer is built on a single-crystal silicon wafer and offers a number of attractive features including: small size and correspondingly precise localization of the sensitive element, a response that can be predicted by relatively simple theory, a large bandwidth and possibility of producing arrays of sensors together with other signal-processing elements in a single processing sequence.

The piezoelectric film (zinc oxide) is sputtered either in one gate region of a field effect transistor (making the "PIFET" structure) or adjacent to the gate electrode of a double-diffused MOS transistor (the "PI-DMOST" structure). The transducer may be excited in various ways: (1) in a thickness mode from the bare silicon surface opposite the piezoelectric-coated region; (2) in a flexural mode caused by bending

the silicon wafer; (3) end excitation by surface motions either normal
or transverse to the edge of the wafer; (4) by surface waves. Various
of these modes are characterized by high sensitivity to strain, low
conversion loss, large bandwidth, and good response at very low or very
high frequencies." (Authors)

1978-459

Alers, G. A., Elsley, R. K., and Flynn, P. L., "Bond Strength Measurements
    by Ultrasonic Spectroscopy," Proc. ARPA/AFML Rev. Prog. Quant. NDE,
    AFML-TR-78-55, 191-197.

"During the current phase of the program, more reliable and
statistically significant mechanical tests for the strength of an
adhesively bonded joint have been developed and more accurate measurement
techniques for extracting quantitative information from both the time
domain and the frequency domain representations of the ultrasonic data
have been found. As a result, measurements of the wave velocity and the
attenuation in FM-400 adhesive sandwiched between sheets of aluminum alloy
and subjected to different degrees of cure successfully predicted the
final cohesive shear strength of the joints. Quantitative measurements
of the standing wave resonant frequencies in aluminum-Chemlok 304
adhesive-aluminum sandwiches showed a strong correlation with the strength
of adhesive joints prepared with weak adhesion at the metal to adhesive
interfaces. By making many adhesive joint samples and testing their
strengths specifically in the region interrogated by the ultrasound, the
reliability of all the correlations between strength and ultrasonic
measurements were greatly improved." (Authors)

1978-460

Evans, A. G., "Overview of Reliability in Structural Ceramics," Proc.
    ARPA/AFML Rev. Prog. Quant. NDE, AFML-TR-78-55, 233-236.

"The failure prediction requirements and the pertinent accept/reject
criteria for structural ceramics are derived, and the available failure
prediction techniques are examinedd, vis-a-vis the failure prediction
relations, in order to highlight the capabilities and limitations of
each technique. The need for additional techniques is thereby demonstrated. The capabilities of the ultrasonic technique are extensively
evaluated in order to determine its ability to satisfy the deficiencies
in the existing failure prediction repertoire. The prospects are shown
to be very encouraging, but the results of several key studies must be
awaited before defining the ultimate role of ultrasonic failure prediction techniques." (Author)

1978-461

Kino, G. S., Khuri-Yakub, B. T., and Tittmann, B. R., "High Frequency
    Ultrasonics," Proc. ARPA/AFML Rev. Prog. Quant. NDE, AFML-TR-78-55,
    237-240.

"A high frequency 250 MHz A-scan system has been used for flaw
detection. We have been able to detect 25-500 μm defects of different
types (C, Si, SiC, BN, Fe, WC) in a Si_3N_4 plate. Since it is difficult
to determine the defect type and size from the amplitude of the backscattered signal, so we have carried out Fourier transforms of the backscattered signal to obtain reflectivity as a function of frequency, and
used that information to characterize the size and type of defect. Our
early experiments have been with voids in glass and Si_3N_4 and we are
able to predict the size of the defects we detect." (Authors), 1 ref.

1978-462

White, R. M., "Programmable Filter for Ultrasonic NDE Systems," Proc.
     ARPA/AFML Rev. Prog. Quant. NDE, AFML-TR-78-55, 307-311.

"Transversal filters based on the charge-coupled device (CCD)
technology may be applied to processing received signals in an ultrasonic
NDE system. Filters having a fixed response (fixed tap weightings) could
be used to compensate for the characteristics of a given ultrasonic
transducer. A more flexible arrangement allowing programmability employs
a CCD delay line whose tap electrodes are accessible externally for weight-
ing. The response of this device can be altered by changing a set of
resistors mounted on a printed circuit board which can be plugged into
a socket connected to the CCD. Proper response for a given ultrasonic
transducer is obtained by plugging in the proper circuit board. Because
present commercial CCD's allowing programmability have clock frequencies
too low for direct processing of ultrasonic NDE signals, the filter is
preceded by a CCD video delay line which quickly stores the return signal
and then outputs it more slowly for processing during the time between
successive excitations of the ultrasonic transducer." (Author)

1978-463

Furgason, E. S., Twyman, R. E., and Newhouse, V. L., "Deconvolution
     Processing for Flaw Signatures," Proc. ARPA/AFML Rev. Prog. Quant.
     NDE, AFML-TR-78-55, 312-318.

"The ultimate resolution of all ultrasonic flaw detection systems
is limited by transducer response. The system output actually contains
detailed information about the target structure that has been masked
by the transducer characteristics. This type of problem is common in
all remote sensing applications and has been attacked by workers in
several other fields.

The output of any linear system can be modeled as the convolution
of the target response and transducer response. For such a system the
convolution process can be easily reversed to remove the effects of the
transducer, yielding an accurate detailed image of the target. Unfortunately, it is well known that real systems have noisy outputs and that
the presence of even relatively small amounts of noise make this deconvolution process impossible.

Recently, ultrasonic flaw detection systems have been demonstrated
which have extremely high output signal-to-noise ratios. For these

systems it is possible to use estimation techniques in the deconvolution process to achieve a good approximation to the actual target response. Results are presented that demonstrate these techniques applied to both computer generated and experimental data. Coupling the deconvolution processing with feature extraction routines is shown to yield an order of magnitude increase in range resolution." (Authors), 3 refs.

1978-464

Lakin, K. M., "Characterization Facility for NDE Transducers," Proc. ARPA/AFML Rev. Prog. Quant. NDE, AFML-TR-78-55, 319-325.

"The characterization of NDE transducers involves an assessment of the bidirectional acousto electric conversion between electrical and acoustic terminals and an evaluation of the near and far-field radiation patterns. The internal details of the transducer are largely unknown and, consequently, the traditional techniques for analyzing the structures cannot be applied. Instead, the transducer may be characterized as a hybrid two port network whose parameters may be determined by relatively simple measurements taken at the electrical port when the acoustic loads are known. The radiation pattern involves measurement only in the region exterior to the transducer. There are several techniques which may be used to accomplish this task. Our approach has been to measure the acoustic field in the far-field region and then to reconstruct the field at the face of the transducer. Once we have the field at the transducer, an evaluation of the source may be readily determined from amplitude and phase contour plots or from gray scale or pseudo-color images." (Author), 6 refs.

1978-465

Elsley, R. K., "Computer Aided Interpretation of NDE Signals," Proc. ARPA/AFML Rev. Prog. Quant. NDE, AFML-TR-78-55, 326-330.

"In order to improve NDE reliability, it is important to recover as much as possible of the useful information in NDE waveforms. An on-line minicomputer is ideally suited to both the collection of data and the performance of sophisticated signal processing tasks. Using a variety of signal processing techniques, including windowing, self-normalization (of transducer properties and far-field diffraction effects), transformations (Fourier magnitude and phase transforms, auto-correlations, cepstra), feature extraction and pattern recognition, it has been possible to obtain information about very small defects, strength of adhesive bonds and acoustic emissions which are not available by conventional means. Examples of these various capabilities will be given." (Author)

1978-466

Alers, G. A., Elsley, R. K., and Flynn, P. L., "Measurement of Strength of Adhesive Bonds," Proc. ARPA/AFML Rev. Prog. Quant. NDE, AFML-TR-78-55, 365-370.

"In order to predict the strength of an adhesive bond between two metal sheets, it is necessary to measure the physical state of the adhesive layer that mechanically joins the two pieces of metal. This requires rapidly performing a detailed analysis of the ultrasonic echos reflected from the entire structure when it is immersed in a water bath for a normal ultrasonic pulse-echo inspection. To achieve this result, computer operated ultrasonic inspection systems have been assembled and equipped with special signal processing routines so that particular features of the ultrasonic echo in both the time domain and the frequency domain can be extracted in a time short enough to meet the requirements of a production inspection system. Such features as the relative amplitude of the signals reflected from the top and bottom of the adhesive layer and the frequencies for which standing waves are excited in the adhesive and in the metal adherends are of particular interest for making the strength predictions. It is also important that the interrogating ultrasonic pulse be of very short time duration so that the echos from the various interfaces in the sandwich-like joint can be resolved in the time domain display. This requires the use of special high frequency pulse generators coupled to broad band transducers and amplifiers. Special procedures are also needed to insure the accuracy of the analog to digital conversion at the input to the computer and the subsequent transformations to and from the frequency domain." (Authors)

1978-467

Busse, L. J., Miller, J. G., Yuhas, D. E., Mimbs, J. W., Weiss, A. N., and Sobel, B. E., "Phase Cancellation Effects: A Source of Attenuation Artifacts Eliminated by a CsS Acoustoelectric Receiver," Ultrasound in Medicine, Vol. 3, Denis White, ed. Plenum Press, New York.

"One salient problem associated with experiments designed to determine the attenuation coefficient of an inhomogeneous specimen is that of phase cancellation effects. Phase cancellation effects may occur if inhomogeneities in the tissue distort the ultrasonic phasefronts presented to a spatially extended piezoelectric receiving transducer. When the wavefronts incident upon a conventional piezoelectric receiver are distorted, the generated electrical signal would be expected to be degraded because of the phase sensitive nature of the receiver. Phase cancellation effects can induce artifacts in attenuation data which might be interpreted incorrectly as reflecting only the absorption and scattering properties of a specimen.

The purpose of this report is threefold: i) to demonstrate the existence of phase cancellation effects and to show how these effects manifest themselves in measurements of the ultrasonic attenuation coefficient of tissue, ii) to show how phase cancellation effects associated with piezoelectric receivers can be reduced by appropriate choice of aperture size, and iii) to show how phase cancellation effects can be virtually eliminated by making use of an intensity sensitive ultrasonic receiver based upon the acoustoelectric effect—the coupling of acoustic energy to the system of charge carriers within a semiconducting crystal." (Authors)

1976-468

Kraut, E. A.,"Applications of Elastic Waves to Electronic Devices, Nondestructive Evaluation and Seismology," Proc. 1976 IEEE Ultrasonics
Symposium.

"Interactions between electrical engineers concerned with device
applications of elastic wave propagation and workers in the fields of
nondestructive evaluation, seismology, and mechanics can be mutually
beneficial and rewarding. Recently, the National Science Foundation
sponsored an interdisciplinary workshop to promote such interactions
through discussions of research activities of current interest in each
of these different fields. Areas where results in one field could
contribute to progress in another were identified. Some examples include
the use of surface acoustic wave correlators for flaw detection in NDE,
application of new methods for time series analysis to ultrasonic and
seismic data processing, fracture mechanics analogies in seismic source
theory, and applications of new results in the theory of piezoelectric
plate vibrations to oscillator and filter design. These and other
examples of overlaps will be presented." (Author), 92 refs.

1974-469

Weller, J. F. and Giallorenzi, T. G., "Optical Detection of Acoustic
Surface Waves in Layered Substrates: Theory and Experiment,"
IEEE Trans. Son. Ultrason., SU-21 (3), 196-203.

"Light diffraction by surface acoustic waves in layered media is
theoretically analyzed. Experimental verification of the theory is
presented for an SiO_2 film on $LiNbO_3$ and a Ta_2O_5 film on quarts."
(Author), 18 refs.

1974-470

Hjellen, G. A., Andersen, J., and Sigelmann, R. A., "Computer-Aided Design
of Ultrasonic Transducer Broadband Matching Networks," IEEE Trans.
Son. Ultrason., SU-21 (4), 302-304.

"The application of computer-aided circuit design (CAD) for modifying
the performance of ultrasonic transducers according to certain design
specifications has received limited recognition. To illustrate the power
of CAD for providing better designs, this paper departs from the more
conventional approach (controlling bonding and matching layers) and outlines a design strategy employing appropriate broadband matching networks."
(Authors), 11 refs.

1975-471

Goll, J. H. and Auld, B. A., "Multilayer Impedance Matching Schemes for
Broadbanding of Water Loaded Piezoelectric Transducers and High Q
Electric Resonators," IEEE Trans. Son. Ultrason., SU-22 (1), 52-
53.

"High efficiency, low ripple piezoelectric transduction into a water load has been achieved experimentally over a bandwidth of about 70% by using a two-layer acoustic impedance matching transformer. Similar performance is predicted for properly designed transducers in a frequency range of 1 to 40 MHz." (Authors), 3 refs.

1977-472

Myrick, R. J. and Arthur, R. M., "Real-Time Digital Echocardiography Using Burst Analog Sampling," IEEE Trans. Son. Ultrason., SU-24 (1), 19-22.

"Analog domain samples, acquired at high speed during the period of the echo from an ultrasonic pulse were stored in a series of sample-and-hold circuits. Analog samples were read out slowly for analog-to-digital (A/D) conversion during the subsequent interval before the next ultrasonic pulse. Burst analog sampling circuitry was combined with a conventional 9-bit A/D converter and a minicomputer to form a digital echocardiograph. An effective sample rate of 7 MHz was obtained with an actual A/D rate of 70 KHz. Gain could be altered under processor control for automatic depth compensation. The A/D rate could be varied by the processor to make analysis context dependent. The system operated in real time at 100 ultrasonic pulses/s. It was tested in A-mode and time-motion studies of cardiac structures." (Authors), 5 refs.

1977-473

Weinert, R. W., "Very High-Frequency Piezoelectric Transducers," IEEE Trans. Son. Ultrason., SU-24 (1), 48-54.

"A discussion of some general problems associated with high-frequency bulk-wave transducer-delay line systems is presented. Design curves concerning the radiation resistance of M-ZnO-M-delay line systems are given, where M represents the metal electrodes and is taken as Au, Ag, or Al for calculation purposes. Some data concerning a 0, π, 0, π beam steering mosaic transducer, which has application in high-frequency broad-band acoustooptic devices, is presented." (Author), 13 refs.

1977-474

Milsom, R. F., Reilly, N. H. C. and Redwood, M., "Analysis of Generation and Detection of Surface and Bulk Acoustic Waves by Interdigital Transducers," IEEE Trans. Son. Ultrason., SU-24 (3), 147-166.

"A method of analysis which uses a combination of analytical and numerical techniques has been developed to obtain an accurate solution to the coupled electromagnetic and acoustic fields set up by an inter-digital transducer on the surface of a piezoelectric substrate. Full account is taken of the coupling to bulk modes as well as surface modes, and the solution for the charge on the electrodes includes both electro-static charge and piezoelectrically regenerated charge. Programs have been written for interdigital arrays with uniform aperture but varying electrode width and pitch and arbitrary electrical connections. The theory is also valid for arbitrary crystal orientations. Generation and detection may be analyzed separately with information being provided on the partition of power into the various acoustic modes and the external load impedance, and the bulk wave radiation patterns are also computed. The program may also be used to find the insertion loss of a p ir of transducers. Results are presented for the Bleustein-Gulyaev orientation of PZT-4 ceramic and the YZ and 41° rotated YX orientations of lithium niobate." (Authors), 31 refs.

1978-475

Linzer, M., Shideler, R. S., and Parks, S. I., "Ultrafast Signal Averaging and Pulse Compression Techniques for Sensitivity Enhancement," Proc. 1st Int. Symp. on Ultrason. Mat. Char., Gaithersburg (June, 1978).

"A signal averager and pulse compression system has been developed for sensitivity enhancement in ultrasonic diagnosis.

The averager is capable of real-time (unbuffered) averaging at 50 MHz rates. . . .

The pulse compression circuit incorporates a surface acoustic wave (SAW) 'chirp' filter. Pulse compression ratios of 30 il and 8 il have been obtained in the case of 8 MHz and 3 MHz filter bandwidths, respectively." (Authors)

1978-476

von Gutfeld, R. J., "MHz Acoustic Waves from Thermoelastic Expansion of Thin Film-Liquid Interfaces," Proc. 1st Int. Symp. on Ultrason. Mat. Char., Gaithersburg (June, 1978).

"We have extended the work on laser induced thermoelastic expansions obtained from acoustically clamped thin metal films in the MHz range. In the work . . . reported here, we study the elastic waves generated from a structure consisting of a clamped metal film in contact with a liquid. . . . A 5 nsec pulsed edge laser was used as the excitation source together with piezoelectric receivers . . .

Based on new results we show some simple structures which use high thermal expansion liquids, such as acetone, to obtain strain amplitudes in liquids considerably larger than any previously observed by us . . ." (Authors), 2 refs.

1978-477

Rocha, H. A. F., Griffen, P. M., and Thomas, C. E., "Opto-Acoustic and Acousto-Electric Wideband Transducers," Proc. 1st Int. Symp. on Ultrason. Mat. Char., Gaithersburg (June, 1978).

"Two major problems are considered: 1) generation of fast acoustic pulses; and 2) detection and amplification of the corresponding echoes without appreciable waveform degradation.

The first problem is solved by the use of a short optical pulse which, when suitably absorbed, produces an acoustic stress-wave.

The second problem required the design and construction of electro-static transducers with a bandwidth compatible with that of the generated acoustic stress waves. . . .

The combination laser transmitter/electrostatic receiver has many advantages over conventional piezoelectric elements. For instance:

   a) A simple lens can be used to change the size of the optical beam falling on the absorber. This is equivalent to having a transmitter with a continuously variable diameter.
   b) Small spot sizes can generate very wide acoustic beams. Beam widths (at 50 percent points) of 160 degrees have been measured in aluminum.
   c) Cylindrical lenses can be used to produce the equivalent of a transducer with a continuously variable shape.
   d) Electro-optic deflectors can produce one- and two-dimensional scanning arrays." (Authors)

1978-478

Resch, M. T., Khuri-Yakub, B. T., Kino, G. S., and Shyne, J. C., "Stress
    Intensity Factor Measurement of Surface Cracks," Proc. 1st Int.
    Symp. on Surface Cracks," Proc. 1st Int. Symp. on Ultrason. Mat.
    Char., Gaithersburg (June, 1978).

"It was shown by Budiansky and Rice that the maximum stress intensity factor of an internal crack can be determined by measurements of the acoustic wave scattering from a crack. We have carried out a simple theory to predict the reflection coefficient of a surface acoustic wave in terms of the stress intensity factor of a surface crack, and predicted from the acoustic measurement the breaking stress of the material." (Authors)

1978-479

Varadan, V. K., and Varadan, V. V., "Characterization of Dynamic Shear
    Modulus in Inhomogeneous Media Using Ultrasonic Waves," Proc. 1st
    Int. Symp. on Ultrason. Mat. Char., Gaithersburg (June, 1978).

"We propose to examine this problem from a scattering theory approach. The model that we will study is a two-dimensional one, namely, a medium containing a random or periodic array of infinitely long cylinders of different material. Shear (SH-) waves are assumed to propagate normal to the axis of the cylinders. The wave undergoes multiple scattering between the cylinders and the effective wave number in the medium is complex. An ensemble average of the displacement field will be performed for a random distribution of the scatterers and expressions will be obtained for the average stress and the displacement field which are related by the average elastic modulus. The scattering properties of each scatterer will be characterized by the T-matrix which depends on the geometry and nature of the scatterer. Using this approach, we have already obtained the phase velocity and attenuation of SH-waves in a medium containing cylinders of elliptical cross section for a wide range of frequencies.

Our approach for relating the average stress to strain using scattering theory bears some resemblance to the treatment given by Bedeaux and Mazur for studying the macroscopic dielectric tensor of an inhomogeneous medium." (Authors)

1978-480

Lloyd, E. A. and Wadhwani, D. S., "Ultrasonic Spectroscopy and the Detection of Hydrothermal Degradation in Adhesive Bonds," Proc. 1st Int. Symp. on Ultrason. Mat. Char., Gaithersburg (June 1978).

"Analytical methods have been developed; determined largely by the geometry of the adhesives and adherent layers under investigation. The use recently of adherent aluminum alloy layers some 4 mm in thickness and supported film adhesives (BSL 312/5) giving adhesive layers some 0.3 mm thick has proved to be a particularly favorable geometry for study. . . .

The second signal returned is reflected from the adhesive bond itself with a spectral window leading to 15 MHz, one, and in many instances, two reflectivity minima are recorded. The position of these minima are inversely related to the transit time in the adhesive layer itself and the depth of modulation is a measure of the two interfacial reflectivities between the adherent and adhesive layers. The sensitivity of the test can be significantly increased by examining the relationship between the third signal returned and the second signal cited. This third signal represents a double reflection from the adhesive layer plus a double transmission through the bond. . . .

When an adhesively bonded joint is subject to a combination of high humidity and temperature both reversible and irreversible degradation, manifest as a reduction of strength, occurs. Within the reversible regime the bonds tend to fail cohesively. This regime, although not yet fully investigated is manifest by a reduction in modulation and a corresponding drop in the frequency at which the bond-line absorption minima occurs." (Authors)

1978-481

Shankar, R., Mucciardi, N., and Lawrie, W. E., "Application of Adaptive Learning Networks to Ultrasonic Signal Processing; Detecting Cracks in Stainless Steel Pipe Welds," Proc. 1st Int. Symp. on Ultrason. Mat. Char., Gaithersburg (June, 1978).

"Unambiguous discrimination has been achieved between cracks and geometrical (benign) reflectors in sample welded sections on 304 Stain-

less Steel (SS) pipes using nonlinear signal processing of ultrasonic
pulse echoes via the Adaptive Learning Network (ALN) methodology. . . .

For the preliminary scan, the ALN classifier was synthesized from
a small set of spectral parameters computed from the UT waveform. The
parameters included the fractional spectral content in frequency bins
descriptive of the low-(0 to 1 MHz), mid-(1 to 2 MHz), and high-(2 to 4
MHz) frequency ranges, and higher order moments of the UT power spectrum.
The synthesized ALN classifier "false dismissal" rate (the percentage of
cracks falsely dismissed as geometrical reflectors) was zero percent,
while the false alarm rate (the percentage of nondefects called cracks)
was 33 percent.

For the more detailed scan, the ALN classifier was synthesized from
parameters related to spectral shifts in an ensemble of pulse-echo waveforms collected around the vicinity of a region tagged as suspicious during
the preliminary scan. The waveform ensemble was obtained by angulating
the transducer over ± 11°, in increments of 5.5°, around the nominal
normal to the assumed crack plane. The ALN classifier reduced the false-
alarm rate to zero and, therefore, discriminating unambiguously between
cracks and benign geometrical reflectors." (Authors)

1978-482

Goebbels, K. and Höller, P., "Quantitative Determination of Grain Size
     and Detection of Inhomogeneities in Steel by Ultrasonic Backscattering
     Methods," Proc. 1st Int. Symp. on Ultrason. Mat. Char., Gaithersburg
     (June 1978).

"Ultrasonic waves in steel are attenuated by absorption and
scattering. Scattering from grains or phase boundaries can be measured
and in backscattering experiments as a function of time of flight
scattering signals are detected and evaluated. Two physical conditions
determine grain scattering:

    1. From the different regions Rayleigh, stochastic and diffuse
scattering in practice only the first and the second one are of interest.
    2. Scattering is only stimulated if there is a change in the wave-
resistance at grain boundaries (because of the elastical anisotropy) or
phase boundaries (because of change in density and/or velocity). . . .
    Several methods allow to evaluate the mean grain size of the
material under test by backscattering experiments:

    a. Measurements with two frequencies under the condition, that the
        frequency dependence of the absorption coefficients is known.
    b. Measurements at two samples with one frequency under the con-
        dition, that alpha is the same in the two samples.
    c. One measurement with a frequency high enough, that multiple
        scattering is generated (without assumptions concerning
        alpha)." (Authors)

1978-483

Heyman, J. S., Cantrell, J. H., Jr., and Whitcomb, J. D., "A Phase
    Insensitive Transducer—Theory and Application," Proc. 1st Int. Symp.
    on Ultrason. Mat. Char., Gaithersburg (June, 1978).

"Materials characterization with ultrasonics is examined from the
viewpoint of the receiving transducer. Signal artifacts are shown to
exist for conventional phase sensitive piezoelectric transducers. The
signal artifacts are always present when differing phases (and/or frequencies) are present on any phase sensitive receiver whose dimensions are
greater than an acoustic wavelength. Attenuation measurements with this
class of receivers under the above circumstances can lead to values of
attenuation in error by orders of magnitude.

These phase cancellation artifacts are eliminated with the use of
an acousto-electric transducer (AET) since it is insensitive to phase
variations across its face. Characteristics of the AET are contrasted
with conventional transducers and are shown to be significantly superior
where phase variations exist in the acoustic wave front." (Authors)

1978-484

Klinman, R., Marsh, F. J., Stephenson, E. T., and Webster, G. R.,
    "Ultrasonic Prediction of Grain Size, Strength and Toughness in
    Plain Carbon Steel," Proc. 1st Int. Symp. on Ultrason. Mat. Char.,
    Gaithersburg (June, 1978).

"It is well known that the mean ferrite grain size and mechanical
properties of steel are closely related. Metallographic measurement of
grain size is both time-consuming and destructive to the product tested.
The ultrasonic literature demonstrated that the contribution of scattering
to ultrasonic attenuation can be used to determine the mean grain size.
However, little work has been carried out to apply ultrasonically estimated
grain size to directly predicting the mechanical properties of steel.
Use of ultrasonic measurements for this purpose is potentially attractive,
because they are rapid and nondestructive and could conceivably be
applied to the on-line measurement of properties during primary
processing. . . .

Our data agree with published results relating grain size to
ultrasonic attenuation. They also agree with accepted Hall-Petch relationships between: a) grain size and yield strength, and b) grain size and
Charpy transition temperature. Multiple regression analysis of our data
showed that the lower yield strength can be estimated to ± 3.5 ksi and
transition temperature to ± 32 F at a 95 percent confidence level by the
combined use of: a) the ultrasonically measured grain size in the optimum
part of the Rayleigh region, and b) knowledge of the chemical composition.
This work supports the potential for development of a nondestructive test
for predicting yield strength, but much work remains before a production
test can be implemented." (Authors)

1968-485

Lee, R. E. and White, R. M., "Excitation of Surface Elastic Waves by
    Transient Surface Heating," Appl. Phys. Lett. 12 (1), 12-14.

    "The generation of surface elastic waves by the transient surface
heating of piezoelectric and nonpiezoelectric solids is described. A Q-
switched ruby laser produces the surface heating; the frequencies of the
resultant surface waves are Fourier components of the laser waveform.
The use of a spatially periodic illumination is shown to increase the
effectiveness of generation at a selected frequency. This method of
generating surface waves appears suitable for microwave frequency
operation as well as operation at high wave amplitudes at low frequencies."
(Authors), 10 refs.

1973-486

Bunkin, F. V. and Komissarov, V. M., "Optical Excitation of Sound Waves,"
    Sov. Phys. Acoust. 19 (3), 203-211.

    "When the powerful optical radiation emitted by present-day lasers
interacts with a medium, sound waves can be excited in the latter. The
physical mechanism of the excitation can be one of several. Below we
discuss only those which are unrelated to nonlinear optical effects,
i.e., for which nonlinear interaction between the radiation and matter
is insignificant. We therefore exclude, for example, the sound
generation mechanism due to induced Brillouin scattering. It is also
important to note that even within the framework of the investigated
mechanisms the amplitude of the excited sound can be very large. Nonlinear
acoustical effects become significant in this case (in particular, shock
waves can be generated)." (Authors), 36 refs.

1962-487

Papoulis, A., The Fourier Integral and Its Applications. McGraw-Hill,
    New York, 1962.

1978-488

Vary, A., "Use of an Ultrasonic-Acoustic Technique for Nondestructive
    Evaluation of Fiber Composite Strength," NASA-TM-73813.

    "This report describes the ultrasonic-acoustic technique used to
measure a "Stress Wave Factor." In a previous study this factor was
found effective in evaluating the interlaminar shear strength of fiber-
reinforced composites. Details of the method used to measure the stress
wave factor are described. In addition, frequency spectra of the stress
waves are analyzed in order to clarify the nature of the wave phenomena
involved. The stress wave factor can be measured with simple contact probes
requiring only one-side access to the part. This is beneficial in nondestructive evaluations because the probes can run parallel to fiber
directions and thus measure material properties in directions assumed by

actual loads. Moreover, the technique can be applied where conventional through transmission techniques are impractical or where more quantitative data are required. The stress wave factor was measured for a series of graphite polymide composite panels and results obtained are compared with through transmission immersion ultrasonic tests." (Author), 5 refs.

1978-489

Vary, A., "Correlations Among Ultrasonic Propagation Factors and Fracture Toughness Properties of Metallic Materials," Mat. Eval., June, 1978, 55-64.

"Empirical evidence was developed to show that a close relation exists among fracture toughness, yield strength, and ultrasonic attenuation properties of metallic materials. The evidence was obtained by ultrasonic probing of specimens of two maraging steels and a titanium alloy. It was concluded that nondestructive ultrasonic methods can be used to indirectly evaluate fracture-related material properties. The results suggest that these nondestructive ultrasonic measurements can also serve as an adjunct to destructive testing, measurement, and analysis of fracture properties." (Author), 19 refs.

1968-490

Papadakis, E. P., "Ultrasonic Attenuation Caused by Scattering in Poly-crystalline Media," Chap. 15 of Physical Acoustics, Vol. IV, Part B, W. P. Mason, ed., Academic Press (New York), 269-329.

85 refs.

1969-491

Youshaw, R. A., Criscuolo, E. L., and Dyer, C. H., "The Magnetic Tape Recording of Ultrasonic Test Information," Mat. Eval., February, 1969, 34-36 and 41.

"This paper describes a method of recording primary ultrasonic test information. A videotape recorder has been converted into a wide band instrumentation recorder. The "A" scan from the ultrasonic tester is directly recorded, together with the operator's voice giving the location, transducer position and interpretation of test data. An oscilloscope is used for the playback. The circuitry necessary to couple the output of the ultrasonic tester to the tape recorder is described." (Author), 6 refs.

1978-492

Kline, R. A., Green, R. E., Jr., and Palmer, C. H., "A Comparison of Optically and Piezoelectrically Sensed Acoustic Emission Signals," J. Acoust. Soc. Am., 64 (6), 1633-1639.

"The ususal sensor for acoustic emission is the piezoelectric transducer. Although this transducer is readily available, reasonably inexpensive, and very sensitive to ultrasonic transients, it has several serious drawbacks as a transducer: It distorts the signals being measured, it exhibits resonances, it has limited bandwidth, it responds differently to surface acoustic waves and bulk waves (because of its large sensitive area), and its calibration is a matter of considerable uncertainty. Essentially, it is a qualitative transducer. Furthermore, it cannot measure local effects within a millimeter of an emission source, where the mechanisms causing the ultrasonic transient are presumably most clearly distinguishable. Optical transducers, on the other hand, have the great advantage of providing accurate, quantitative, highly localized information; they do not disturb the waves being measured and are not limited by frequency response." (Authors), 18 refs.

1970-493

Elliott, B. J., "System for Precise Observations of Repetitive Picosecond Pulse Waveforms," IEEE Trans. Instrum. Msmt., November, 1970, 391-395.

"Conventional sampling techniques yield minimum risetime in the oscillography of repetitive electrical waveforms. However, system timing uncertainties introduce drift and jitter errors, which are typically comparable in magnitude to the cathode-ray-oscilloscope risetime. By using two sampling oscilloscopes in cascade it is possible to reduce the drift by a factor of 10^{-3}, down to a level of 10^{-14} seconds (during an averaging and recording time interval of 2 minutes). Successive sampling also allows accurate jitter filtering. With the aid of a tunnel-diode step generator the total system has a 10-90 percent risetime of 25×10^{-12} second, a step response that is closely integral Gaussian, a time-measurement uncertainty of about 10^{-13} second, and amplitude accuracy of 0.2 percent. Absolute time calibration is possible with a 3×10^{-14}-second resetting capability. Applications include the measurement of the impulse-response functions of coaxial two-port networks in the 10^{-11}-second range. Finally, it is shown that amplitude averaging at a sampled waveform with time jitter causes convolution error and loss of resolution." (Author), 5 refs.

1973-494

Lees, S., Gerhard, F. B., Barber, F. E., and Cheney, S. P., "DONAR: A Computer Processing System to Extend Ultrasonic Pulse-Echo Testing," Ultrasonics, July, 165-173.

"A dedicated general purpose digital computer has been built on the principle of a sampled-data system to run an ultrasonic subsystem under programmed control. A most significant application is the ability to extract a signal from an interfering background. As illustrated in the paper, a 1 mm diameter transducer was used to measure the diameter of a 2.5 mm OD plastics tube with 0.4 mm wall thickness. Echoes from all four surfaces were displayed and the measurements indicated an uncertainty of less than 0.1 mm." (Authors), 5 refs.

1976-495

Fay, B., Brendel, K., and Ludwig, G., "Studies of Inhomogeneous Substances by Ultrasonic Back-Scattering," Ultrasound in Med. and Biol. 2 (3), 195-198.

"A method is outlined by which ultrasonic back-scattering measurements may reveal information concerning both the scattering and absorption properties of inhomogeneous substances. After a description of the principle of the measuring method, experimental studies of a sample consisting of four layers with different scattering properties are discussed. To carry out measurements on substances, the acoustic properties of which are similar to those of biological tissues, inhomogeneous gelatine gels are investigated. The gels are produced using appropriate ethanol-glycerine mixtures. The inhomogeneities were introduced by adding tiny plastic spheres to the gel. It is shown, that the ultrasonic back-scattering method allows separation of the attenuation into scattering and absorption as functions of the location. In this way recognition of the inhomogeneities is possible. This fact should be helpful in the field of medical diagnostics." (Authors), 9 refs.

1973-496

Fay, B., "Theoretische Betrachtungen zur Ultraschall-rückstreuung," Acustica 28, 354-357.

"Theoretical Considerations of Ultrasound Back-Scatter"

"Nondestructive structure testing with ultrasound using the impulse-echo method on materials with localized structure only yields the attenuation constant averaged over the sound path. Using theoretical considerations it will be shown that by the method of ultrasonic back-scatter one can calculate the change of back-scatter and absorption coefficient in materials." (Author), 7 refs.

1978-497

Sigrist, M. W. and Kneubühl, F. K., "Laser-Generated Stress Waves in Liquids," J. Acoust. Soc. Am. 64 (6), 1652-1663.

"The generation of laser-induced stress waves in liquids by the vaporization process and the thermoelastic effect was studied experimentally. A high-speed camera and special high-sensitivity stress transducers with a response time of a few nanoseconds have been used for these investigations. The experimental results obtained for water, n-heptane, and carbon tetrachloride are discussed. For the first time, the individual contribution of vaporization and the thermoelastic effect on stress generation are separated. In addition, tunable high-frequency acoustic waves, with frequencies up to 60 MHz, have been generated in water by the impact of a laser pulse exhibiting longitudinal mode beating. Since existing theories on the thermoelastic generation of acoustic waves do not yield satisfactory agreement with our experimental data, a new spherical model is proposed, where the transient heating caused by the

laser impact, is represented by the three-dimensional heat pole. This
solution of the equation of heat conduction corresponds to a Gaussian
distribution of the excessive temperature in space, and thus to the TEM_{00}
mode of the incident laser beam. An analytical solution of the thermoelastic pressure wave is derived for this case of temperature distribution.
Its good agreement and the experiment is discussed for various liquids
and for two different laser characteristics." (Authors), 64 refs.

1973-498+

Canella, G., "Resonances and Effects of Couplant Layers in Ultrasonic
    Contact Testing," Mat. Eval., 61-66 (April).

"A theoretical and experimental examination has been carried out on
the variation of echo amplitude as a function of the thickness of the
liquid couplant layer in direct-contact ultrasonic testing, using compression waves. Four liquids were used and compared, namely water,
glycerine, lubricating oil, and oil for penetrant liquids. The
theoretical resonances were confirmed by experimental measurements.
Amplitude differences of up to 20 dB were measured for variations in
liquid thickness of 0.1 mm. The echo amplitude was also measured as a
function of thickness of the protective layers of rubber and it was found
that the maximum efficiency of a probe is attained at a fixed thickness
of these layers." (Author), 5 refs.

1978-499+

Childers, D. G. (editor), Modern Spectrum Analysis. IEEE Press, New
    York.

"Power spectrum estimation has progressed through several stages
since the turn of the century. Perhaps the first estimator to be used
extensively was the periodogram which, although it is known to be a poor
estimator because the variance of the estimate exceeds its mean value,
is still used today. Twenty years ago in 1958 Blackman and Tukey published
their autocorrelative method for power spectrum estimation, the steps of
which include estimating the autocorrelation function from the observed
data, windowing the autocorrelation estimate in an appropriate manner and
then Fourier transforming the windowed autocorrelation function to finally
obtain the estimated power spectrum. With the advent of the fast Fourier
transform (FFT) algorithm in 1965, the direct method for power spectrum
estimation became popular. This approach generally uses the magnitude
squared of the transform of the windowed data record as the power
spectrum estimate. More recently, the maximum entropy method (MEM) for
spectral estimation has attracted the attention of scientists and
engineers. This procedure is equivalent to the linear predictive and
autoregressive methods and is being applied to a wide range of data
because of its potential to achieve increased spectral resolution. Another
recent technique is the maximum likelihood method (MLM) which has been
used for high resolution frequency wavenumber spectral estimation. The
MEM and MLM are related as we shall discuss later and as one of the
reprints (Burg, 1972) shows. This collection of papers concentrates on
the MEM and MLM but we do provide two background papers as well as a

brief tutorial review of the other major spectral estimation procedures."
(Editor), numerous references.

1975-500+

Ramirez, R. W., The FFT: Fundamentals and Concepts," Tektronix, Inc.,
    Beaverton, OR, Part 070-1754-00.

1973-501+

Auld, B. A., Acoustic Fields and Waves in Solids, Wiley, New York.

1979-502

Rose, J. H. and Krumhansl, J. A., "A Technique for Determining Flaw
    Characteristics from Ultrasonic Scattering Amplitudes," Proc.
    ARPA/AFML Rev. of Prog. in Quant. NDE, AFML-TR-78-205, 368-372.

   "We report an approximate technique for determining the characteristics of flaws in elastic media from a knowledge of their ultrasonic
scattering amplitudes. The technique is rigorously valid in the weak
scattering limit. Good results have been obtained for strongly scattering
flaws. In particular, we tested the technique for a 2-1 oblate spheroidal
void in Ti, and for various strongly scattering spherical defects. For
these tests the technique yields good results for the volume of the flaws.
In the case of the oblate spheroid, satisfactory results were obtained
for the calculated ratio of major to minor axis, indicating that the
technique is sensitive to the shape of the flaw." (Authors), 10 refs.

1979-503

Richardson, J. M.,'Direct and Inverse Problems Pertaining to the Scattering
    of Elastic Waves in the Rayleigh (Long Wavelength) Regime," Proc.
    ARPA/AFML Rev. of Prog. in Quant. NDE, AFML-TR-78-205, 332-340.

   "It is well known that in the scattering of elastic waves from
localized inhomogeneities the scattering amplitude A is proportional
to the square of the frequency ω in the Rayleigh (long wavelength)
regime, i.e., $A = A_2\omega^2 + \ldots$ This talk deals with the problem of
(1) extracting A_2 from experimental scattering data, (2) calculating A_2
for an assumed scatterer and (3) deducing the properties of the scatterer
from a set of values of A_2 measured for various transducer configurations.
A review of experimental and theoretical results for A_2 will be presented
for the case of spheroidal voids and the remaining discrepancies between
the two kinds of results will be discussed. The inverse problem (i.e.,
deducing the scatterer properties from the scattering measurements) will
be discussed in detail. The probabilistic inverse problem, which provides
the appropriate framework for the interpretation of real data, will be
covered at greater length. In the case in which it is assumed that the

scatterer is an ellipsoid void, whose size, shape and orientation are unknown a priori, a number of computational results involving best estimates and associated measures of significance will be given. Analogous results will be derived for parameters related to fracture mechanics." (Author), 12 refs.

1979-504

Achenbach, J. D., Adler, L., Lewis, D. K. and McMaken, H., "Diffraction of Ultrasonic Waves by Penny-Shaped Cracks in Metals: Theory and Experiment" (accepted for publication in J. Acoust. Soc. of Am., 1979.

"In this paper an analytical solution to the diffraction of elastic waves by penny-shaped cracks in metals is compared with experimental observations. The analysis, which is based on elastodynamic ray theory, is valid for the region of ka > 1. A digitized spectrum analysis system is described which measures the frequency components of the waves diffracted from a 2500 μ radius crack in diffusion bonded titanium. The amplitude spectra show good agreement between experiment and theory. The theoretically predicted periodicity of the diffracted spectra provides a simple formula for the inverse problem. Application of this formula to the experimental measurements determines the crack size with excellent accuracy." (Authors) 16 refs.

1979-505

Adler, L. and Achenbach, J. D., "Elastic Wave Diffraction from Elliptical Cracks: Theory and Experiment" (will be published in J. of Nondestr. Eval.).

1980-506

Crane, R. L. and Nayfeh, A., "Fundamental Considerations in the Nondestructive Measurement of Adhesive Bond Properties," Proc. of ASNT Spring Conf., Philadelphia (March 1980).

"Recently there has been a great deal of research with a goal of the predicting of the mechanical response of adhesively bonded joints. Several techniques have been developed that purportedly correlate an ultrasonic parameter such as wave speed, attenuation, or an acoustic spectral feature with lap shear strength. These techniques are fundamentally based on the empirical observation that there is a weak correlation between stiffness and ultimate tensile strength for many polymeric materials. If the stress distribution in the adhesive joint is taken into account, then it can be shown that such correlations depend critically on two assumptions that do not usually occur in service: (1) the uniformity of the adhesive and (2) the absence of defects. A fracture mechanics/acoustic model is presented which shows what measurements need to be made before predictions of the mechanical response of the adhesive joint can be made. Supporting data will also be presented." (Authors)

Subject Index

This subject index lists the references found in Chapter 5, Abstracted Bibliography. The entries are listed here according to the sequence of section headings given in Chapters 1, 2, and 3. At the end of this index is a *miscellaneous* section.

One should note that many more references are listed in this index than were mentioned in the text of Chapters 1–3. Unfortunately, time did not permit inclusion of all pertinent references in the text.

CHAPTER 1

INTRODUCTION

Background and General Information

Surveys of Ultrasonic Spectroscopy

CHAPTER 2

ULTRASONIC SPECTROSCOPIC SYSTEMS

Introduction

System Model

Transmitter

Continuous-Wave Sinusoid and Sinusoidal Burst

1966-050	1970-064	1974-360	1976-082
	066		
	268		

Ideal Broadband Pulse

Rectangular Pulse

1963-006	1964-061	1965-008	1966-007	1968-062	1970-064
			116		

Other Pulse Shapes

1972-157

Single Transition

1972-168   1973-155

Exponential Pulse

1972-239   1974-192

Arbitrary Pulse Shapes

1977-240   1978-257

Electrical Coupling Network, I

Coaxial Cable

1972-168

Response Equalization Networks

1971-354	1974-470	1977-238
		355

Transmitting Transducer

1970-064	1971-286	1975-368	1976-407	1977-419	1978-320
		398		436	458
				473	464

Frequency Response

1942-273	1964-147	1966-050	1969-284	1972-283	1974-191
		274			
		282			
1975-089	1976-102	1977-236	1978-230		
223		355			

Response Equalization Networks

1975-223   1977-355

Producing an Ultrasonic Pulse Having an Arbitrarily Chosen Spectrum

1975-190   1977-240   1978-257

Laser Generation of Ultrasonic Waves

1973-486

Bulk Waves in Liquids

1966-364   1973-486   1978-497

Bulk Waves in Solids

1973-486   1974-382   1977-365   1978-476
                                             477

Surface Waves in Solids

1968-485

Radiation Coupling, I

Mechanical Coupling Layers

1958-224	1960-296	1969-284	1970-064	1974-470	1976-082
1977-195					

Acoustic Impedance Matching

| 1966-282 | 1971-354 | 1973-183 | 1974-191 | 1975-471 | 1977-195 |

Intrinsic Energy Losses - Scattering and Absorption

1969-251

Geometrical Energy Losses - The Ultrasonic Field

1956-015	1964-277	1969-005	1970-278	1971-231	1972-276
			280	294	
				295	
1973-016	1974-281	1975-235	1976-082	1977-198	1978-230
194	292	343	102		320
			259		

Material Under Investigation

Refer to the specific application in Chapter 3

Radiation Coupling, II

Transfer Function for Disk-to-Disk Coupling

1977- 265

Receiving Transducer

Piezoelectric Receivers

1961-148	1963-297	1964-147	1966-050	1971-286	1975-398
214				294	
1976-407	1977-236	1978-230			
	419	262			
	428	320			
	436	458			
		464			
		492			

Open-Circuit Operation

1961-148	1963-297	1975-223	1977-236	1978-230
			355	

Short-Circuit Operation

1963-297

Acoustoelectric Transducer

1977-185	1978-363
443	467
	483

Capacitive Receiver

1966-352	1976-082	1977-353

Optical Receiver

1974-469	1978-492

Electrical Coupling Network, II

Differentiator

Low-Pass Filter

Amplifier

1971-122	1972-239

Analog Gate

1969-005	1970-270	1975-086

Sampling and Digitization

Sampling

1969-491	1970-493	1972-009	1973-001	1974-146	1975-110
			123	227	161
			494	384	184
					293
					371

1976-099	1977-219	1978-298
299	220	300
	226	465
	473	475

Digitization

1973-123	1974-146	1975-110	1976-099	1977-219	1978-298
	227	161	248		465
	384	184	299		475
		371			
		401			

Transient Recorders

1976-118	1977-266	1978-298
	448	300
		475

Charge-Transfer Devices

1976-099	1977-420	1978-462

Equivalent-Time Sampling

1970-134	1974-146	1975-033	1978-298
493	384	086	
		371	

Receiving Subsystem

Response Equalization Networks

1970-010	1977-238
	355
	425

Equalization Network Incorporating a Transmission Line

1975-223

Surface Acoustic Wave (SAW) Filter

1975-370	1976-099	1977-441	1978-475
392	408		
	468		

Programmable Charge Coupled Device (CCD) Filter

1976-099	1977-420	1978-462

Spectrum Analysis

Fourier Analysis

1974-013	1976-002

Additional Analysis Techniques

Multiple Signals

1974-013	1976-002	
	152	

The Cepstrum

1963-347	1970-269	1977-189	1978-454
		206	
		222	

Other Transformations

1974-092

Maximum Entropy and Maximum Likelihood Spectrum Analysis

1976-404	1978-499
468	

Pattern Recognition

1974-092	1975-093	1975-394	1976-106	1977-187	1978-454
141	094		410	426	481
380	126			447	
	228				
	393				

Miscellaneous Techniques

1970-010	1974-092	1975-088
134		086
269		

Complete Ultrasonic Spectroscopic System

Continuous-Wave Systems

Continuous-Wave System - Single Frequency (Narrowband)

1976-082

Continuous-Wave System - Swept Frequency

1970-064	1974-360	1975-229	1976-082	1977-434	1978-253

Pulsed Systems

Pulsed System - "Single" Frequency RF Burst

1976-082
098

Pulsed System - Swept Frequency

1966-050	1970-064	1976-082
	268	

CHAPTER 3

APPLICATIONS OF ULTRASONIC SPECTROSCOPY TO MATERIALS EVALUATION

Defect Characterization

Inversion Techniques

1974-380	1975-094	1976-119	1977-426	1978-454	1979-502
	126		427	455	503
			447		

Adhesive Bonds

1973-366	1974-210	1975-367	1976-244	1977-418	1978-449
			405	431	

Models and Theoretical Developments

1960-296	1966-051	1973-256	1974-026	1975-034	176-106
			054	048	107
			132	120	108
			141	129	118
			150	400	177
			237		416
			383		417

1977-153	1978-218	1980-506
200		
432		
433		

Experimental Investigations

1966-051	1973-020	1974-026	1974-237	1975-033	1975-377
	053	054	384	048	378
	256	132		120	399
		141		129	400
		150		180	401

1976-028	1976-415	1977-153	1977-433	1978-218
097	416	200	434	442
108		429		459
118		430		466
177		432		480

Surface Properties

Surface Roughness

1975-137	1976-139
138	246

Periodic Structure

1975-138	1976-139	1977-140	1978-339

Corrosion, Deformation, and Fatigue

1967-159	1975-179	1978-442

Surface-Breaking Cracks

1970-163	1974-083	1975-164	1976-097	1977-341	1978-205
	085		109		339
			119		
			196		
			345		

Miscellaneous

Thickness Measurement

1967-128	1970-064	1971-294	1974-188	1975-044
159		362	199	361

Stress Analysis

1975-402	1976-152	1978-456

Acoustic Emission

1974-124	1975-089	1976-173	1978-492
	144		

Diffusion Bond Testing

1968-135   1973-121

Low-Frequency Excitation

1970-163   1971-165   1975-164

Material Properties

1973-127   1974-130

Holographic Spectroscopy

1975-032
   217

Harmonic Generation

1976-100

Completeness of Sintering

1976-143

Relaxation Spectroscopy

1975-197

Interferometer

1968-213

Resonance Testing

1974-237

Viscosity

1971-294

Author Index

Achenbach, J. D.
  1975-337
  1977-323
  1978-324
  1979-504, 505

Adler, L.
  1970-036, 037
  1971-012, 038
  1972-039, 040, 041
  1973-042, 043
  1975-044, 045, 372,
    391
  1976-145, 242, 413
  1977-046, 047, 069,
    424, 446
  1978-156, 322, 452
  1979-504, 505

Ahlberg, L. A.
  1976-175
  1977-428

Aldridge, E. E.
  1970-270

Alers, G. A.
  1974-090
  1975-361, 377, 399
  1976-415
  1977-200, 429, 432
  1978-459, 466

Anderson, J.
  1974-470

Ang, D. D.
  1964-308, 309, 310,
    311

Arthur, R. M.
  1977-472

Auld, B. A.
  1973-501
  1975-471

Avioli, M.
  1977-448

Baboux, J. C.
  1975-089, 223
  1977-236

Bantz, W. J.
  1968-049

Barber, F. E.
  1973-494

Barrus, L.
  1971-165

Beaver, W. L.
  1974-292

Bednar, J. B.
  1975-190

Behravesh, M.
  1977-198

Bell, J. C.
  1978-442

Bell, J. R.
  1975-371
  1976-177
  1978-218

Belyi, V. E.
  1975-104

Berlincourt, D. A.
  1964-274

Bess, L.
  1974-281

Bifulco, F.
  1975-151

Birchak, J. R.
  1976-109

Biswas, R.
  1977-428

Bleistein, N.
  1977-427
  1978-455

Blinka, J.
  1976-152

Bogert, B. P.
  1963-347

Boore, D. M.
  1972-331

Bray, D. E.
  1978-254

Breazeale, M. A.
  1966-352
  1977-353

Brekhovskikh, L. M.
  1960-296

Brendel, K.
  1976-495

Brockelman, R. H.
  1973-121